Stochastic Processes and Related Topics

STOCHASTICS MONOGRAPHS
Theory and Applications of Stochastic Processes
A series of books edited by Saul Jacka, Department of Statistics,
University of Warwick, UK

Stochastic Processes and Related Topics

Proceedings of the 12th Winter School
Siegmundsburg, Germany

27 February – 4 March 2000

Edited by

Rainer Buckdahn
Université de Bretagne Occidentale
Brest, France

Hans-Jürgen Engelbert
Friedrich-Schiller-Universität
Jena, Germany

and

Marc Yor
Université Pierre et Marie Curie
Paris, France

CRC Press
Taylor & Francis Group
Boca Raton London New York

CRC Press is an imprint of the
Taylor & Francis Group, an informa business
A TAYLOR & FRANCIS BOOK

CRC Press
Taylor & Francis Group
6000 Broken Sound Parkway NW, Suite 300
Boca Raton, FL 33487-2742

First issued in paperback 2019

ISBN-13: 978-0-415-29883-4 (hbk)
ISBN-13: 978-0-367-39614-5 (pbk)

Visit the Taylor & Francis Web site at
http://www.taylorandfrancis.com

and the CRC Press Web site at
http://www.crcpress.com

Publisher's note
This book has been prepared from camera-ready copy
supplied by the authors.

British Library Cataloguing in Publication Data
A catalogue record for this book is available from the British Library

Library of Congress Cataloging in Publication Data
A catalog record for this book has been requested.

CONTENTS

PREFACE

This volume contains a significant part of the contributions held at the 12th Winter School on Stochastic Processes and their Applications, February 27 – March 4, 2000, in Siegmundsburg, a little village in the midst of the Forest of Thuringia, near to the famous *Rennsteig*, the high-level road over the ridge of hills, sixty miles south-west of Jena (Federal Republic of Germany). The conference was organized by the Friedrich Schiller University of Jena as part of a series of regular meetings in the field of stochastic processes and their applications which began in 1976.

The Winter School was attended by thirty-six participants, among them eighteen from abroad. The scientific programme included ten 90-minutes and twenty 30–45-minutes lectures.

The scope of the conference covered aspects of the following topics:

- Stochastic analysis,
- Mathematical Finance,
- Stochastic differential equations and, in particular,
- Backward stochastic differential equations,
- Stochastic partial differential equations.

The Winter School included four series of invited lectures:

R. BUCKDAHN (Brest)
Backward Stochastic Differential Equations. Semilinear PDE and SPDE,

YU. KABANOV (Besançon)
Arbitrage Theory,

W.M. SCHMIDT (Frankfurt a.M.)
Credit Derivatives and Models for Correlated Defaults,

M. YOR (Paris)
Three Intertwined Brownian Topics: Exponential Functionals, Winding Numbers, and Ray–Knight Theorems on Local Times.

The organizers would like to take the opportunity to thank Rainer Buckdahn, Yuri Kabanov, Wolfgang Schmidt and Marc Yor heartily for their lively and interesting lectures, which had a major influence on the success of the conference.

We hope that the volume will give a representative insight into the work of the Winter School and that it will encourage the worldwide cooperation of probabilists working in the field of stochastic processes and their applications.

We would like to thank all speakers and participants for their contribution to the success of the meeting. Particular thanks are due to Jochen Wolf and Gunar Tittel for their participation in the organization of the Winter School and the preparation of this volume. We are also indebted to Nicole Serfling for her valuable assistence in organizing the Winter School.

The Winter School was supported by the Deutsche Forschungsgemeinschaft (DFG) through the Graduation College *Analytical and Stochastic Structures and Systems*. It is a pleasure for us to express our gratitude to the DFG for this financial support without which the conference would not have taken place. Research presented at the Winter School was also partly supported by INTAS-97-30204 which is gratefully acknowledged.

For the Organizers
Jena, in March 2001   H.-J. Engelbert

BACKWARD STOCHASTIC DIFFERENTIAL EQUATIONS AND VISCOSITY SOLUTIONS OF SEMILINEAR PARABOLIC DETERMINISTIC AND STOCHASTIC PDE OF SECOND ORDER

RAINER BUCKDAHN

Université de Bretagne Occidentale
Département de Mathématiques
6, Av. le Gorgeu, B.P. 809, F-29285 Brest

Abstract The paper, written in form of a survey but including the proofs of the key results, introduces the reader into the theory of backward stochastic differential equations and backward doubly stochastic differential equations. For the Markovian case the connection between these both types of equations and viscosity solutions of deterministic and stochastic parabolic partial differential equations is developed.

Key Words Backward stochastic differential equation, backward doubly stochastic differential equation, parabolic PDE, stochastic PDE, Feynman-Kac formula, viscosity solution.

0. INTRODUCTION

The aim of this article is, in a first step, to present to the reader an introduction to the theory of backward stochastic differential equations (in short, BSDEs) and then to develop the theory of BSDEs in a Markovian framework and its connection with viscosity solutions of semilinear second order partial differential equations of parabolic type. Afterwards, this connection is generalized to that between backward doubly stochastic differential equations (BDSDEs) and stochastic viscosity solutions of semilinear stochastic second order partial differential equations (SPDEs, for short) of parabolic type.

1

The first paper on nonlinear BSDEs (Pardoux and Peng [36]) appeared in 1990. Since then, motivated by the vast field of applications, in particular in finance, stochastic control and PDEs, the theory of BSDEs has developed continually. We refer the reader more interested in the theory of BSDEs to Pardoux and Peng [36], [37] and [38], to the book edited by El Karoui and Mazliak [17], the references therein and to Hamadène [21], [22]. The reader more interested in applications of BSDEs in finance should consult El Karoui, Peng and Quenez [18], El Karoui and Quenez [19] and also the references therein; he is also referred to Duffie and Epstein [13], [14], Duffie and Lions [15] and Duffie, Ma and Yong [16]. The reader whose interest concerns particularly the applications to stochastic control can find more information in El Karoui and Mazliak [17], in Peng [41], [42] and Hamadène and Lepeltier [23], [24]. More details about the connection between BSDEs and second order PDEs will be found by the reader in Peng [39], [41], Pardoux and Peng [37], [38]. A nice survey which includes also recent works on homogenization of PDEs whith BSDEs is given by Pardoux in [35]. Finally, the reader who wants to learn more about the connection between BDSDEs and SPDEs can consult Pardoux and Peng [38] and, for the case of stochastic viscosity solutions, Buckdahn and Ma [7], [8].

In Section 1 we will motivate and introduce BSDEs. By respecting the chronological order of the development of the theory of BSDE we will begin with a rather easy problem of characterization of optimal control processes for a controlled system of stochastic differential equations. The adjoint equation to this problem will be interpreted as linear BSDEs. Such linear BSDEs related to the stochastic version of Pontryagin's maximum principle had been studied long time before the pioneering work on general nonlinear BSDEs by Pardoux and Peng [36]: Bismut [5], in 1973, was the first to introduce such a linear BSDE; other works on the maximum principle in stochastic control were written by Kabanov [26] in 1978 and Arkin and Saksonov [1] in 1979. In 1978, Bismut [6] introduced an BSDE with nonlinearity (Riccati equation) for which he proved existence and uniqueness of the solution.

This motivation of BSDEs will be followed, in Section 2, by a survey on some fundamental results of BSDEs; this survey includes the proof of the main results. In Section 3 we will restrict to BSDEs in the Markovian framework and develop their connection with viscosity solutions of semilinear elliptic PDEs. In the Sections 4 – 7 we present the generalization by Buckdahn and Ma ([7], [8]) of this connection to that between BDSDEs and stochastic second order parabolic PDEs: Section 4 is devoted to the introduction of the notion of stochastic viscosity solution of SPDEs; in Section 5 we consider BDSDEs and study a comparison theorem for such a type of equations; in Section 6, with the help of a Doss–Sussmann transformation for SPDEs and BDSDEs, existence of stochastic viscosity solutions is shown and, finally, Section 7 is devoted to the study of uniqueness of the stochastic viscosity solution.

Acknowledgement It is my pleasure to thank the organizer of the Winter School at Jena for inviting me and giving me the opportunity to deliver a small series of lectures on BSDEs and their connection to viscosity and stochastic viscosity solutions.

1. THE ADJOINT EQUATION AS LINEAR BSDE

In this section we consider the problem of characterizing an optimal control process for a controlled stochastic differential equation (SDE, for short). This discussion will lead us in a natural way to the notion of a BSDE.

Let $(\Omega, \mathcal{F}, P)$ be a complete probability space on which a k-dimensional standard Brownian motion $W = (W_t)_{t \in [0,T]}$ is defined; $T > 0$ is some fixed finite time horizon. By $\mathbf{F} = (\mathcal{F}_t)_{t \in [0,T]}$ we denote the natural filtration generated by W and augmented by all the P-null sets of $\mathcal{F}$. We consider the m-dimensional Euclidean space R^m as control state space and by $\mathcal{A} = L^2(\mathbf{F}, [0,T]; R^m)$ we denote the linear space of all $\mathbf{F}$-progressively measurable R^m-valued processes $v : \Omega \times [0,T] \to R^m$ such that $E[\int_0^T |v_t|^2 dt] < +\infty$. A process $v \in \mathcal{A}$ is called admissible control.

The control problem we shall study is governed by the SDE

$$dX_t^v = b(X_t^v, v_t)dt + \sigma(X_t^v, v_t)dW_t, \; t \in [0,T]; \quad X_0^v = x, \tag{1.1}$$

with initial value $x \in R^n$ and $v \in \mathcal{A}$. To this controlled equation we associate the cost functional

$$J(v) = E\left[\int_0^T g(X_t^v, v_t)dt + h(X_T^v)\right], \quad v \in \mathcal{A}. \tag{1.2}$$

Our aim is to minimize this cost functional over $\mathcal{A}$ and to derive a necessary condition for the optimality of an admissible control $\bar{v}$. This necessary condition will lead us in a natural way to the notion of a linear BSDE.

Let us suppose that the coefficients

$$b : R^n \times R^m \to R^n, \quad \sigma : R^n \times R^m \to R^{n \times k},$$
$$g : R^n \times R^m \to R, \quad h : R^n \to R$$

of SDE (1.1) and cost functional (1.2), respectively, are of class C^1 with bounded derivatives in x, u. We observe that then, for each $v \in \mathcal{A}$, SDE (1.1) has a unique continuous solution $X^v \in L^2(\mathbf{F}, [0,T]; R^n)$ and J takes all its values in R.

Let us make also the assumption that there is some admissible control $\bar{v} \in \mathcal{A}$ which realizes the minimum of the cost functional J. To characterize this optimal control we should study the behavior of J under small perturbations of the control. For this we fix any $v \in \mathcal{A}$ and put

$$v(\theta) = \bar{v} + \theta v, \quad \theta \in R.$$

Clearly, also $v(\theta)$ is an admissible control and

$$J(v(\theta)) \geq J(\bar{v}), \quad \theta \in R. \tag{1.3}$$

Let us now show that $\theta \mapsto J(v(\theta))$ is differentiable at 0. For this we recall the following well-known result which is based on standard estimates for $X^{v(\theta)}$, $\theta \in R$ and on Kolmogorov's continuity criterion.

Lemma 1.1 *The process $\{X_t^{v(\theta)}, (t, \theta) \in [0,T] \times R\}$ possesses a version which is continuous in (t, θ) and differentiable with respect to θ. Moreover, if $Y = (Y_t)_{t \in [0,T]} \in L^2(\mathbf{F}, [0,T]; R^n)$ denotes the unique continuous solution to the SDE*

$$
\begin{aligned}
dY_t &= \left(Y_t \nabla_x b(X_t^{\bar{v}}, \bar{v}_t) + v_t \nabla_u b(X_t^{\bar{v}}, \bar{v}_t)\right) dt + \left(Y_t \nabla_x \sigma(X_t^{\bar{v}}, \bar{v}_t)\right.\\
&\quad \left. + v_t \nabla_u \sigma(X_t^{\bar{v}}, \bar{v}_t)\right) dW_t, \quad t \in [0,T],\\
Y_0 &= 0,
\end{aligned}
$$

then

$$
\sup_{t \in [0,T]} \left| Y_t - \frac{1}{\theta}\left(X_t^{v(\theta)} - X_t^{\bar{v}}\right) \right| \longrightarrow 0, \quad as \ \theta \to 0,
$$

in L^2 and P-a.e.

Thus, with the notation $X_t^{\theta, \varsigma} = X_t^{\bar{v}} + \varsigma(X_t^{v(\theta)} - X_t^{\bar{v}})$, $\varsigma \in [0,1]$, we have

$$
\begin{aligned}
&\frac{1}{\theta}\{J(v(\theta)) - J(\bar{v})\}\\
&= \int_0^1 E\left[\int_0^T \left(\nabla_x g(X_t^{\theta, \varsigma}, v_t(\theta\varsigma)) \frac{X_t^{v(\theta)} - X_t^{\bar{v}}}{\theta} + \nabla_u g(X_t^{\theta, \varsigma}, v_t(\theta\varsigma)) v_t\right) dt\right.\\
&\quad \left. + \nabla_x h(X_T^{\theta, \varsigma}) \frac{X_T^{v(\theta)} - X_T^{\bar{v}}}{\theta}\right] d\varsigma\\
&\longrightarrow E\left[\int_0^T \left(\nabla_x g(X_t^{\bar{v}}, \bar{v}_t) Y_t + \nabla_u g(X_t^{\bar{v}}, \bar{v}_t) v_t\right) dt + \nabla_x h(X_T^{\bar{v}}) Y_T\right]
\end{aligned}
$$

as $\theta \to 0$. Consequently, $\theta \mapsto J(v(\theta))$ is differentiable at θ and from (1.3) we see that

$$
E\left[\int_0^T \left(\nabla_x g(X_t^{\bar{v}}, \bar{v}_t) Y_t + \nabla_u g(X_t^{\bar{v}}, \bar{v}_t) v_t\right) dt + \nabla_x h(X_T^{\bar{v}}) Y_T\right] = 0, \ v \in \mathcal{A}. \quad (1.4)
$$

Our aim is now to transform (1.4) into an expression of the form

$$
E\left[\int_0^T \Gamma_t v_t dt\right] = 0
$$

where the process $\Gamma \in L^2(\mathbf{F}, [0,T]; R^k)$ should depend only on the processes $X^{\bar{v}}, Y$ and $\bar{v}$, but not on $v \in \mathcal{A}$. This will lead us to the necessary condition $\Gamma_t = \Gamma_t(X^{\bar{v}}, Y, \bar{v}) = 0$, dtdP-a.e.

To this end we consider an R^n-valued Itô process given by the following equation with terminal condition

$$
dp_t = \alpha_t dt + K_t dW_t, \ t \in [0,T]; \quad p_T = -\nabla_x h(X_T^{\bar{v}}).
$$

While the drift coefficient $\alpha \in L^2(\mathbf{F}, [0,T]; R^n)$ will be specified later, the diffusion coefficient $K \in L^2(\mathbf{F}, [0,T]; R^{n \times k})$ is completely determined by the martingale representation

$$
L^2(\Omega, \mathcal{F}_T, P; R^n) \ni \nabla_x h(X_T^{\bar{v}}) - \int_0^T \alpha_t dt = E\left[\nabla_x h(X_T^{\bar{v}}) - \int_0^T \alpha_t dt\right] + \int_0^T K_t dW_t.
$$

We recall that the filtration $\mathbf{F} = (\mathcal{F}_t)_{t \in [0,T]}$ is generated by the Brownian motion W.

Applying Itô's formula to the product $p_t Y_t$ we obtain

$$
\begin{aligned}
p_T Y_T \;=\; & \int_0^T d(p_t Y_t) && (1.5) \\
=\; & \int_0^T \Big\{ Y_t \left(\nabla_x b(X_t^{\bar{v}}, \bar{v}_t) p_t + \nabla_x \left(\mathrm{tr}\{K_t^* \sigma\} \right) (X_t^{\bar{v}}, \bar{v}_t) + \alpha_t \right) \\
& \quad + v_t \left(\nabla_u b(X_t^{\bar{v}}, \bar{v}_t) p_t + \nabla_u \left(\mathrm{tr}\{K_t^* \sigma\} \right) (X_t^{\bar{v}}, \bar{v}_t) \right) \Big\} dt + M_T,
\end{aligned}
$$

where M_T is the increment of an integrable martingale over the time interval $[0, T]$. Hence, by setting

$$
\alpha_t = - \left(\nabla_x g(X_t^{\bar{v}}, \bar{v}_t) + \nabla_x b(X_t^{\bar{v}}, \bar{v}_t) p_t + \nabla_x \left(\mathrm{tr}\{K_t^* \sigma\} \right) (X_t^{\bar{v}}, \bar{v}_t) \right), \qquad (1.6)
$$

and taking the expectation in (1.5) we reduce equation (1.4) to the relation

$$
E\left[\int_0^T v_t \left(\nabla_u g(X_t^{\bar{v}}, \bar{v}_t) + \nabla_u b(X_t^{\bar{v}}, \bar{v}_t) p_t + \nabla_u \left(\mathrm{tr}\{K_t^* \sigma\} \right) (X_t^{\bar{v}}, \bar{v}_t) \right) dt \right] = 0.
$$

Since neither the process α nor p depend on $v \in \mathcal{A}$, it follows that

$$
\nabla_u g(X_t^{\bar{v}}, \bar{v}_t) + \nabla_u b(X_t^{\bar{v}}, \bar{v}_t) p_t + \nabla_u \left(\mathrm{tr}\{K_t \sigma\} \right) (X_t^{\bar{v}}, \bar{v}_t) = 0 \quad dt dP\text{-a.e.} \qquad (1.7)
$$

Summarizing our above calculus we see that, with the system of equations formed by (1.1) and (1.5) – (1.7), we have got a necessary condition for the optimality of the cost function $J : \mathcal{A} \to R$ at some $\bar{v} \in \mathcal{A}$,

$$
\left\{
\begin{aligned}
dX_t^{\bar{v}} \;=\;& b(X_t^{\bar{v}}, \bar{v}_t) dt + \sigma(X_t^{\bar{v}}, \bar{v}_t) dW_t, \\
dp_t \;=\;& - \left(\nabla_x g(X_t^{\bar{v}}, \bar{v}_t) + \nabla_x b(X_t^{\bar{v}}, \bar{v}_t) p_t + \nabla_x \left(\mathrm{tr}\{K_t \sigma\} \right) (X_t^{\bar{v}}, \bar{v}_t) \right) dt \\
& + K_t dW_t, \\
X_0^{\bar{v}} \;=\;& x, \quad p_T = -\nabla_x h(X_T^{\bar{v}}), \\
0 \;=\;& \nabla_u g(X_t^{\bar{v}}, \bar{v}_t) + \nabla_u b(X_t^{\bar{v}}, \bar{v}_t) p_t + \nabla_u \left(\mathrm{tr}\{K_t \sigma\} \right) (X_t^{\bar{v}}, \bar{v}_t), \\
& \hspace{6cm} dt dP\text{-a.e.}
\end{aligned}
\right.
$$

The second equation in the above system is an equation in p, K and with terminal condition; while this terminal condition can depend on the whole trajectory of the Brownian motion, the processes p, K are assumed to be adapted to the filtration $\mathbf{F}$ generated by the underlying Brownian motion. That is why this equation cannot be interpreted as a time inversed forward SDE but should be interpreted as backward SDE which will be discussed in the next section.

2. BACKWARD SDE: EXISTENCE AND UNIQUENESS

As in Section 1, let $(\Omega, \mathcal{F}, \mathbf{F}, P)$ be a filtered probability space associated to an underlying k-dimensional Brownian motion W. We recall that $T > 0$ denotes a finite time horizon and $L^2(\mathbf{F}, [0, T]; R^m)$ the space of all $\mathbf{F}$-measurable R^m-valued processes X : $\Omega \times [0, T] \to R^m$ such that $E[\int_0^T |X_t|^2 dt] < +\infty$. Moreover, $S^2(\mathbf{F}, [0, T]; R^m)$ denotes the subspace of continuous semimartingales Y which satisfy $E[\sup_{t \in [0,T]} |Y_t|^2] < +\infty$. If $m = 1$ we will simply write $S^2(\mathbf{F}, [0, T])$.

We are given a final condition $\xi \in L^2(\Omega, \mathcal{F}_T, P; R^m)$ and a random function

$$f : \Omega \times [0, T] \times R^m \times R^{m \times k} \to R^m$$

which is such that, for some $\gamma \in L^2(\mathbf{F}, [0, T], R_+)$ and some real constants $\mu, K > 0$:

H.i) $f(., ., y, z)$ is $\mathbf{F}$-progressively measurable, for all (y, z);
H.ii) $|f(t, y, z)| \leq \gamma_t + K(|y| + |z|)$, for all (t, y, z), P-a.e.;
H.iii) $|f(t, y, z) - f(t, y, z')| \leq K|z - z'|$, for all t, y, z, z', P-a.e.;
H.iv) $(y - y')(f(t, y, z) - f(t, y', z)) \leq \mu|y - y'|^2$, for all t, y, y', z, P-a.e.;
H.v) $y \longrightarrow f(t, y, z)$ is continuous, for all (t, z), P-a.e.

A pair of processes $(Y, Z) \in S^2(\mathbf{F}, [0, T]; R^m) \times L^2(\mathbf{F}, [0, T], R^{m \times k})$ is said to be a solution of BSDE(ξ, f) if

$$Y_t = \xi + \int_t^T f(s, Y_s, Z_s) ds - \int_t^T Z_s dW_s, \quad t \in [0, T].$$

The random function f is called "driver" of the BSDE. We observe that the progressive measurability together with the continuity of Y implies, in particular, that Y_0 is deterministic.

Theorem 2.1 (cf. [37]) *Under the assumptions H.i) – H.v), BSDE(ξ, f) has a unique solution $(Y, Z) \in S^2(\mathbf{F}, [0, T]; R^m) \times L^2(\mathbf{F}, [0, T], R^{m \times k})$.*

For our purposes it will be sufficient to consider and to prove a special case of Theorem 2.1:

Proposition 2.2 (cf. [36]) *Assume H.i) – H.iii) and*
   H.iv') $|f(t, y, z) - f(t, y', z)| \leq K|y - y'|$, *for all t, y, y', z, P-a.e.*
Then BSDE(ξ, f) has a unique solution

$$(Y, Z) \in S^2(\mathbf{F}, [0, T]; R^m) \times L^2(\mathbf{F}, [0, T]; R^{m \times k}).$$

Proof. Let $\mathcal{B}^2 = L^2(\mathbf{F}, [0, T]; R^m) \times L^2(\mathbf{F}, [0, T]; R^{m \times k})$. Given $(U, V) \in \mathcal{B}^2$, we put $\varphi(U, V) = (Y, Z)$, where

$$Y_t = E\left[\xi + \int_0^T f(s, U_s, V_s) ds \Big| \mathcal{F}_t\right] - \int_0^t f(s, U_s, V_s) ds, \quad t \in [0, T],$$

and $Z \in L^2(\mathbf{F}, [0, T]; R^{m \times k})$ is defined by Itô's martingale representation theorem applied to

$$\xi + \int_0^T f(s, U_s, V_s) ds \in L^2(\Omega, \mathcal{F}_T, P; R^m).$$

We observe that the so defined (Y, Z) belongs to

$$S^2(\mathbf{F}, [0, T]; R^m) \times L^2(\mathbf{F}, [0, T]; R^{m \times d}) \, (\subset \mathcal{B}^2)$$

and satisfies the relation

$$Y_t = \xi + \int_t^T f(s, U_s, V_s)ds - \int_t^T Z_s dW_s, \quad t \in [0, T].$$

Consequently, (Y, Z) is a fixed point of $\varphi : \mathcal{B}^2 \longrightarrow \mathcal{B}^2$ if and only if it solves the equation BSDE(ξ, f). Let us now define a norm on $\mathcal{B}^2$ under which φ is a contraction. For this let $(U^i, V^i) \in \mathcal{B}^2$, $i = 1, 2$, and set

$$
\begin{array}{rcl}
(Y^i, Z^i) & = & \varphi(U^i, V^i), \\
(\overline{U}, \overline{V}) & = & (U^1 - U^2, V^1 - V^2), \\
(\overline{Y}, \overline{Z}) & = & (Y^1 - Y^2, Z^1 - Z^2).
\end{array}
$$

From Itô's formula we get that, for any real γ,

$$
\begin{aligned}
& e^{\gamma t} E[|Y_t|^2] + \gamma E\left[\int_t^T e^{\gamma s} |Y_s|^2 ds\right] + E\left[\int_t^T e^{\gamma s} |Z_s|^2 ds\right] \\
& \leq \ 2KE\left[\int_t^T e^{\gamma s} |\overline{Y}_s|(|\overline{U}_s| + |\overline{V}_s|)ds\right] \\
& \leq \ 4K^2 E\left[\int_t^T e^{\gamma s} |\overline{Y}_s|^2 ds\right] + \frac{1}{2}E\left[\int_t^T e^{\gamma s}(|\overline{U}_s|^2 + |\overline{V}_s|^2)ds\right].
\end{aligned}
$$

Hence, with the choice $\gamma = 4K^2 + 1$,

$$E\left[\int_0^T e^{\gamma s}(|\overline{Y}_s|^2 + |\overline{Z}_s|^2)ds\right] \leq \frac{1}{2}E\left[\int_0^T e^{\gamma s}(|\overline{U}_s|^2 + |\overline{V}_s|^2)ds\right].$$

This shows that $\varphi : \mathcal{B}^2 \longrightarrow \mathcal{B}^2$ is a contraction on $\mathcal{B}^2$ under the norm

$$\|(Y, Z)\| = \left(E\left[\int_0^T e^{(1+4K^2)s}(|Y_s|^2 + |Z_s|^2)ds\right]\right)^{1/2}, \quad (Y, Z) \in \mathcal{B}^2.$$

Consequently, there is a unique fixed point $(Y, Z) \in \mathcal{B}^2$ of φ and this fixed point $(Y, Z) = \varphi(Y, Z) \in S^2(\mathbf{F}, [0, T]; R^m) \times L^2(\mathbf{F}, [0, T]; R^{m \times k}) \, (\subset \mathcal{B}^2)$ is the unique solution of BSDE(ξ, f). This completes the proof. $\qquad\square$

We still give some estimates for the solutions of BSDEs and a comparison result, cf. [37].

Lemma 2.3 *For any fixed reals $\mu, K > 0$ there exists a real constant $c = c(T, \mu, K)$ such that,*

(i) *for every* $\xi \in L^2(\Omega, \mathcal{F}_T, P; R^m)$ *and every driver* f *with* H.i) – H.v), *it holds: If* $(Y, Z) \in \mathcal{S}^2(\mathbf{F}, [0, T]; R^m) \times L^2(\mathbf{F}, [0, T]; R^{m \times k})$ *is a solution of the* BSDE(ξ, f) *then*

$$E\left[\sup_{t \in [0,T]} |Y_t|^2 + \int_0^T |Z_t|^2 dt\right] \leq cE\left[|\xi|^2 + \int_0^T |f(t,0,0)|^2 dt\right];$$

(ii) *for all* $\xi^1, \xi^2 \in L^2(\Omega, \mathcal{F}_T, P; R^m)$ *and all drivers* f^1, f^2 *with* H.i) – H.v), *one has: If* $(Y^i, Z^i) \in \mathcal{S}^2(\mathbf{F}, [0, T]; R^m) \times L^2(\mathbf{F}, [0, T]; R^{m \times k})$ *is a solution of the equation* BSDE(ξ^i, f^i), $i = 1, 2$, *then*

$$E\left[\sup_{t \in [0,T]} |Y_t^1 - Y_t^2|^2 + \int_0^T |Z_t^1 - Z_t^2|^2 dt\right]$$
$$\leq cE\left[|\xi^1 - \xi^2|^2 + \int_0^T |f^1(s, Y_s^1, Z_s^1) - f^2(s, Y_s^1, Z_s^1)|^2 ds\right].$$

Proof. (i) Let $(Y, Z) \in \mathcal{S}^2(\mathbf{F}, [0, T]; R^m) \times L^2(\mathbf{F}, [0, T], R^{m \times k})$ be a solution of BSDE(ξ, f). Then, from Itô's formula we have

$$|Y_t|^2 + \int_t^T |Z_s|^2 ds \tag{2.1}$$
$$= |\xi|^2 + 2\int_t^T Y_s f(s, Y_s, Z_s) ds - 2\int_t^T Y_s Z_s dW_s$$
$$\leq |\xi|^2 + 2\int_t^T \{|Y_s||f(s,0,0)| + \mu|Y_s|^2 + K|Y_s||Z_s|\} ds - 2\int_t^T Y_s Z_s dW_s$$
$$\leq \left(|\xi|^2 + \int_0^T |f(s,0,0)|^2 ds\right) + (1 + 2\mu + 2K^2)\int_t^T |Y_s|^2 ds$$
$$+ \frac{1}{2}\int_t^T |Z_s|^2 ds - 2\int_t^T Y_s Z_s dW_s.$$

Note that from the Burkholder–Davis–Gundy inequality

$$E\left[\sup_{t \in [0,T]} \left|\int_0^t Y_s Z_s dW_s\right|\right] \leq 2\sqrt{2} E\left[\left(\int_0^T |Y_s Z_s|^2 ds\right)^{1/2}\right] \tag{2.2}$$
$$\leq 2\sqrt{2}\left(E\left[\sup_{s \in [0,T]} |Y_s|^2\right]\right)^{1/2} \left(E\left[\int_0^T |Z_s|^2 ds\right]\right)^{1/2} < +\infty.$$

Consequently, $\int_0^T Y_s Z_s dW_s$ is the increment of a martingale. Thus, taking the expectation in (2.1) and applying Gronwall's inequality we obtain

$$\sup_{t \in [0,T]} E\left[|Y_t|^2\right] + E\left[\int_0^T |Z_s|^2 ds\right] \tag{2.3}$$
$$\leq 3\exp\{(1 + 2\mu + 2K^2)T\} E\left[|\xi|^2 + \int_0^T |f(s,0,0)|^2 ds\right].$$

Finally, the relations (2.1), (2.2) and (2.3) allow to estimate $E[\sup_{t\in[0,T]}|Y_t|^2]$.

(ii) For $i = 1, 2$, let $(Y^i, Z^i) \in S^2(\mathbf{F}, [0, T]; R^m) \times L^2(\mathbf{F}, [0, T]; R^{m\times k})$ be a solution of BSDE(ξ^i, f^i). Then $(\overline{Y}, \overline{Z}) = (Y^1 - Y^2, Z^1 - Z^2)$ solves BSDE($\overline{\xi}, \overline{f}$) with terminal condition $\overline{\xi} = \xi^1 - \xi^2 \in L^2(\Omega, \mathcal{F}_T, P; R^m)$ and driver

$$\overline{f}(s, y, z) = f^1(s, Y_s^1, Z_s^1) - f^2(s, Y_s^1 - y, Z_s^1 - z), \quad (s, y, z) \in [0, T] \times R^m \times R^{m\times k}.$$

We remark that the driver $\overline{f}$ satisfies H.i) – H.v). This allows to apply statement (i), from which we obtain the desired result. □

We continue with the above setting, now restricting ourselves to the case $m = 1$ and prove the comparison theorem, cf. [37], [18].

Theorem 2.4 *Let $\xi^1, \xi^2 \in L^2(\Omega, \mathcal{F}_T, P)$ and f^1, f^2 be two drivers satisfying H.i) – H.v). By $(Y^i, Z^i) \in S^2(\mathbf{F}, [0, T]) \times L^2(\mathbf{F}, [0, T]; R^k)$ we denote the solution of BSDE(ξ^i, f^i), $i = 1, 2$, and we suppose that*

$$\xi^1 \geq \xi^2, \quad P\text{-a.e.},$$
$$f^1(t, Y_t^j, Z_t^j) \geq f^2(t, Y_t^j, Z_t^j), \quad dtdP\text{-a.e.},$$

for some $j \in \{1, 2\}$. Then

(i) $Y_t^1 \geq Y_t^2, \quad t \in [0, T], P\text{-a.e.}$
(ii) *If moreover $Y_0^1 = Y_0^2$, then $Y_t^1 = Y_t^2, \quad t \in [0, T], P\text{-a.e.}$*

Proof. Without loss of generality for the method of the proof we suppose that the above relation between f^1 and f^2 is satisfied for $j = 1$. We set

$$\overline{\xi} = \xi^1 - \xi^2, \quad \overline{Y} = Y^1 - Y^2,$$
$$\gamma_t = f^1(t, Y_t^1, Z_t^1) - f^2(t, Y_t^1, Z_t^1),$$
$$\overline{Z} = (\overline{Z}^1, \ldots, \overline{Z}^k) = Z^1 - Z^2 = (Z^{1,1} - Z^{2,1}, \ldots, Z^{1,k} - Z^{2,k}).$$

Moreover, we shall define the processes

$$\alpha_t = \overline{Y}_t^\oplus \left(f^2(t, Y_t^1, Z_t^1) - f^2(t, Y_t^2, Z_t^1)\right),$$
$$\beta_t^i = \overline{Z}_t^{i\oplus} \big(f(t, Y_t^2, (Z_t^{2,1}, \ldots, Z_t^{2,i-1}, Z_t^{1,i}, \ldots, Z_t^{1,k}))$$
$$- f(t, Y_t^2, (Z_t^{2,1}, \ldots, Z_t^{2,i}, Z_t^{1,i+1}, \ldots, Z_t^{1,k}))\big),$$

for $t \in [0, T]$, $1 \leq i \leq k$. We recall that, for $a \in R$, $a^\oplus = a^{-1}$ if $a \neq 0$, and $a^\oplus = 0$ otherwise. We observe that the processes α and $\beta = (\beta^1, \ldots, \beta^k)$ are progressively measurable, $\alpha_t \leq \mu$ and $|\beta^i| \leq K$, $1 \leq i \leq k$. Obviously, $(\overline{Y}, \overline{Z})$ solves the BSDE

$$\overline{Y}_t = \overline{\xi} + \int_t^T (\gamma_s + \alpha_s \overline{Y}_s + \beta_s \overline{Z}_s) \, ds - \int_t^T \overline{Z}_s \, dW_s, \quad t \in [0, T],$$

For $0 \leq s \leq t \leq T$, let

$$\mathcal{E}_{s,t} = \exp\left\{\int_s^t \beta_r dW_r + \int_s^t \left(\alpha_r - \frac{1}{2}|\beta_r|^2\right) dr\right\}.$$

One checks easily that, for all $0 \leq s \leq t \leq T$,

$$Y_s = \mathcal{E}_{s,t}\overline{Y}_t + \int_s^t \mathcal{E}_{s,r}\gamma_r dr - \int_s^t \mathcal{E}_{s,r}(\overline{Y}_r\beta_r + \overline{Z}_r)dW_r,$$

and hence,

$$\overline{Y}_s = E\left[\mathcal{E}_{s,t}\overline{Y}_t + \int_s^t \mathcal{E}_{s,r}\gamma_r dr \Big| \mathcal{F}_s\right].$$

We first set $t = T$. Then, from the nonnegativity of $\overline{\xi}$ and γ we get $Y_s^1 - Y_s^2 = \overline{Y}_s \geq 0$, $s \in [0, T]$, P-a.e. If moreover $\overline{Y}_0 = Y_0^1 - Y_0^2 = 0$, then we choose $s = 0 \leq t \leq T$ and from $\mathcal{E}_{0,t} > 0$ and $\mathcal{E}_{0,r}\gamma_r \geq 0$, $r \in [0, t]$, we can conclude that $Y_t^1 - Y_t^2 = \overline{Y}_t = 0$, P-a.e. The proof is complete now.   □

3.   BACKWARD SDE AND SEMILINEAR PARABOLIC PDE

In this section we put the BSDEs in a Markovian framework in order to associate them to semilinear parabolic PDEs. Let $\sigma : R^{n\times k} \longrightarrow R^n$ and $\beta : R^n \longrightarrow R^n$ two Lipschitz continuous functions with a Lipschitz constant $K > 0$. For any $(t, x) \in [0, T] \times R^n$, let $X^t(x) = (X_s^t(x))_{s\in[t,T]} \in \mathcal{S}^2(\mathbf{F}, [t, T]; R^n)$ be the unique solution of the SDE

$$X_s^t(x) = x + \int_t^s \sigma(X_r^t(x))dW_r + \int_t^s \beta(X_r^t(x))dr, \quad s \in [t, T]. \tag{3.1}$$

To this SDE we associate the BSDE

$$Y_s^t(x) = h(X_T^t(x)) + \int_s^T f(r, X_r^t(x), Y_r^t(x), Z_r^t(x))dr - \int_s^T Z_r^t(x)dW_r \tag{3.2}$$

for $s \in [t, T]$. Here $h : R^n \to R$ and $f : [0, T] \times R^n \times R \times R^k \longrightarrow R$ are continuous functions such that for some $K, p > 0$,
   G.i) $|h(x)| \leq K(1 + |x|^p)$,
   G.ii) $|f(t, x, 0, 0)| \leq K(1 + |x|^p)$,
   G.iii) $|f(t, x, y, z) - f(t, x, y', z')| \leq K(|y - y'| + |z - z'|)$,
for all t, x, y, y', z, z'.

Lemma 3.1 *Assume G.i) – G.iii). Then, for each $(t, x) \in [0, T] \times R^n$, BSDE (3.2) has a unique solution*

$$(Y^t(x), Z^t(x)) \in \mathcal{S}^2(\mathbf{F}, [t, T]) \times L^2(\mathbf{F}, [t, T]; R^k).$$

Moreover, the following properties are true:
   (i) *For each $(t, x) \in [0, T) \times R^n$, there exists a version of $X^t(x)$ such that $(t, s, x) \to X_{s\vee t}^t(x)$ is continuous and $(s, x) \to X_{s\vee t}^t(x)$ is locally Hölder-$C^{\alpha, 2\alpha}$, for some $\alpha \in (0, 1/2)$ and for such a version, it holds that*

$$X_s^t(x) = X_s^r\left(X_r^t(x)\right); \quad 0 \leq t \leq r \leq s \leq T, x \in R^n;$$

(ii) *for any $q \geq 2$, there exists a real $M_q > 0$, such that for $t \in [0, T]$ and $x, x' \in R^n$,*

$$\left. \begin{aligned} E\left[\sup_{r \in [t,s]} |X_r^t(x) - x|^q \right] &\leq M_q(s-t)^{q/2}(1 + |x|^q), \\ E\left[\sup_{r \in [t,s]} |(X_r^t(x) - X_r^t(x')) - (x - x')|^q \right] &\leq M_q(s-t)^{q/2}|x - x'|^q; \end{aligned} \right\} \quad (3.3)$$

(iii) *for any $0 \leq t \leq r \leq T$ and $x \in R^n$, we have*

$$Y_s^t(x) = Y_s^r(X_r^t(x)), \quad Z_s^t(x) = Z_s^r(X_r^t(x)), \quad ds\text{-}a.e. \text{ in } [r, T], \ P\text{-}a.e.;$$

(iv) *for any $q \geq 2$, there exists a real $C_{p,q}(T) > 0$ such that for all $t, t' \in [0, T]$ and $x, x' \in R^n$, it holds that*

$$\left. \begin{aligned} &E\left[\sup_{r \in [t,T]} |Y_r^t(x)|^q + \left(\int_t^T |Z_s^t(x)|^2 ds \right)^{q/2} \right] \leq C_{p,q}(T)(1 + |x|^{pq}), \\ &E\left[\sup_{r \in [t \lor t',T]} |Y_r^t(x) - Y_r^{t'}(x')|^q + \left(\int_{t \lor t'}^T |Z_s^t(x) - Z_s^{t'}(x')|^2 ds \right)^{q/2} \right] \\ &\leq C_{p,q}(T) E\left[|h(X_T^t(x)) - h(X_T^{t'}(x'))|^q \right. \\ &\quad + \left. \left(\int_{t \lor t'}^T |f(s, X_s^t(x), Y_s^t(x), Z_s^t(x)) - f(s, X_s^{t'}(x'), Y_s^t(x), Z_s^t(x))|^2 \right)^{q/2} \right]. \end{aligned} \right\} \quad (3.4)$$

Proof. The results given in (i) and (ii) are classic. We remark that from (3.3) and G.i) – G.iii) it follows that, for all $(t, x) \in [0, T] \times R^n$, the random variable $h(X_T^t(x)) \in L^2(\Omega, \mathcal{F}_T, P)$ and $f(s, X_s^t(x), y, z)$ satisfies the assumptions H.i) – H.iii) and H.iv'). Hence, BSDE (3.2) possesses a unique solution $(Y^t(x), Z^t(x))$. Then, from the uniqueness of the solution of BSDE (3.2) and the Markov property (i) of $X = (X_s^t(x))$ we obtain (iii). For $q = 2$, estimate (iv) is a direct consequence of the estimates for BSDE$(h(X_T^t(x)), f(., X.^t(x), ., .))$ given by Lemma 2.3 and those in (ii). For $q > 2$ the reader is referred to [37].

Lemma 3.2 *Suppose G.i) – G.iii). Then the function*

$$u(t, x) = Y_t^t(x), \quad (t, x) \in [0, T] \times R^n$$

is deterministic and continuous with at most polynomial growth,

$$|u(t, x)| \leq C_{p,2}(T)(1 + |x|^p), \quad (t, x) \in [0, T] \times R^n.$$

If, moreover, h is bounded and $|f(t, x, 0, 0)| \leq C$, $(t, x) \in [0, T] \times R^n$, for some real $C > 0$, then u is even bounded.

Proof. Clearly, for each $t \leq s \leq T$, $Y_s^t(x)$ is

$$\mathcal{F}_{t,s} = \sigma\{W_r - W_t, \ r \in [t, s]\} \lor \mathcal{N} \text{ -measurable,}$$

where $\mathcal{N}$ denotes the class of P-null sets of $\mathcal{F}$. Hence $u(t, x) = Y_t^t(x)$ is P-a.e. constant, i.e., deterministic.

Let us now prove the continuity of u. For $(t, x), (t', x') \in [0, T] \times R^n$ with $t \leq t'$, we have

$$u(t, x) - u(t', x')$$
$$= h(X_T^t(x)) - h(X_T^{t'}(x')) + \int_t^{t'} f(s, X_s^t(x), Y_s^t(x), Z_s^t(x)) ds$$
$$+ \int_{t'}^T \left(f(s, X_s^t(x), Y_s^t(x), Z_s^t(x)) - f(s, X_s^{t'}(x'), Y_s^{t'}(x'), Z_s^{t'}(x')) \right) ds$$
$$- \int_t^T Z_s^t(x) dW_s + \int_{t'}^T Z_s^{t'}(x') dW_s,$$

and, consequently,

$$u(t, x) - u(t', x')$$
$$= E\left[h(X_T^t(x)) - h(X_T^{t'}(x')) \right] + E\left[\int_t^{t'} f(s, X_s^t(x), Y_s^t(x), Z_s^t(x)) ds \right]$$
$$+ E\left[\int_{t'}^T \left(f(s, X_s^t(x), Y_s^{t'}(X_{t'}^t(x)), Z_s^{t'}(X_{t'}^t(x))) \right) \right.$$
$$\left. - f(s, X_s^{t'}(x'), Y_s^{t'}(x'), Z_s^{t'}(x')) \right) ds \Big].$$

Then, from Lemma 3.1, for some constant $c_2 > 0$,

$$|u(t, x) - u(t', x')|^2$$
$$\leq c_2 \Big\{ E\left[|h(X_T^t(x)) - h(X_T^{t'}(x'))|^2 \right] + |t' - t|(1 + |x|^{2p} + |x'|^{2p})$$
$$+ E\left[\int_{t \lor t'}^T |f(s, X_s^t(x), Y_s^t(x), Z_s^t(x)) - f(s, X_s^{t'}(x'), Y_s^t(x), Z_s^t(x))|^2 ds \right] \Big\}.$$

We note that, for $t \geq t'$, we get the same estimate. We fix now any point (t, x) and take the limit in the above estimate, as (t', x') tends to (t, x). From Lemma 3.1 we see that the right-hand side of the above estimate converges to zero. This proves the continuity of u in (t, x) and hence in $[0, T] \times R^n$. Finally, the estimate of the growth of u follows immediately from the Lemmata 2.3 and 3.1. The proof is complete now. $\square$

We now want to identify the function $u(t, x) = Y_t^t(x)$ as solution of some PDE. By

$$\mathcal{A} = \frac{1}{2} \sum_{i,j=1}^n \sum_{\ell=1}^k \sigma^{i,\ell} = (x)\sigma^{j,\ell}(x)\partial_{x_i x_j}^2 + \sum_{i=1}^n \beta^i(x)\partial_{x_i}$$

we denote the infinitesimal generator of the Markov process $X^t(x)$ and we consider the following (backward) semilinear PDE:

$$\partial_t u(t, x) + \mathcal{A}u(t, x) + f(t, x, u(t, x), (\nabla u\sigma)(t, x)) = 0, \quad (t, x) \in (0, T) \times R^n,$$
$$u(T, x) = h(x), \quad x \in R^n. \tag{3.5}$$

Pardoux and Peng [37] have shown that, under some additional smoothness assumptions on the coefficients σ, β, h and f, the function $u(t,x) = Y_t^t(x)$ belongs to $C^{1,2}([0,T] \times R^n)$ and is the unique classic solution to the semilinear parabolic PDE (3.5). Later, in order to avoid restrictive additional assumptions on the coefficients, Pardoux and Peng have considered the PDE in viscosity sense.

Definition 3.3 (cf., e.g., [10]) Let $v : [0,T] \times R^n \longrightarrow R$ be a continuous function.

(i) The function v is called a viscosity subsolution (resp. supersolution) of PDE (3.5) if $v(T,x) \leq$ (resp. $\geq$)$h(x)$, for all $x \in R^n$; and if moreover for any $(t,x) \in (0,T) \times R^n$ and any test function $\varphi \in C^{1,2}((0,T) \times R^n)$ satisfying

$$v(t',x') - \varphi(t',x') \leq \text{ (resp. } \geq) \, 0 = v(t,x) - \varphi(t,x),$$

for all (t',x') in a neighbourhood of (t,x), it holds that

$$\mathcal{A}\varphi(t,x) + f(t,x,v(t,x),(\nabla v\sigma)(t,x)) \geq \text{ (resp. } \leq) \, -\partial_t\varphi(t,x). \qquad (3.6)$$

(ii) The function v is called a viscosity solution of PDE (3.5) if it is both a viscosity subsolution and a supersolution.

We now give the main result of this section.

Theorem 3.4 *Under the above assumptions G.i) – G.iii), the function*

$$u(t,x) = Y_t^t(x), \, (t,x) \in [0,T] \times R^n,$$

is a viscosity solution of PDE (3.5). If, moreover, for all $R > 0$ there is some increasing function $m_R : R_+ \to R_+$ with $\lim_{s\to 0+} m_R(s) = 0$ such that
G.iv) $|f(t,x,y,z) - f(t',x',y,z)| \leq m_R\left((|t-t'| + |x-x'|)(1 + |z|)\right),$
for all $t,t' \in [0,T], |x|^2, |x'|^2, |y|^2 \leq R, z \in R^k$, is satisfied then u is unique in the class of viscosity solutions of (3.5) which are of at most polynomial growth.

We first prove that u is a viscosity solution of (3.5). The proof of uniqueness which is more delicate and uses purely analytic arguments will be treated separately.

Proof of the existence. Let us prove that u is a viscosity subsolution of (3.5); the proof that u is also a viscosity supersolution is analogous. We recall that u is continuous and that $u(T,x) = Y_T^T(x) = h(x)$, $x \in R^n$. Let $(t,x) \in (0,T) \times R^n$ and $\varphi \in C^{1,2}((0,T) \times R^n)$ be a test function such that

$$u(t',x') - \varphi(t',x') \leq 0 = v(t,x) - \varphi(t,x),$$

for all (t',x') in a neighbourhood of (t,x). We suppose that

$$\partial_t\varphi(t,x) + \mathcal{A}\varphi(t,x) + f(t,x,u(t,x),(\nabla\varphi\sigma)(t,x)) < -\delta,$$

for some $\delta > 0$ and we will construct a contradiction. To this end we fix $\varepsilon \in (0,T-t)$ such that, for all $t' \in [t,t+\varepsilon], |x-x'| \leq \varepsilon,$
i) $u(t',x') \leq \varphi(t',x'),$

ii) $(\partial_t \varphi + \mathcal{A}\varphi)(t', x') + f(t', x', u(t', x'), (\nabla \varphi \sigma)(t', x')) \le -\delta$;
and we introduce the stopping time

$$\tau = \inf\{s \ge t \mid |X_s^t(x) - x| \ge \varepsilon \text{ or } s \ge t + \varepsilon\}.$$

Clearly, $t < \tau \le t + \varepsilon$, P-a.e. Let $\zeta \in L^2(\mathbf{F}, [t, T]; R^k)$ be such that

$$\tau = E[\tau] + \int_t^T \zeta_s dW_s, \quad \text{P-a.e.}$$

We notice that $\zeta_s = 0$, dsdP-a.e. on $[\tau, T]$. We now define, for $s \in [t, T]$,

$$\begin{cases} (\overline{Y}_s, \overline{Z}_s) &= (Y_{s\wedge\tau}^t(x), I_{\{s \le \tau\}} Z_s^t(x)), \\ (\widehat{Y}_s, \widehat{Z}_s) &= (\varphi(s, X_{s\wedge\tau}^t(x)) \\ & \quad - \delta E[\tau - \tau \wedge s | \mathcal{F}_s], I_{\{s \le \tau\}} (\nabla \varphi \sigma)(s, X_s^t(x)) - \delta\zeta_s). \end{cases}$$

Obviously, $(\overline{Y}, \overline{Z})$, $(\widehat{Y}, \widehat{Z})$ are in $\mathcal{S}^2(\mathbf{F}, [0, T]) \times L^2(\mathbf{F}, [0, T]; R^k)$ and solve the BSDE

$$\overline{Y}_s = u(\tau, X_\tau^t(x)) + \int_s^T I_{\{r \le \tau\}} f(r, X_r^t(x), u(r, X_r^t(x)), \overline{Z}_r) dr - \int_s^T \overline{Z}_r dW_r,$$

and

$$\widehat{Y}_s = \varphi(\tau, X_\tau^t(x)) - \int_s^T I_{\{r \le \tau\}} \{(\partial_t \varphi + \mathcal{A}\varphi)(r, X_r^t(x)) + \delta\} dr - \int_s^T \widehat{Z}_r dW_r,$$

for $s \in [t, T]$, respectively. From i), ii) and the choice of the stopping time τ, with the help of the comparison theorem (Theorem 2.4) we conclude that

$$\overline{Y}_s \le \widehat{Y}_s, \ s \in [t, T], \ \text{P-a.e.}$$

In particular, we have

$$u(t, x) = Y_t^t(x) = \overline{Y}_t \le \widehat{Y}_t = \varphi(t, x) - \delta E[\tau - t] < \varphi(t, x).$$

This contradicts our assumption. Consequently,

$$\partial_t \varphi(t, x) + \mathcal{A}\varphi(t, x) + f(t, x, u(t, x), (\nabla \varphi \sigma)(t, x)) \ge 0.$$

This proves that u is a viscosity subsolution of (3.5). Since a symmetric argument shows that u is also a supersolution, u must be a viscosity solution of PDE (3.5).

Proof of the uniqueness. We recall that if the functions h and $f(., ., 0, 0)$ are bounded then also the viscosity solution u of (3.5) is bounded. We will restrict ourselves to the proof of the uniqueness inside the class of bounded continuous functions in order to better prepare some arguments which will be translated later into the framework of the stochastic viscosity solution. The reader interested in the proof of the general case is referred to [37]. In order to adapt our approach to that one finds in the literature, we make a time inversion $v(t, x) = u(T - t, x)$, $g(t, x, y, z) = f(T - t, x, y, z)$. This transforms (3.5) into the PDE with initial condition

$$\begin{aligned} -\partial_t v(t, x) + \mathcal{A}v(t, x) + g(t, x, v(t, x), (\nabla v \sigma)(t, x)) &= 0, \ (t, x) \in (0, T) \times R^n, \\ v(0, x) &= h(x), \ x \in R^n, \end{aligned} \tag{3.7}$$

and the definition of the viscosity solution of (3.7) takes the following form.

Definition 3.5 (i) The function $v \in C([0, T] \times R^n)$ is a viscosity subsolution (resp., supersolution) of PDE (3.7) if $v(0, x) \leq$ (resp. $\geq$) $h(x)$, for all $x \in R^n$; and if moreover for any $(t, x) \in (0, T) \times R^n$ and any test function $\varphi \in C^{1,2}((0, T) \times R^n)$ satisfying

$$v(t', x') - \varphi(t', x') \leq \text{(resp. } \geq) \ 0 = v(t, x) - \varphi(t, x),$$

for all (t', x') in a neighbourhood of (t, x), it holds that

$$\mathcal{A}\varphi(t, x) + g(t, x, v(t, x), (\nabla v \sigma)(t, x)) \geq \text{(resp. } \leq) \ \partial_t \varphi(t, x). \tag{3.8}$$

(ii) The function v is called a viscosity solution of PDE (3.7) if it is both a viscosity subsolution and a supersolution.

The uniqueness of the viscosity solution of (3.7), and hence also of (3.5), in the class of bounded continuous functions is an immediate consequence of the following comparison theorem:

Theorem 3.6 *Assume G.i) – G.iv). Suppose that $v_1 \in C_b([0, T] \times R^n)$ is a viscosity subsolution of (3.7) and $v_2 \in C_b([0, T] \times R^n)$ is a viscosity supersolution of (3.7). Then it holds that*
$$v_1(t, x) \leq v_2(t, x), \quad \text{for all } (t, x) \in [0, T] \times R^n.$$

Proof. We split the proof in several steps. To begin with, we define

$$\zeta = \sup_{(t,x) \in [0,T] \times R^n} (v_1(t, x) - v_2(t, x)),$$

and we suppose that $\zeta > 0$, on which a contradiction will be drawn.

1st Step For $\delta_1, \delta_2 > 0$, we introduce the test function

$$\varphi_{\delta_1, \delta_2}(t, x, t', x') = \frac{1}{2\delta_1} \left(|t - t'|^2 + |x - x'|^2\right) + \frac{\delta_2}{2}\left(|x|^2 + |x'|^2 + \frac{1}{T - t} + \frac{1}{T - t'}\right),$$

and the function

$$\psi(t, x, t', x') = (v_1(t, x) - v_2(t', x')) - \varphi_{\delta_1, \delta_2}(t, x, t', x'),$$

for $(t, x), (t', x') \in [0, T) \times R^n$. We observe that, since v_1 and v_2 are bounded, the function $\psi_{\delta_1, \delta_2}$ attains its maximum at some point $(\hat{t}, \hat{x}, \hat{t}', \hat{x}') \in [0, T) \times R^n \times [0, T) \times R^n$. For simplicity of the notation we don't indicate explicitly the dependence of this point on δ_1, δ_2. A simple straightforward calculus proves the following properties of $\psi_{\delta_1, \delta_2}$, cf. [10]:

Lemma 3.7 *Under the above assumptions that v_1, v_2 are continuous and bounded by some real $C > 0$, we have: For every $\varepsilon > 0$ there exists some $\delta_2(\varepsilon) > 0$ such that*
(i) *for all $\delta_1 > 0$, $\delta_2 \in (0, \delta_2(\varepsilon)]$,*

$$\psi_{\delta_1, \delta_2}(\hat{t}, \hat{x}, \hat{t}', \hat{x}') \geq \zeta - \varepsilon,$$

$$|\widehat{t} - \widehat{t'}|^2 + |\widehat{x} - \widehat{x'}|^2 \le 4C\delta_1, \quad |\widehat{x}|^2 + |\widehat{x'}|^2 \le \frac{4C}{\delta_2} \quad \text{and} \quad \widehat{t}, \widehat{t'} \le T - \frac{\delta_2}{4C};$$

(ii) *for all $\delta_2 \in (0, \delta_2(\varepsilon)]$ there is a $\delta_1(\delta_2) > 0$ such that, for all $\delta_1 \in (0, \delta_1(\delta_2)]$,*

$$\widehat{t} > 0, \ \widehat{t'} > 0;$$

(iii) *for all $\delta_2 \in (0, \delta_2(\varepsilon)]$,*

$$\liminf_{\delta_1 \downarrow 0} \frac{1}{2\delta_1} \left(|\widehat{t} - \widehat{t'}|^2 + |\widehat{x} - \widehat{x'}|^2 \right) \le 0, \quad \liminf_{\delta_1 \downarrow 0} \frac{\delta_2}{2} \left(|\widehat{x}|^2 + |\widehat{x'}|^2 \right) \le \varepsilon.$$

2nd Step   In this step we shall apply Theorem 3.2 of Crandall, Ishii and Lions [10]. For this we first recall some notions in [10]. Let $\mathcal{S}^n$ denote the class of symmetric matrices of $R^{n \times n}$. For given $v \in C([0, T] \times R^n)$ and $(\bar{t}, \bar{x}) \in (0, T) \times R^n$, a triplet $(a, p, X) \in R \times R^n \times \mathcal{S}^n$ is called a *parabolic superjet* of v at $(\bar{t}, \bar{x})$ if, for any (t, x) in a neighbourhood of $(\bar{t}, \bar{x})$, it holds that

$$\begin{aligned} v(t, x) \ &\le \ v(\bar{t}, \bar{x}) + a(t - \bar{t}) + p(x - \bar{x}) + \frac{1}{2} \left(X(x - \bar{x}) \right) (x - \bar{x}) \qquad (3.9) \\ &+ o(|t - \bar{t}|) + o(|x - \bar{x}|^2). \end{aligned}$$

We denote the set of all parabolic superjets of v at $(\bar{t}, \bar{x})$ by $\mathcal{P}^{1,2,+} v(\bar{t}, \bar{x})$. The set of *parabolic subjets*, defined in a similar way by reversing the direction of the inequality in (3.9), is denoted by $\mathcal{P}^{1,2,-} v(\bar{t}, \bar{x})$. Further, if the function v does not depend on t, then the set of the second order superjets of v at $\bar{x}$, denoted by $\mathcal{P}^{2,+} v(\bar{x})$, are those couples (p, X) for which (3.9) holds with the obvious modifications. The set $\mathcal{P}^{2,-} v(\bar{x})$ is defined likewise. Finally, the closure of $\mathcal{P}^{1,2,+} v(\bar{t}, \bar{x})$ is denoted by $\overline{\mathcal{P}}^{1,2,+} v(\bar{t}, \bar{x})$. The closure of the sets of other "jets" are defined similarly.

Lemma 3.8 (cf. Proposition V.4.1 [20]) *Let $v \in C([0, T] \times R^n)$. Then it holds that*

   (i)   *v is a viscosity subsolution of (3.7) if and only if $v(0, x) \le h(x)$, for all $x \in R^n$ and for all $(\bar{t}, \bar{x}) \in (0, T) \times R^n$,*

$$\frac{1}{2} \mathrm{tr}(\sigma\sigma^*(\bar{x}) X) + \beta(\bar{x}) p + g(\bar{t}, \bar{x}, v(\bar{t}, \bar{x}), p\sigma(\bar{x})) \ge a, \quad \forall (a, p, X) \in \overline{\mathcal{P}}^{1,2,+} v(\bar{t}, \bar{x});$$

   (ii)   *v is a viscosity supersolution of (3.7) if and only if $v(0, x) \ge h(x)$, for all $x \in R^n$, and for all $(\bar{t}, \bar{x}) \in (0, T) \times R^n$,*

$$\frac{1}{2} \mathrm{tr}(\sigma\sigma^*(\bar{x}) X) + \beta(\bar{x}) p + g(\bar{t}, \bar{x}, v(\bar{t}, \bar{x}), p\sigma(\bar{x})) \le a, \quad \forall (a, p, X) \in \overline{\mathcal{P}}^{1,2,-} v(\bar{t}, \bar{x}).$$

The following statement is an adaptation of Theorem 3.2 [10] to our framework.

Theorem 3.9 (cf. [10]) *Assume G.i) – G.iii). Let $\delta_1, \delta_2 > 0$ and suppose that the maximizer $(\widehat{t}, \widehat{x}, \widehat{t'}, \widehat{x'})$ of $\psi_{\delta_1, \delta_2}$ is in $(0, T) \times R^n \times (0, T) \times R^n$. Then, for any $\gamma > 0$, there exist two matrices $\mathcal{X}, \mathcal{Y} \in \mathcal{S}^n$ such that*

   (i) $\left((\partial_t \varphi, \nabla_x \varphi)(\widehat{t}, \widehat{x}, \widehat{t'}, \widehat{x'}), \mathcal{X} \right)$

$$\Big(= \Big(\frac{\hat{t} - \check{t}}{\delta_1} + \frac{\delta_2}{2(T - \hat{t})^2}, \frac{1}{\delta_1}(\hat{x} - \hat{x}') + \delta_2 \hat{x}, \mathcal{X} \Big) \Big) \in \overline{\mathcal{P}}^{1,2,+} v(\bar{t}, \bar{x});$$

(ii) $\ (-(\partial_{t'}\varphi, \nabla_{x'}\varphi)(\hat{t}, \hat{x}, \check{t}, \hat{x}'), \mathcal{Y})$

$$\Big(= \Big(\frac{\hat{t} - \check{t}}{\delta_1} - \frac{\delta_2}{2(T - \check{t})^2}, \frac{1}{\delta_1}(\hat{x} - \hat{x}') - \delta_2 \hat{x}', \mathcal{Y} \Big) \Big) \in \overline{\mathcal{P}}^{1,2,-} v(\bar{t}, \bar{x});$$

(iii) *with*

$$B = \begin{pmatrix} D^2_{xx}\varphi & D^2_{xx'}\varphi \\ D^2_{xx'}\varphi & D^2_{x'x'}\varphi \end{pmatrix} (\hat{t}, \hat{x}, \check{t}, \hat{x}') \ \Big(= \frac{1}{\delta_1} \begin{pmatrix} I & -I \\ -I & I \end{pmatrix} + \delta_2 \begin{pmatrix} I & 0 \\ 0 & I \end{pmatrix} \Big),$$

it holds

$$\begin{pmatrix} \mathcal{X} & 0 \\ 0 & -\mathcal{Y} \end{pmatrix} \le B + \gamma \Big\{ B^2 + \begin{pmatrix} I & 0 \\ 0 & I \end{pmatrix} \Big\}. \tag{3.10}$$

3rd Step We are now ready to prove the uniqueness by drawing a contradiction from the hypothesis $\zeta > 0$. To begin with, we first claim that under the assumption G.iii) we can assume without loss of generality that there exists a constant $\mu > 0$ such that for all $r > 0$ it holds that

$$g(t, x, y + r, Z) - g(t, x, y, z) \le -\mu r, \quad \forall (t, x, y, z). \tag{3.11}$$

In fact, thanks to G.iii), the function

$$\hat{g}(t, x, y, z) = e^{-(\mu + K)t} g(t, x, e^{(\mu + K)t} y, e^{(\mu + K)t} z) - (\mu + K)y$$

satisfies (3.11). Furthermore, one shows easily that v is a viscosity subsolution (resp., supersolution) of (3.7) if and only if $\hat{v} = (e^{-(\mu+K)t}v(t, x))$ is a viscosity subsolution (resp., supersolution) to the PDE:

$$-\partial_t \hat{v}(t, x) + A\hat{v}(t, x) + \hat{g}(t, x, \hat{v}(t, x), (\nabla \hat{v}\sigma)(t, x)) = 0, \quad (t, x) \in (0, T) \times R^n,$$
$$\hat{v}(0, x) = h(x), \ x \in R^n.$$

Let now $\varepsilon \in (0, \zeta/4]$, $\delta_2 \in (0, \delta_2(\varepsilon)]$ and $\delta_1 \in (0, \delta_1(\delta_2)]$. Then, with the choice of $\gamma = \min(\frac{\delta_1}{2(1 + \delta_1 \delta_2)}, \frac{\delta_2}{(1 + \delta_2^2)})$ in Theorem 3.9, we have

$$\frac{1}{2}\mathrm{tr}(\sigma\sigma^*(\hat{x})\mathcal{X}) - \frac{1}{2}\mathrm{tr}(\sigma\sigma^*(\hat{x}')\mathcal{Y}) \le \frac{1}{\delta_1}|\sigma(\hat{x}) - \sigma(\hat{x}')|^2 + \delta_2 (|\sigma(\hat{x})|^2 + |\sigma(\hat{x}')|^2). \tag{3.12}$$

Moreover, since v_1 is a viscosity subsolution and v_2 a supersolution of (3.7), we can conclude from the Lemmata 3.7 and 3.8 and from Theorem 3.9 that

$$\left. \begin{aligned} \frac{1}{\delta_1}(\hat{t} - \check{t}) + \delta_2 \frac{1}{(T - \hat{t})^2} &\le \frac{1}{2}\mathrm{tr}(\sigma\sigma^*(\hat{x})\mathcal{X}) + \beta(\hat{x})(\frac{1}{\delta_1}(\hat{x} - \hat{x}') + \delta_2 \hat{x}) \\ &\quad + f(\hat{t}, \hat{x}, v_1(\hat{t}, \hat{x}), (\frac{1}{\delta_1}(\hat{x} - \hat{x}') + \delta_2 \hat{x})\sigma(\hat{x})); \\ \frac{1}{\delta_1}(\hat{t} - \check{t}) - \delta_2 \frac{1}{(T - \check{t})^2} &\ge \frac{1}{2}\mathrm{tr}(\sigma\sigma^*(\hat{x}')\mathcal{Y}) + \beta(\hat{x}')(\frac{1}{\delta_1}(\hat{x} - \hat{x}') - \delta_2 \hat{x}') \\ &\quad + f(\check{t}, \hat{x}', v_2(\check{t}, \hat{x}'), (\frac{1}{\delta_1}(\hat{x} - \hat{x}') - \delta_2 \hat{x}')\sigma(\hat{x}')). \end{aligned} \right\} \tag{3.13}$$

Consequently, from Lemma 3.7 (i), (3.11), (3.12), (3.13) and assumption G.iv), for $\varepsilon \in (0, \zeta/4]$ and for all $\delta_2 \in (0, \delta_2(\varepsilon)], \delta_1 \in (0, \delta_1(\delta_2)]$,

$$
\begin{aligned}
\mu\zeta/2 \;\leq\; & \mu\psi_{\delta_1,\delta_2}(\hat{t},\hat{x},\hat{t}',\hat{x}') \leq \mu(v_1(\hat{t},\hat{x}) - v_2(\hat{t}',\hat{x}')) \\
\leq\; & -\mu\Big\{ f\big(\hat{t},\hat{x}, v_1(\hat{t},\hat{x}), (\tfrac{1}{\delta_1}(\hat{x}-\hat{x}') + \delta_2\hat{x})\sigma(\hat{x})\big) \\
& \quad - f\big(\hat{t},\hat{x}, v_2(\hat{t}',\hat{x}'), (\tfrac{1}{\delta_1}(\hat{x}-\hat{x}') + \delta_2\hat{x})\sigma(\hat{x})\big)\Big\} \\
\leq\; & \mu\Big(\big\{\tfrac{1}{2}\mathrm{tr}(\sigma\sigma^*(\hat{x})\mathcal{X}) + \beta(\hat{x})[\tfrac{1}{\delta_1}(\hat{x}-\hat{x}') + \delta_2\hat{x}] \\
& \quad - [\tfrac{1}{\delta_1}(\hat{t}-\hat{t}') + \delta_2\tfrac{1}{(T-\hat{t})^2}]\big\} \\
& \quad - \big\{\tfrac{1}{2}\mathrm{tr}(\sigma\sigma^*(\hat{x}')\mathcal{Y}) + \beta(\hat{x}')[\tfrac{1}{\delta_1}(\hat{x}-\hat{x}') - \delta_2\hat{x}'] \\
& \quad - [\tfrac{1}{\delta_1}(\hat{t}-\hat{t}') - \delta_2\tfrac{1}{(T-\hat{t}')^2}]\big\}\Big) \\
& - \mu\Big\{ f\big(\hat{t}',\hat{x}', v_2(\hat{t}',\hat{x}'), (\tfrac{1}{\delta_1}(\hat{x}-\hat{x}') - \delta_2\hat{x}')\sigma(\hat{x}')\big) \\
& \quad - f\big(\hat{t},\hat{x}, v_2(\hat{t}',\hat{x}'), (\tfrac{1}{\delta_1}(\hat{x}-\hat{x}') + \delta_2\hat{x})\sigma(\hat{x})\big)\Big\} \\
\leq\; & \mu\Big\{\tfrac{1}{\delta_1}|\sigma(\hat{x}) - \sigma(\hat{x}')|^2 + \delta_2(|\sigma(\hat{x})|^2 + |\sigma(\hat{x}')|^2) \\
& \quad + \tfrac{1}{\delta_1}|\beta(\hat{x}) - \beta(\hat{x}')||\hat{x}-\hat{x}'| + \delta_2(|\beta(\hat{x})||\hat{x}| + |\beta(\hat{x}')||\hat{x}'|) \\
& \quad + K\delta_2(|\hat{x}||\sigma(\hat{x})| + |\hat{x}'||\sigma(\hat{x}')|) \\
& \quad + m_R\big((|\hat{t}-\hat{t}'| + |\hat{x}-\hat{x}'|)(1 + [\tfrac{1}{\delta_1}|\hat{x}-\hat{x}'| + \delta_2|\hat{x}|]|\sigma(\hat{x})|)\big)\Big\},
\end{aligned}
$$

where $R > 0$ is a sufficiently great constant which depends only on δ_2, the bound $C > 0$ of v_1 and v_2, cf. Lemma 3.7 (i). Therefore, for some constant $C > 0$,

$$
\begin{aligned}
\mu\zeta/2 \;\leq\; & C\Big(\frac{|\hat{x}-\hat{x}'|^2}{\delta_1} + \delta_2(1 + |\hat{x}|^2 + |\hat{x}'|^2)\Big) \\
& + \mu m_R\Big(C\big(\tfrac{1}{\delta_1}[|\hat{t}-\hat{t}'|^2 + |\hat{x}-\hat{x}'|^2] + \delta_1\big)(1 + |\hat{x}| + \delta_2|\hat{x}|^2)\Big).
\end{aligned}
$$

Hence, taking the "lim inf" as $\delta_1 \to 0$, with the help of Lemma 3.7 (iii) we obtain that

$$
\mu\zeta/2 \leq C(2\varepsilon + \delta_2) + \mu m_R(0+) = C(2\varepsilon + \delta_2).
$$

Finally, first $\delta_2 \to 0$ and then $\varepsilon \to 0$ give $\mu\zeta/2 \leq 0$. This contradicts our hypothesis that $\zeta > 0$. Therefore, this hypothesis must be wrong. The proof is now complete. □

4. STOCHASTIC SEMILINEAR PARABOLIC PDE AND STOCHASTIC VISCOSITY SOLUTIONS

The objective of the following sections is to generalize the relation between BSDEs and semilinear parabolic PDEs, which we have reviewed before, to that between backward doubly stochastic differential equations (for short, BDSDEs) and stochastic semilinear parabolic PDEs.

Let $(\Omega, \mathcal{F}, P)$ be a complete probability space on which a d-dimensional Brownian motion $B = (B_t)_{t \geq 0}$ is defined which will drive our PDE. Let $\mathbf{F}^B := \{\mathcal{F}_t^B\}_{t \geq 0}$ denote the natural filtration generated by B, augmented by the P-null sets of $\mathcal{F}$; and let $\mathcal{F}^B = \mathcal{F}_T^B$. By $\mathcal{M}_{0,T}^B$ we will denote all the $\mathbf{F}^B$-stopping times τ such that $0 \leq \tau \leq T$, P-a.s., where $T > 0$ is again some fixed time horizon. We shall also introduce some spaces with which we will have often to do in the sequel: Let E denote a generic Euclidean space; in case other Euclidean spaces are needed we shall label them as $E_1, E_2, \cdots$, etc. We denote

$$C^{k,l}(\mathcal{G}, [0, T] \times E; E_1),$$

for any sub-σ-field $\mathcal{G} \subset \mathcal{F}_T^B$, to be the space of all $\mathcal{G} \otimes \mathcal{B}([0, T] \times E)$-measurable random variables $\varphi : \Omega \to C^{k,l}([0, T] \times E; E_1)$;

$$C^{k,l}(\mathbf{F}^B, [0, T] \times E; E_1)$$

to be the space of all random fields $\varphi \in C^{k,l}(\mathcal{F}_T^B, [0, T] \times E; E_1)$ such that, for fixed $x \in E$, the mapping $(\omega, t) \to \varphi(\omega, t, x)$ is $\mathbf{F}^B$-progressively measurable. The following simplification of notations will be frequently used throughout:

$$\begin{aligned}
C^{k,l}(\mathcal{G}, [0, T] \times E) &= C^{k,l}(\mathcal{G}, [0, T] \times E; R); \\
C^{k,l}(\mathbf{F}^B, [0, T] \times E) &= C^{k,l}(\mathbf{F}^B, [0, T] \times E; R); \\
C(\mathcal{G}, [0, T] \times E) &= C^{0,0}(\mathcal{G}, [0, T] \times E; R); \\
C(\mathbf{F}^B, [0, T] \times E) &= C^{0,0}(\mathbf{F}^B, [0, T] \times E; R).
\end{aligned}$$

Throughout the following sections of we shall make use of the following *Standing Assumptions:*

A.i)   The functions $\sigma : R^n \longrightarrow R^{n \times k}$ and $\beta : R^n \longrightarrow R^n$ are Lipschitz continuous, with a common Lipschitz constant $K > 0$.

A.ii) $f : \Omega \times [0, T] \times R^n \times R \times R^k \longrightarrow R$ is a continuous random field such that for fixed (x, y, p), $f(., ., x, y, p\sigma(x))$ is $\mathbf{F}^B$-progressively measurable; and there exists some $K > 0$, such that for P-a.e. $\omega \in \Omega$,

$$|f(\omega, 0, 0, 0, 0)| \leq K,$$
$$|f(\omega, t, x, y, z) - f(\omega, t', x', y', z')| \leq K\big(|t - t'| + |x - x'| + |y - y'| + |z - z'|\big);$$
$$\forall (t, x, y, z), (t', x', y', z') \in [0, T] \times R^n \times R \times R^k.$$

A.iii) $h : R^n \longrightarrow R$ is a continuous function such that, for some constants K, $p > 0$,

$$|h(x)| \leq K(1 + |x|^p), \qquad x \in R^n.$$

Finally, we are also given a function

A.iv) $g \in C_b^{0,2,3}([0, T] \times R^n \times R; R^d)$.

We consider the following semilinear stochastic PDE (SPDE):

$$\left.\begin{aligned}
du(t, x) &= \{Au(t, x) + f(t, x, u(t, x), \nabla_x u(t, x)\sigma(x))\}dt \\
&\qquad\qquad + g(t, x, u(t, x)) \circ dB_t, \\
u(0, x) &= h(x), \ x \in R^d, \quad (t, x) \in (0, T) \times R^n,
\end{aligned}\right\} \tag{4.1}$$

where $\mathcal{A}$ is the second order differential operator introduced in Section 3 and the stochastic integral is interpreted in Stratonovich's sense. For SPDE (4.1) we will also write SPDE(f, g) in the sequel.

Lions and Souganidis have introduced in [30], [31] the "stochastic viscosity solution" for the first time. One of their key ideas is to use the so-called stochastic characteristics to "remove" the stochastic integral from the SPDE, so that the stochastic viscosity solution can be studied ω-wisely. We present the approach of [7], [8] that, although technically different, has the same spirit.

We will relate the definition of a stochastic viscosity solution to the following stochastic flow $\eta \in C(\mathbf{F}^B, [0, T] \times R^n \times R)$, defined as the unique solution of the SDE in Stratonovich's sense:

$$\eta(t, x, y) = y + \int_0^t g(s, x, \eta(s, x, y)) \circ dB_s, \quad t \geq 0. \tag{4.2}$$

We recall that $g \in C_b^{0,2,3}([0, T] \times R^n \times R; R^d)$. Hence, applying Itô's formula to $g(t, x, \eta(t, x, y))$ and using the definition of Stratonovich's integral, one easily shows that the Stratonovich SDE (4.2) is equivalent to the following Itô SDE (with parameter):

$$\eta(t, x, y) = y + \frac{1}{2} \int_0^t g\partial_y g(s, x, \eta(s, x, y))ds + \int_0^t g(s, x, \eta(s, x, y))dB_s. \tag{4.3}$$

We note that under the assumption A.iv), for fixed x the random field $\eta(., x, .)$ is continuously differentiable with respect to y; and the mapping $y \rightarrow \eta(\omega, t, x, y)$ defines a diffeomorphism for all (t, x), P-a.s. We denote the y-inverse of $\eta(t, x, y)$ by $\mathcal{E}(t, x, y)$ in the sequel.

We now introduce the notion of *stochastic viscosity solution* for SPDE(f, g) (4.1).

Definition 4.1 A random field $u \in C(\mathbf{F}^B, [0, T] \times R^n)$ is called a stochastic viscosity subsolution (resp., supersolution) of SPDE(f, g), if $u(0, x) \leq$ (resp., $\geq$) $h(x)$, for all $x \in R^n$; and if for any $\tau \in \mathcal{M}_{0,T}^B$, $\xi \in L^0(\mathcal{F}_\tau^B; R^n)$ and any random test field $\varphi \in C^{1,2}(\mathcal{F}_\tau^B, [0, T] \times R^n)$ satisfying

$$u(t, x) - \eta(t, x, \varphi(t, x)) \leq \text{ (resp. } \geq) 0 = u(\tau, \xi) - \eta(\tau, \xi, \varphi(\tau, \xi)),$$

for all (t, x) in a neighbourhood of (τ, ξ), P-a.e. on the set $\{0 < \tau < T\}$, it holds that

$$\mathcal{A}\psi(\tau, \xi) + f(\tau, \xi, \psi(\tau, \xi), \nabla_x \psi(\tau, \xi)\sigma(\xi)) \tag{4.4}$$
$$\geq \text{ (resp. } \leq) \partial_y \eta(\tau, \xi, \varphi(\tau, \xi))\partial_t \varphi(\tau, \xi),$$

P-a.e. on $\{0 < \tau < T\}$, where $\psi(t, x) = \eta(t, x, \varphi(t, x))$. A random field $u \in C(\mathbf{F}^B, [0, T] \times R^n)$ is called a stochastic viscosity solution of SPDE(f, g), if it is both a stochastic viscosity subsolution and a supersolution.

We observe that if in SPDE(f, g) the function $g \equiv 0$, the flow η becomes $\eta(t, x, y) = y$, $\forall(t, x, y)$, and $\psi(t, x) = \varphi(t, x)$. Thus the definition of a stochastic viscosity solution becomes the same as Definition 3.3 for the deterministic viscosity solution, for each fixed $\omega \in \{0 < \tau < T\}$, modulo the $\mathcal{F}_\tau^B$-measurability requirement on the test function φ. The following notion of a random viscosity solution will be a bridge linking the stochastic viscosity solution and its deterministic counterpart.

Definition 4.2 A random field $u \in C(\mathbf{F}^B, [0, T] \times R^n)$ is called an ω-*wise* viscosity solution (resp., subsolution, supersolution) of SPDE$(f, 0)$ if for P-a.e. $\omega \in \Omega$, $u(\omega, ., .)$ is a (deterministic) viscosity solution (resp., subsolution, supersolution) of the PDE (3.7) with $g(., ., ., .) = f(\omega, ., ., ., .)$.

Remark 4.3 (i) From the Definitions 4.1 and 4.2 one easily sees that every ω-wise viscosity (sub-, super-) solution of SPDE$(f, 0)$ is also a stochastic viscosity (sub-, super-) solution to this equation. In Section 7 we will show that under some additional assumption the notions of the ω-wise and the stochastic viscosity (sub-, super-) solutions of SPDE$(f, 0)$ coincide.

(ii) We should direct the reader's attention also to the fact that in Definition 4.1 the random field φ is only required to belong to $C^{1,2}(\mathcal{F}^B_{\tau}, [0, T] \times R^n)$ instead of $C^{1,2}(\mathbf{F}^B, [0, T] \times R^n)$. In other words, φ is not necessarily progressively measurable! However, if we restrict ourselves to the subclass of test fields $\varphi \in C^{1,2}(\mathbf{F}^B, [0, T] \times R^n)$ then a straightforward computation using the Itô–Ventzell formula shows that the random field $\psi(t, x) = \eta(t, x, \varphi(t, x))$ satisfies

$$d\psi(t, x) = \partial_y \eta(t, x, \varphi(t, x)) \partial_t \varphi(t, x) dt + g(t, x, \psi(t, x)) \circ dB_t, \quad t \in [0, T]. \quad (4.5)$$

Since $g(\tau, \xi, \psi(\tau, \xi)) = g(\tau, \xi, u(\tau, \xi))$ by definition, it seems natural to compare

$$\mathcal{A}\psi(\tau, \xi) + f(\tau, \xi, \psi(\tau, \xi), \nabla_x \psi(\tau, \xi) \sigma(\xi))$$

with $\partial_y \eta(t, x, \varphi(t, x)) \partial_t \varphi(t, x)$ to characterize a viscosity solution of SPDE(f, g), as we did in (4.4). The reason why we can't restrict ourselves to such progressively measurable test fields $\varphi \in C^{1,2}(\mathbf{F}^B, [0, T] \times R^n)$ is that this class of progressively measurable test fields seems to be too poor for the uniqueness of the stochastic viscosity solution.

We shall now introduce the both tools which will allow to prove existence of a stochastic viscosity solution to SPDE(f, g): the Doss–Sussmann transformation which "removes" the martingale term from SPDE(f, g) and converts it to an SPDE$(\tilde{f}, 0)$, where $\tilde{f}$ is some (progressively measurable) random field and, in the next section, the BDSDEs which will play the same role here as the BSDEs for the PDEs.

To begin with the Doss–Sussmann transformation, let us remark that under the assumption A.iv) on g, the stochastic flows $\eta(t, x, y)$ and $\mathcal{E}(t, x, y) = \eta(t, x, .)^{-1}(y)$ are in $C^{0,2,2}(\mathbf{F}^B, [0, T] \times R^n \times R)$. For any random field $\psi : \Omega \times [0, T] \times R^n \longrightarrow R$, we consider the transformation introduced in the Definition 4.1:

$$\varphi(t, x) = \mathcal{E}(t, x, \psi(t, x)), \quad (t, x) \in [0, T] \times R^n, \quad (4.6)$$

or the inverse transformation, $\psi(t, x) = \eta(t, x, \varphi(t, x))$, $(t, x) \in [0, T] \times R^n$. It is easy to check that the function $\psi \in C^{0,p}(\mathcal{F}^B_T, [0, T] \times R^n)$ if and only if the function $\varphi \in C^{0,p} \mathcal{F}^B_T, [0, T] \times R^n)$, for $p = 0, 1, 2$.

Let us define the new random field

$$\tilde{f}(t, x, y, z) = \frac{1}{\partial_y \eta(t, x, y)} \Big(f\big(t, x, \eta(t, x, y), \nabla_x \eta(t, x, y) \sigma(x) + \partial_y \eta(t, x, y) z\big)$$

$$+ \mathcal{A}_x \eta(t, x, y) + \left(\sigma^*(x) \nabla_{xy} \eta(t, x, y)\right) z + \frac{1}{2} \partial^2_{yy} \eta(t, x, y) |z|^2\right),$$

for $(t, x, y, z) \in [0, T] \times R^n \times R \times R^k$. Here $\mathcal{A}_x$ is the same as the operator $\mathcal{A}$, with the emphasis that all the partial derivatives are with respect to x. We shall, however, often omit the subscript x in the sequel when there is no danger of confusion. It is clear that $\widetilde{f}$ is in the space $C(\mathbf{F}^B, [0, T] \times R^n \times R \times R^k)$; and a simple straightforward computation shows that

$$\partial_y \mathcal{E}(t, x, \psi(t, x)) \{ \mathcal{A}\psi(t, x) + f(t, x, \psi(t, x), \nabla_x \psi(t, x))\sigma(x) \} \qquad (4.7)$$
$$= \mathcal{A}\varphi(t, x) + \widetilde{f}(t, x, \varphi(t, x), \nabla_x \varphi(t, x)\sigma(x)),$$

for all $(t, x) \in (0, T) \times R^n$.

The SPDE($\widetilde{f}, 0$) is called the robust form of the SPDE(f, g); and the following result tells us why.

Proposition 4.4 *Assume A.i) – A.iv). A random field u is a stochastic viscosity subsolution (resp., supersolution) to SPDE(f, g) if and only if*

$$v = (v(t, x) = \mathcal{E}(t, x, u(t, x)))$$

is a stochastic viscosity subsolution (resp., supersolution) to SPDE($\widetilde{f}, 0$). Consequently, u is a stochastic viscosity solution of SPDE(f, g) if and only if

$$v = (v(t, x) = \mathcal{E}(t, x, u(t, x)))$$

is a stochastic viscosity solution to SPDE($\widetilde{f}, 0$).

Proof. We only need to prove that if u is a stochastic viscosity subsolution to the equation SPDE(f, g), then v is a stochastic viscosity subsolution to SPDE($\widetilde{f}, 0$). The remaining parts of the proposition can be proved with analogous arguments.

Given a stochastic viscosity subsolution $u \in C(\mathbf{F}^B, [0, T] \times R^n)$ of SPDE(f, g) we set

$$v = (v(t, x) = \mathcal{E}(t, x, u(t, x))).$$

Clearly, $v \in C(\mathbf{F}^B, [0, T] \times R^n)$. In order to show that v is a stochastic viscosity subsolution of SPDE($\widetilde{f}, 0$), we choose arbitrarily $\tau \in \mathcal{M}^B_{0,T}$, $\xi \in L^2(\mathcal{F}^B_\tau, R^n)$ and let $\varphi \in C^{1,2}(\mathcal{F}^B_\tau, [0, T] \times R^n)$ be such that

$$v(\omega, t, x) - \varphi(\omega, t, x) \leq 0 = v(\omega, \tau(\omega), \xi(\omega)) - \varphi(\omega, \tau(\omega), \xi(\omega)),$$

for all (t, x) of some neighbourhood $\mathcal{O}(\omega, \tau(\omega), \xi(\omega))$ of $(\tau(\omega), \xi(\omega))$ and for P-a.e $\omega \in \{0 < \tau < T\}$.

Now let $\psi(t, x) = \eta(t, x, \varphi(t, x))$, $(t, x) \in [0, T] \times R^n$. Since the mapping $y \mapsto \eta(t, x, y)$ is strictly increasing, we get

$$u(t, x) - \psi(t, x) = \eta(t, x, v(t, x)) - \eta(t, x, \varphi(t, x)) \qquad (4.8)$$
$$\leq 0$$

$$= \eta(\tau, \xi, v(\tau, \xi)) - \eta(\tau, \xi, \varphi(\tau, \xi))$$
$$= u(\tau, \xi) - \psi(\tau, \xi),$$

for all $(t, x) \in \mathcal{O}(\tau, \xi)$, P-a.e on $\{0 < \tau < T\}$. Further, since u is a viscosity subsolution of SPDE(f, g), we can conclude that, P-a.e. on $\{0 < \tau < T\}$,

$$\mathcal{A}\psi(\tau, \xi) + f(\tau, \xi, \psi(\tau, \xi), \nabla_x \psi(\tau, \xi)\sigma(\xi)) \geq \partial_y \eta(\tau, \xi, \varphi(\tau, \xi))\partial_t \varphi(\tau, \xi). \quad (4.9)$$

We thus deduce from (4.7) that

$$\mathcal{A}\varphi(\tau, \xi) + \widetilde{f}(\tau, \xi, \varphi(\tau, \xi), \nabla_x \varphi(\tau, \xi)\sigma(\xi)) \geq \partial_t \varphi(\tau, \xi), \quad P\text{-a.e. on } \{0 < \tau < T\}.$$

That is, v is a stochastic viscosity subsolution of SPDE$(\widetilde{f}, 0)$. $\qquad\qquad\square$

In the remaining part of this section we prove the stochastic boundedness of the random fields η and $\mathcal{E}$ which will be needed later to characterize the stochastic viscosity solution of SPDE(f, g) and SPDE$(\widetilde{f}, 0)$.

Definition 4.5 A random field $u \in C(\mathbf{F}^B, [0, T] \times R^n)$ is said to be *stochastically bounded* if there exists a positive real-valued random variable $\theta \in L^0(\mathcal{F}_T^B)$ such that, P-a.e., it holds

$$|u(t, x)| \leq \theta \quad \text{for all } (t, x) \in [0, T] \times R^n.$$

To study the stochastic boundedness of random fields η and $\mathcal{E}$, we shall impose an extra condition on the functions $g = (g_1, \ldots, g_d)$ which we shall call *compatibility condition* in the sequel:

A.iv′) g satisfies A.iv); and, for any $\varepsilon > 0$, there exists a function G^ε from the space $C^{1,2,2,2}([0, T] \times R^d \times R^n \times R)$, such that

$$\partial_t G^\varepsilon(t, q, x, y) = \varepsilon; \quad \nabla_q G^\varepsilon = g(t, x, G^\varepsilon(t, q, x, y)); \quad G^\varepsilon(0, 0, x, y) = y.$$

Remark 4.6 We observe that the existence of such a function G^ε is not trivial. A necessary condition is the following "compatibility" among the components of $g = (g^1, \cdots, g^d)$ (suppressing variables):

$$g^i \partial_y g^j = \partial^2_{q_i q_j} G^\varepsilon = \partial^2_{q_j q_i} G^\varepsilon = g^j \partial_y g^i, \qquad \text{for all } 1 \leq i, j \leq d,$$

which is of course not necessarily true in general. However, the assumption A.iv′) is trivially satisfied in the case if $d = 1$ and g is independent of t and satisfies A.iv). Indeed, in such a case we can choose G to be the solution of the ordinary differential equation with parameter x:

$$\partial_q G(q, x, y) = g(x, G(q, x, y)), \quad G(0, x, y) = y,$$

and then let $G^\varepsilon(t, q, x, y) = \varepsilon t + G(q, x, y), \varepsilon > 0$.

We have the following result.

Proposition 4.7 *Assume A.iv'). Let η be the unique solution to SDE (2.3) and $\mathcal{E}(t, x, y)$ = $\eta(t, x, .)^{-1}(y)$. Then there exists a constant $C > 0$, depending only on the bound of g and its partial derivatives, such that for $\zeta = \eta$, $\mathcal{E}$, it holds that*

$$
\left.
\begin{aligned}
|\zeta(t, x, y)| &\leq |y| + C|B_t|, \\
|\nabla_x \zeta(t, x, y)|, |\partial_y \zeta(t, x, y)|, |D_{xx}^2 \zeta(t, x, y)|, \\
|D_{xy}^2 \zeta(t, x, y)|, |D_{yy}^2 \zeta(t, x, y)| \\
&\leq C \exp\{C|B_t|\},
\end{aligned}
\right\} \tag{4.10}
$$

for all (t, x, y), P-a.e. Consequently, the partial derivatives of the random fields η and $\mathcal{E}$ with respect to x and y, up to the second order, are all stochastically bounded, with the required real-valued random variable in Definition 4.5 being $\theta = C \exp\{C|B|_T^*\}$, where C is some generic constant and $|B|_t^* = \sup_{0 \leq s \leq t} |B_s|$.

The proof of the preceding proposition is the result of a straightforward calculus using ordinary differential equation methods. The reader interested in is referred to [7].

5. A BACKWARD DOUBLY SDE (BDSDE)

In this section we recall the notion of a backward doubly SDE (BDSDE for short) introduced by Pardoux and Peng in [38] in 1994. We refer to the fact that the version of BDSDE presented here is in fact a time reversal of that considered in [38]. We nevertheless use the same name because they are of the same nature.

Let $(\Omega', \mathcal{F}', P')$ be a complete probability space on which a k-dimensional Brownian motion W is defined. For a fixed terminal time $T > 0$ we introduce the backward filtration $\mathbf{F}_T^W := (\mathcal{F}_{t,T}^W)_{0 \leq t \leq T}$ generated by W:

$$
\mathcal{F}_{t,T}^W := \sigma\{W_s - W_T, t \leq s \leq T\} \vee \mathcal{N}', \quad t \in [0, T],
$$

where $\mathcal{N}'$ denotes the collection of P'-null sets in $\mathcal{F}$. Next, we define the product space

$$
(\overline{\Omega} = \Omega \times \Omega'; \overline{\mathcal{F}} = \mathcal{F} \otimes \mathcal{F}'; \overline{P} = P \times P')
$$

and endow it with the family of sub-σ-fields $\overline{\mathbf{F}} := \{\overline{\mathcal{F}}_t = \mathcal{F}_t^B \otimes \mathcal{F}_{t,T}^W\}_{0 \leq t \leq T}$. Observe that $\overline{\mathbf{F}}$ is not a filtration! Further, random variables $\xi(\omega)$, $\omega \in \Omega$ and $\eta(\omega')$, $\omega' \in \Omega'$, are interpreted as random variables in $\overline{\Omega}$ by the identification $\xi(\overline{\omega}) = \xi(\omega)$, $\eta(\overline{\omega}) = \eta(\omega')$, $\overline{\omega} := (\omega, \omega')$.

By $L^2(\overline{\mathbf{F}}, [0, T]; R^n)$ let us denote the set of $\overline{\mathcal{F}} \otimes \mathcal{B}([0, T])$-measurable random processes $h = (h_t)_{t \in [0,T]}$ which satisfy

(i) $\overline{E}\left[\int_0^T |h_t|^2 dt\right] < \infty$;

(ii) h_t is $\overline{\mathcal{F}}_t$-measurable, dt-a.e.

We will also have to do with the space

$$
S^2(\overline{\mathbf{F}}, [0, T]; R^m) = \left\{h \in C(\overline{\mathbf{F}}, [0, T]; R^m) \,\middle|\, \overline{E}\left[\sup_{0 \leq t \leq T} |h_t|^2\right] < +\infty\right\}.
$$

For $H \in L^2(\overline{\mathbf{F}}, [0, T]; R^k)$, we denote the Itô "backward" integral of H against W on an interval $[s, t]$ by $\int_s^t H_r \downarrow dW_r$. Recall that, if H is $\mathbf{F}^W$-adapted, then we can "reverse" the time and consider such an integral as a standard Itô integral from t to s. For every $(t, x) \in [0, T] \times R^n$, we now consider the following SDE:

$$X_s^t(x) = x + \int_s^t b(X_r^t(x)) dr + \int_s^t \sigma(X_r^t(x)) \downarrow dW_r, \quad 0 \le s \le t. \tag{5.1}$$

This equation should be interpreted as going from t to 0; $X_0^t(x)$ depends on the whole Brownian path on $[0, t]$ and is to understand as "terminal" value of the solution process X. It is clear that under our standard assumptions on the coefficients b and σ the above SDE has a unique solution $X \in \mathcal{S}^2(\overline{\mathbf{F}}, [0, T]; R^n)$ which has the same law as the time-reversed solution of the forward SDE (3.1).

To the above SDE we associate the following equation: for $(t, x) \in [0, T] \times R^n$,

$$Y_s^t(x) = h(X_0^t(x)) + \int_0^s f(r, X_r^t(x), Y_r^t(x), Z_r^t(x)) dr \tag{5.2}$$

$$+ \int_0^s g(r, X_r^t(x), Y_r^t(x)) \circ dB_r - \int_0^s Z_r^t(x) \downarrow dW_r, \quad 0 \le s \le t,$$

where h satisfies A.iii).

We observe that although SDE (5.2) looks like a "forward" SDE, it is indeed a "backward" one into direction of W because a "terminal" condition in form of an $\mathcal{F}_{0,t}^W$-measurable random variable is given at time $t = 0$: $Y_0^t = u_0(X_0^t(x))$. We also remark that, on the other hand, the equation is a forward one in direction of the second Brownian source B, since the initial condition $u_0(X_0^t(x))$ is independent of B. Equations with two independent Brownian sources W and B, backward with respect to the one and progressive with respect to the other, were studied by Pardoux and Peng [38] in 1994 and have been called by them "backward doubly stochastic differential equation". Let us mention that here, in contrast to the BDSDEs considered by Pardoux and Peng, the integral with respect to dB_s is in the Stratonovich form and not in Itô form. The Stratonovich integral in (5.2) is needed to make this equation compatible to SPDE(f, g) (4.1). On the other hand, it is not possible to consider the integral with respect to W also in Stratonovich sense since, in general, the process $Z^t(x)$ does not have much regularity, much less a semimartingale.

As direct consequence of the paper of Pardoux and Peng [38] translated into our context we have the following result:

Lemma 5.1 *Assume* Ai) – Aiv). *Then, for each* $(t, x) \in [0, T] \times R^n$, *BDSDE* (5.2) *has a unique solution* $(Y^t(x), Z^t(x)) \in \mathcal{S}^2(\overline{\mathbf{F}}, [0, t]) \times L^2(\overline{\mathbf{F}}, [0, t]; R^k)$. *Moreover, let* $X^t(x)$ *denote the unique solution of* (5.1). *Then*

(i) *For each* $t > 0$, *there exists a version of* $X^t(x) = \{X_s^t(x), 0 \le s \le t\}$ *such that* $(\mathfrak{n}, \mathfrak{m}) \mapsto X_s^t(\mathfrak{m})$ *is locally Hölder* $C^{\alpha, 2\alpha}$, *for some* $\alpha \in (0, 1/2)$ *and for such a version, it holds*

$$X_s^t(x) = X_s^r(X_r^t(x)), \quad 0 \le s \le r \le t \le T, \ x \in R^n$$

and

(ii) *for any $q \geq 2$, there exists a real $M_q > 0$ such that for $t \in [0, T]$ and $x, x' \in R^n$*

$$\left. \begin{array}{c} \overline{E}\left[\sup_{s \leq r \leq t} |X_r^t(x) - x|^q\right] \leq M_q(t-s)^{q/2}(1 + |x|^q), \\ \overline{E}\left[\sup_{s \leq r \leq t} |(X_r^t(x) - X_r^t(x')) - (x - x')|^q\right] \leq M_q(t-s)^{q/2}|x - x'|^q; \end{array} \right\} \quad (5.3)$$

(iii) *for any $0 \leq r \leq t \leq T$ and $x \in R^n$, one has*

$$Y_s^t(x) = Y_s^r(X_r^t(x)), \quad Z_s^t(x) = Z_s^r(X_r^t(x)), \quad ds\text{-a.e. in } [0, r], \; \overline{P}\text{-a.s.};$$

(iv) *for any $q \geq 2$, there exists a real $C_{p,q}(T) > 0$ such that, for $(t, x) \in [0, T] \times R^n$, it holds that*

$$\overline{E}\left[\left(\sup_{0 \leq s \leq t} |Y_s^t(x)|^2 + \int_0^t |Z_s^t(x)|^2 ds\right)^{q/2}\right] \leq C_{p,q}(T)(1 + |x|^{pq}). \quad (5.4)$$

While the statements (i) and (ii) of Lemma 5.1 are a direct consequence of Lemma 3.1 and the fact that the solution of SDE (5.1) has the same law as the time-reversed solution of SDE (3.1), the proof of (iii) and (iv) retakes the BSDE arguments developed in the proof of Lemma 3.1 and combines them with SDE arguments. For the detailed proof the reader is referred to [38].

Next, we present a generalized version of the Itô–Ventzell formula that combines the generalized Itô formula of Pardoux and Peng [38] and the Itô–Ventzell formula of Ocone and Pardoux [33] (regarding all of our stochastic integrals as Skorohod integrals). Its proof is based on the Itô formula of [38] and adopts the argument of the proof of the Itô–Ventzell formula given in [33]. This generalized version of the Itô–Ventzell formula will be a main device for the transformation of BDSDEs into BSDEs.

Theorem 5.2 ([7], Theorem 4.2) *Suppose that $F \in C^{0,2}(\overline{F}, [0, T] \times R^\ell)$ is a semimartingale with spatial parameter $x \in R^\ell$:*

$$F(t, x) = F(0, x) + \int_0^t G(s, x)ds + \int_0^t H(s, x)dB_s + \int_0^t K(s, x) \downarrow dW_s, \; t \in [0, T],$$

where $G, H \in C^{0,2}(F^B, [0, T] \times R^\ell)$ and $K \in C^{0,2}(F^W; [0, T] \times R^\ell)$. Moreover, let $\alpha \in C(\overline{F}, [0, T]; R^\ell)$ be a process of the form

$$\alpha_t = \alpha_0 + A_t + \int_0^t \gamma_s dB_s + \int_0^t \delta_s \downarrow W_s, \quad t \in [0, T],$$

where $\gamma, \delta \in L^2(\overline{F}, [0, T]; R^\ell)$ and A is a continuous, $\overline{F}$-adapted R^ℓ-valued process with paths of locally bounded variation. Then, P-a.s., it holds for all $0 \leq t \leq T$

$$\begin{aligned} F(t, \alpha_t) &= F(0, \alpha_0) + \int_0^t G(s, \alpha_s)ds + \int_0^t H(s, \alpha_s)dB_s \\ &+ \int_0^t K(s, \alpha_s) \downarrow dW_s + \int_0^t \nabla_x F(s, \alpha_s)dA_s \end{aligned}$$

$$+ \int_0^t (\nabla_x F(s, \alpha_s)\gamma_s) \, dB_s + \int_0^t (\nabla_x F(s, \alpha_s)\delta_s) \downarrow dW_s$$

$$+ \frac{1}{2} \int_0^t D^2_{xx} F(s, \alpha_s)\gamma_s \gamma_s^* ds - \frac{1}{2} \int_0^t D^2_{xx} F(s, \alpha_s)\delta_s \delta_s^* ds$$

$$+ \int_0^t \nabla_x H(s, \alpha_s)\gamma_s ds - \int_0^t \nabla_x K(s, \alpha_s)\delta_s ds.$$

This generalized Itô–Ventzell formula allows to prove the following comparison theorem for BDSDEs in the same manner as the Itô formula allows to prove the comparison theorem for BSDEs, cf. [37], [18]:

Theorem 5.3 *Assume A.i) – A.iv). For $(t, x) \in [0, T] \times R^n$, let*

$$(Y^{t,i}(x), Z^{t,i}(x)) \in \mathcal{S}^2(\overline{\mathbf{F}}, [0, T] \times R^n) \times L^2(\overline{\mathbf{F}}, [0, T] \times R^n), \ i = 1, 2,$$

be solutions of the BDSDEs:

$$Y_s^{t,i}(x) = \varphi_i\left(X_0^t(x)\right) + \int_0^s f_i(r, X_r^t(x), Y_r^{t,i}(x), Z_r^{t,i}(x))dr \qquad (5.5)$$

$$+ \int_0^s g(r, X_r^t(x), Y_r^{t,i}(x))dB_r - \int_0^s Z_r^{t,i}(x) \downarrow W_r, \quad s \in [0, t],$$

where f_1, f_2 satisfy A.ii); φ_1, φ_2 satisfy A.iii) and g satisfies A.iv). Suppose that
1) $\varphi_1(x) \le \varphi_2(x)$, *for all $x \in R^n$;*
2) *For either $i = 1$ or $i = 2$, it holds that, for all $(t, x) \in [0, T] \times R^d$,*

$$f_1\left(s, X_s^t(x), Y_s^{t,i}(x), Z_s^{t,i}(x)\right) \le f_2\left(s, X_s^t(x), Y_s^{t,i}(x), Z_s^{t,i}(x)\right),$$
$$ds \times dP\text{-a.e. on } [0, t] \times \Omega.$$

Then, one has $Y_s^{t,1}(x) \le Y_s^{t,2}(x)$, for all $0 \le s \le t \le T$, $x \in R^n$, P-a.e.

6. EXISTENCE OF THE STOCHASTIC VISCOSITY SOLUTION

In this section we show the existence of the stochastic viscosity solution to SPDE(f, g) (4.1). The main idea is to apply the "Doss-transformation" to the BDSDE (5.2) in order to let disappear the stochastic integral with respect to B in the resulting BSDE. Thus, in light of Theorem 3.4, we will get then an ω-wise viscosity solution to the SPDE and hence a stochastic viscosity solution, thanks to Remark 4.3 (i) and Proposition 4.4. To substantiate the above idea we define the "Doss-transform" of the solution of BSDE (4.2):

$$\left. \begin{array}{l} U_s^t(x) = \mathcal{E}(s, X_s^t(x), Y_s^t(x)), \ 0 \le s \le t \le T, \ x \in R^n; \\ V_s^t(x) = \partial_y \mathcal{E}(s, X_s^t(x), Y_s^t(x))Z_s^t(x) + \nabla_x \mathcal{E}(s, X_s^t(x), Y_s^t(x))\sigma(X_s^t(x)). \end{array} \right\} \ (6.1)$$

From the estimates of Proposition 4.7 and Lemma 5.1 we deduce that

$$(U^t(x), V^t(x)) \in \mathcal{S}^2(\overline{\mathbf{F}}; [0, T]) \times L^2(\overline{\mathbf{F}}; [0, T]; R^k).$$

We now give the main result of this section.

Theorem 6.1 *For each $(t, x) \in [0, T] \times R^n$, the couple $(U^t(x), V^t(x))$ is the unique solution of the BSDE: for $0 \le s \le t$,*

$$U_s^t(x) = h(X_0^t(x)) + \int_0^s \widetilde{f}(r, X_r^t(x), U_r^t(x), V_r^t(x)) dr - \int_0^s V_r^t(x) \downarrow dW_r. \quad (6.2)$$

Let us note that the existence and uniqueness of BDSDE (6.2) does not follow directly from the standard theory, because of the quadratic growth in the variable z. BSDEs with quadratic growth in z were studied recently by Kobylanski [27]. However, using (6.1) we give a direct proof.

Proof. Obviously, the mapping

$$(X^t(x), Y^t(x), Z^t(x)) \mapsto (X^t(x), U^t(x), V^t(x))$$

is invertible; the inverse transformation is given by:

$$\left. \begin{array}{ll} Y_s^t(x) = & \eta(s, X_s^t(x), U_s^t(x)); \\ Z_s^t(x) = & \partial_y \eta(s, X_s^t(x), U_s^t(x)) V_s^t(x) + \nabla_x \eta(s, X_s^t(x), U_s^t(x)) \sigma(X_s^t(x)). \end{array} \right\} \quad (6.3)$$

Consequently, thanks to (6.1) and (6.3), the uniqueness of (6.2) follows from that of BDSDE (5.2), once existence is proved. Thus it remains only to show that (U, V) is a solution of the BSDE (6.2). To prove this it suffices to apply the generalized Itô–Ventzell theorem (Theorem 5.2) to $\mathcal{E}(s, X_s^t(x), Y_s^t(x))$, $s \in [0, t]$ with $F(t, (x, y), \omega) = \mathcal{E}(s, x, y, \omega)$. The calculus, straightforward but a little laborious, yields the desired result. $\square$

The next statement concerns the stability of the stochastical boundedness property of the solutions to the BDSDE (5.2) under the "Doss-transformation". Such a property will lead to the same type of boundedness of the stochastic viscosity solution, which will be essential for the study of the uniqueness of stochastic viscosity solutions. Recall that Proposition 4.7 gives only the stochastically boundedness of the derivatives of the random fields η and $\mathcal{E}$, but not for η and $\mathcal{E}$ themselves. Thus the stochastic boundedness of the processes U or Y are not clear. We should emphasize that the proof of the following statement borrows some ideas of Kobylanski [27].

Lemma 6.2 *Suppose A.i) – A.iii) and A.iv). Assume also that the functions σ, β in A.i) and h in A.iii) are bounded and that the random function f in A.ii) is such that for some real constant $C > 0$, $|f(t, x, 0, 0)| \le C$, for all (t, x), uniformly on Ω. Then, there exists an $\mathbf{F}^B$-adapted increasing real-valued process Θ, such that for the solution $(X^t(x), Y^t(x), Z^t(x))$ of the SDEs (5.1) and (5.2), it holds that, $\overline{P}$-a.e.,*

$$\left. \begin{array}{rcl} |Y_s^t(x)| & \le & \Theta_s, \\ |\mathcal{E}(s, X_s^t(x), Y_s^t(x))| & \le & \Theta_s, \quad 0 \le s \le t \le T, \, x \in R^n. \end{array} \right\} \quad (6.4)$$

Let us sketch the proof.

Proof. We put

$$f_1(\omega, t, x, y, z) = f(\omega, t, x, y, z) - \frac{1}{2}(g\partial_y g)(t, x, y).$$

By assumption A.ii) and the extra boundedness condition we made on f, there exists a constant $C_1 > 0$, such that

$$f_1(\omega, t, x, y, z) \leq C(1 + |y| + |z|) - \tfrac{1}{2}(g\partial_y g)(t, x, y) =: f_2(\omega, t, x, y, z),$$
$$(\omega, t, x, y, z) \in \Omega \times [0, T] \times R^n \times R \times R^k.$$

Now let $(Y^{t,1}(x), Z^{t,1}(x)) = (Y^t(x), Z^t(x))$ and $(Y^{t,2}(x), Z^{t,2}(x))$ be the unique solution of BDSDE (5.5) with generator f_2 and terminal condition $Y_0^{t,2}(x) = C_0$; C_0 denotes the bound for h. Then, applying the comparison theorem (Theorem 5.3) we see that

$$Y_s^t(x) \leq Y_s^{t,2}(x), \quad \text{for all } s \in [0, t], \ \overline{P}\text{-a.e.}$$

Next, let $(U^{t,2}(x), V^{t,2}(x))$ be the Doss-transformation of $(Y^{t,2}(x), Z^{t,2}(x))$ evaluated via (6.1). From Theorem 6.1 we see that $(U^{t,2}(x), V^{t,2}(x))$ solves the BSDE:

$$U_s^{t,2}(x) = C_0 + \int_0^s \widetilde{f}_2(r, X_r^t(x), U_r^{t,2}(x), V_r^{t,2}(x))dr \qquad (6.5)$$
$$- \int_0^s V_r^{t,2}(x) \downarrow dW_r, \quad s \in [0, t],$$

where

$$\widetilde{f}_2(\omega, t, x, y, z) = (\partial_y \eta(t, x, y, \omega))^{-1} \{ C(1 + |\eta(t, x, y, \omega)| + |\nabla_x \eta(t, x, y, \omega)\sigma(x) + \partial_y \eta(t, x, y, \omega)z|) + \mathcal{A}_x \eta(t, x, y, \omega) + (\sigma^*(x)D_{xy}^2 \eta(t, x, y, \omega))z + \tfrac{1}{2}D_{yy}^2 \eta(t, x, y, \omega)|z|^2 \}.$$

By Proposition 4.7 there exists a constant $L > 0$ such that

$$|\widetilde{f}_2(t, \omega, x, y, z)| \leq L \exp\{L|B_t(\omega)|\}(1 + |y| + |z|^2) \qquad (6.6)$$
$$\leq a(t, \omega)(1 + |y|) + A(\omega)|z|^2, \ \forall (t, x, y, z), \ P\text{-a.e.}$$

where

$$a(t, \omega) := L \exp\{L|B_t(\omega)|\};$$
$$A(\omega) := L \exp\{L|B|_T^*(\omega)|\},$$
$$|B|_t^*(\omega) := \sup_{0 \leq s \leq t} |B_s(\omega)|.$$

Clearly, the process a defined in this way is $\mathbf{F}^B$-adapted, therefore the solution $\theta(t, \omega)$ to the ω-wise differential equation

$$\frac{d\theta}{dt}(t, \omega) = a(t, \omega)(1 + \theta(t, \omega)); \quad \theta(0, \omega) = C_0 \ (> 0), \qquad (6.7)$$

is a continuous, positive, increasing and $\mathbf{F}^B$-adapted solution.

Recall that P-a.e. on Ω and for every $t \in [0, T]$, the process $(U^{t,2}(x, \omega, .)$ and $V^{t,2}(x, \omega, .))$ is an $(\mathcal{F}^W_{s,t})_{s \in [0,t]}$-adapted solution to the BSDE (6.5), defined on the probability space $(\Omega', \mathcal{F}', P')$. For notational simplicity we will often suppress (x, ω) and write $U = U^{t,2}(x, \omega)$, $V = V^{t,2}(x, \omega)$. Let us first show that $|U_s| \leq \theta_s$, for all $s \in [0, t]$, P'-a.e. For this purpose, for $M > 0$ we introduce the random function

$$\widetilde{f}_2^M(\omega, t, x, y, z) := \widetilde{f}_2(\omega, t, x, y, z)\chi_M(y),$$

where $\chi_M \in C^\infty$ such that $0 \leq \chi_M(y) \leq 1$, for all $y \in R$, $\chi_M(y) = 0$ for $|y| \geq M + 1$ and $\chi_M(y) = 1$ for $|y| \leq M$. Clearly, for each M,

$$|\widetilde{f}_2^M(t, x, y, z)| \leq a(t)\,(1 + \chi_M(y)|y|) + A|z|^2 \leq K_M(1 + |z|^2).$$

Thus, by Kobylanski [27], there exists a unique solution (U^M, V^M) of BSDE (6.5) with $\widetilde{f}_2^M$ instead of $\widetilde{f}_2$. Furthermore, by the stability result in [27] we know that as $M \to \infty$, possibly along a subsequence, one has

$$\sup_{s \in [0,t]} |U^M_s - U_s| \xrightarrow{\text{P}} 0 \text{ and } V^M \xrightarrow{L^2} V,$$

provided that $\{U^M; M > 0\}$ is bounded uniformly with respect to M. Therefore, it suffices to prove that for all $M > 0$ and P-a.e. $\omega \in \Omega$, it holds that $|U^M_s(\omega)| \leq \theta(s, \omega)$, for all $s \in [0, t]$, P'-a.e., where θ is the increasing process defined by (6.7).

To show this we first apply Tanaka's formula to $|U^M_s|$. This yields

$$|U^M_s| = C_0 + \int_0^s \text{sign}(U^M_s)\varphi_M(U^M_r)\widetilde{f}_2(r, \omega, X^t_r(x), U^M_r, V^M_r)dr \qquad (6.8)$$
$$- \int_0^s \text{sign}(U^M_r)V^M_r \downarrow dW_r + (K_s - K_0), \quad 0 \leq s \leq t,$$

where K is an $(\mathcal{F}^W_{s,t})_{s \in [0,t]}$-adapted, continuous decreasing process with $K_t = 0$ and

$$K_0 = -\int_0^t \mathbf{I}_{\{U^M_r = 0\}}dK_r, \quad 0 \leq s \leq t, \; P'\text{-a.e.}$$

Next, for fixed $\omega \in \Omega$ we put:

$$\Phi(\rho, \omega) = \begin{cases} \sum_{k=3}^\infty \frac{(2A(\omega)\rho)^k}{k!} & = \exp\{2A(\omega)\rho\} - 1 \\ & \quad -2A(\omega)\rho - 2A(\omega)^2\rho^2, \quad \rho > 0; \\ 0 & \rho \leq 0. \end{cases} \qquad (6.9)$$

Then it is easy to check that $\Phi(\cdot, \omega)$ is twice continuously differentiable, nonnegative and $\Phi(\rho, \omega) > 0$ if and only if $\rho > 0$. Again, we will suppress ω in Φ if there is no risk of confusion.

Now let us put $\Delta^M_s = |U^M_s| - \theta(s)$, $0 \leq s \leq t$, and apply Itô's formula to $\Phi(\Delta^M_s)$. This gives

$$\Phi(\Delta^M_s) = \int_0^s \Phi'(\Delta^M_r)\{\text{sign}(U^M_r)\varphi_M(U^M_r)\widetilde{f}_2(r, X_r, U^M_r, V^M_r) \qquad (6.10)$$

$$- a(r)(1 + \theta(r))\} dr$$

$$+ \int_0^s \text{sign}(U_r^M) V_r^M \downarrow dW_r + \int_0^s \Phi'(\Delta_r^M) dK_r$$

$$- \frac{1}{2} \int_0^s \Phi''(\Delta_r^M) |V_r^M|^2 dr.$$

Since $\Phi'(u) \equiv 0$ for $u < 0$ and $\theta(\cdot) \geq 0$, we have

$$\int_0^s \Phi'(\Delta_r^M) dK_r = \int_0^s \Phi'(\Delta_r^M) \mathbf{I}_{\{U_r^M = 0\}} dK_r = \int_0^s \Phi'(-\theta(r)) dK_r = 0. \quad (6.11)$$

Moreover, since $\Phi' \geq 0$, it holds

$$\int_0^s \Phi'(\Delta_r^M) \{\text{sign}(U_r^M) \varphi_M(U_r^M) \tilde{f}_2(r, X_r, U_r^M, V_r^M) - a(r)(1 + \theta(r))\} dr$$

$$\leq \int_0^s \Phi'(\Delta_r^M) \{\varphi_M(U_r^M) |\tilde{f}_2(r, X_r, U_r^M, V_r^M)| - a(r)(1 + \theta(r))\} dr$$

$$\leq \int_0^s \Phi'(\Delta_r^M) \Big[a(r) \{\varphi_M(U_r^M) |U_r^M| - \theta(r)\} + A |V_r^M|^2 \Big] dr. \quad (6.12)$$

We remark that $\Phi''(\rho) - 2A\Phi'(\rho) \geq 0$, for all ρ and recall that $a(r) \leq A$, for all r. Combining (6.10) – (6.12) we have

$$\Phi(\Delta_s^M)$$

$$\leq \int_0^s \Big[-\frac{1}{2}\Phi'' + A\Phi' \Big] (\Delta_r^M) |V_r^M|^2 dr$$

$$+ \int_0^s A\Phi'(\Delta_r^M) \big[\varphi_M(U_r^M) |U_r^M| - \theta(r) \big] dr + \int_0^s \text{sign}(U_r^M) V_r^M \downarrow dW_r$$

$$\leq A \int_0^s \Phi'(\Delta_r^M) \big[\varphi_M(U_r^M) |U_r^M| - \theta(r) \big] dr + \int_0^s \text{sign}(U_r^M) V_r^M \downarrow dW_r.$$

Hence, for every $s \leq v \leq t$ (and for fixed ω) it holds

$$E_{P'} \big[\Phi(\Delta_s^M) | \mathcal{F}_{v,t}^W \big] \leq A E_{P'} \Big[\int_0^s \Phi'(\Delta_r^M) \big[\varphi_M(U_r^M) |U_r^M| - \theta(r) \big] dr \Big| \mathcal{F}_{v,t}^W \Big]. \quad (6.13)$$

From definition (6.9) one deduces easily that $\Phi'(\rho) = 2A\big[\Phi(\rho) + 2A^2 \rho^2 \big]$ and $\Phi'(\rho) = 0$ for $\rho < 0$. Consequently, we have

$$\Phi'(\Delta_r^M) \big[\varphi_M(U_r^M) |U_r^M| - \theta(r) \big] \leq 2A \Phi(\Delta_r^M) \big[\varphi_M(U_r^M) |U_r^M| - \theta(r) \big] + 4A^3 (\Delta_r^M)^3. \quad (6.14)$$

But since

$$\lim_{\rho \to 0} \rho^3 / \Phi(\rho) = \frac{3}{4A^3} \quad \text{and} \quad \lim_{\rho \to \infty} \rho^3 / \Phi(\rho) = 0,$$

there exists some constant $\tilde{A} > 0$ such that $\rho^3 \leq \tilde{A} \Phi(\rho)$ for all ρ. Thus, if we put

$$\psi_s^{t,v} := E_{P'} \big[\Phi(\Delta_s^M) | \mathcal{F}_{v,t}^W \big], \quad 0 \leq s \leq v \leq t,$$

then (6.13) and (6.14) gives

$$\psi_s^{t,v} \le K \int_0^s \psi_r^{t,v} dr, \ s \in [0, v],$$

where $K = 2A(M + 1 + 2A^2 \tilde{A})$. Then, Gronwall's inequality yields that $\psi_s^{t,v} = 0$, for all $s \in [0, v]$, P'-a.e. Finally, since Δ_v^M is $\mathcal{F}_{v,t}^W$-measurable, we can conclude that $\Phi(\Delta_v^M) = 0$, $\forall v \in [0, t]$, P'-a.e.; and from the construction of Φ we see that it must hold $\Delta_r^M \le 0$, P'-a.e. Namely, $|U_s^M| \le \theta(s)$, P'-a.e., proving the claim.

The above result now allows to complete the proof. Since both processes η and $\mathcal{E}$ are increasing in y, for all (t, x) we have

$$
\begin{aligned}
U_s^t(x) &= \mathcal{E}(s, X_s^t(x), Y_s^t(x)) \le \mathcal{E}(s, X_s^t(x), Y_s^{t,2}(x)) \\
&= U_s^{t,2}(x) \le \theta_s, \quad \forall s \in [0, t], \ P\text{-a.e.,}
\end{aligned}
$$

and by Proposition 4.7, for all $0 \le s \le t$,

$$
\begin{aligned}
Y_s^t(x) &= \eta(s, X_s^t(x), U_s^t(x)) \le \eta(s, X_s^t(x), U_s^{t,2}(x)) \\
&\le \eta(s, X_s^t(x), \theta_s) \le C|B|_s^* + \theta_s := \Theta_s.
\end{aligned}
$$

We recall that θ is an $\mathbf{F}^B$-adapted increasing process, so is Θ. The proof is now complete. $\square$

We are now ready to prove the existence of the stochastic viscosity solution. Following the approach by Pardoux and Peng [37] we introduce for each $(t, x) \in [0, T] \times R^n$ two random fields

$$u(t, x) = Y_t^t(x); \qquad v(t, x) = U_t^t(x), \tag{6.15}$$

where Y, U are the solutions to the BDSDEs (5.2) and (6.2), respectively. By (6.1) and (6.3) these random fields are related as follows: for $(\omega, t, x) \in \Omega \times [0, T] \times R^n$,

$$u(\omega, t, x) = \eta(\omega, t, x, v(\omega, t, x)); \ v(\omega, t, x) = \mathcal{E}(\omega, t, x, u(\omega, t, x)). \tag{6.16}$$

Proposition 4.4 reduces the proof of the existence of the viscosity solution to the equation SPDE(f, g) to the proof that the random field v is a stochastic viscosity solution to the SPDE($\tilde{f}, 0$). However, from Remark 4.3 (i) we know that an $\mathbf{F}^B$-progressively measurable, ω-wise viscosity solution is automatically a stochastic viscosity solution. Thus the following theorem cannot surprise.

Theorem 6.3 *Assume A.i) – A.iii) and A.iv'). Then the random field v is a stochastic viscosity solution of* SPDE($\tilde{f}, 0$); *and u is a stochastic viscosity solution to* SPDE(f, g), *respectively. Furthermore, if in addition we assume that,*

$$|f(\omega, t, x, 0, 0)| \le C, \ \forall (\omega, t, x) \in \Omega \times [0, T] \times R^n,$$

for some real constant $C > 0$, then the random fields u and v are stochastically bounded.

Proof. Since the equation (5.2) is a time-reversed version of the BDSDE considered by Pardoux and Peng [20], it follows from the proof of Theorem 2.1 in [38], together with Lemma 5.1, that there exists a continuous version of the random field $(s, t, x) \mapsto Y^t_{s \wedge t}(x)$, for $(s, t, x) \in [0, T]^2 \times R^n$. Taking this continuous version from now on, also the random field $u(t, x) = Y^t_t(x)$ is continuous and hence jointly measurable on $\overline{\Omega} \times [0, T] \times R^n$.

Next, by Lemma 5.1 we know that $Y^t_s(x)$ is $\mathcal{F}^B_s \otimes \mathcal{F}^W_{s,t}$-measurable. Hence, $u(t, x)$ is $\mathcal{F}^B_t \otimes \mathcal{F}^W_{t,t}$-measurable. But since W is a Brownian motion on the probability space $(\Omega', \mathcal{F}', P')$, we conclude from Blumenthal's 0-1 law that u is independent of $\omega' \in \Omega'$. Therefore, the random field u can be viewed as one that is defined on $\Omega \times [0, T] \times R^n$ and is $\mathcal{F}^B_t$-measurable for each $t \in [0, T]$. In other words, $u \in C(\mathbf{F}^B, [0, T] \times R^n)$. Consequently, from (6.1), $v \in C(\mathbf{F}^B, [0, T] \times R^n)$ as well.

To complete the proof it still remains to verify that v is a stochastic viscosity solution to SPDE$(\tilde{f}, 0)$. For fixed $\omega \in \Omega$ we put

$$\overline{U}^\omega_s(x)(\omega') = U^t_s(x)(\omega, \omega'); \quad \overline{V}^\omega_s(x)(\omega') = V^t_s(x)(\omega, \omega').$$

Since for fixed $\omega \in \Omega$ we can interpret (6.2) as a time-reversed version of a standard BSDE (3.2) defined on the probability space $(\Omega', \mathcal{F}', P')$, with unique solution $(\overline{U}^\omega, \overline{V}^\omega)$, we get from Theorem 3.4 that $\overline{v}(\omega, t, x) := \overline{U}^\omega_t(x)$ is a viscosity solution to a semilinear PDE with coefficient $\tilde{f}(\omega, ., ., ., .)$. By Blumenthal's 0–1 law again we have $\overline{P}(\overline{U}^\omega_t(x) = U^t_t(x)(\omega, \omega')) = 1$, hence $\overline{v}(t, x) \equiv v(t, x)$, for all (t, x), P-a.e.

Since $v \in C(\mathbf{F}^B, [0, T] \times R^n)$ and $v(\omega, ., .)$ is a viscosity solution to SPDE$(\tilde{f}, 0)$ for (P-almost) each fixed $\omega \in \Omega$, it is by definition an ω-wise viscosity solution, hence a stochastic viscosity solution to SPDE$(\tilde{f}, 0)$. The first conclusion of the theorem now follows from Proposition 4.4, while the last statement of the theorem follows from Lemma 5.1 and Theorem 5.3. The proof is now complete. □

Remark 6.4 A direct consequence of Proposition 4.7 and Theorem 6.3 is that the stochastic viscosity solution constructed above is stochastically bounded.

7. UNIQUENESS OF THE STOCHASTIC VISCOSITY SOLUTION

In this section we will give a comparison theorem for stochastic viscosity solutions. This comparison theorem is the key for the proof of uniqueness of the stochastic viscosity solution for SPDE(f, g) (4.1).

Let us begin with a reduction of the problem. We note that, in light of Proposition 4.4, we only need to prove the uniqueness of the stochastic viscosity solution to SPDE$(\tilde{f}, 0)$:

$$\partial_t v(t, x) = \mathcal{A}v(t, x) + \tilde{f}(t, x, v(t, x), \nabla_x v(t, x)\sigma(x)), \ (t, x) \in [0, T] \times R^n. \quad (7.1)$$

We should point out that, although SPDE (3.1) does not have a martingale term, the definitions of a stochastic viscosity solution and an ω-wise stochastic viscosity solution to (7.1) are still different! However, since the proof of the uniqueness for the ω-wise stochastic viscosity solution is essentially parallel to that of Theorem 3.6, we shall establish the uniqueness by identifying the two definitions 4.1 and 4.3 in the case of (7.1).

To do this, we shall make a technical assumptions on the random field $\widetilde{f}$:

A.v) There exists an $\mathbf{F}^B$-adapted, increasing real-valued process

$$\Theta = (\Theta_t)_{t \in [0,T]}$$

such that

$$|\widetilde{f}(t, x, y + r, z) - \widetilde{f}(t, x, y, z)| \le \Theta_t r, \quad \forall r > 0, (t, x, y, z), \ P\text{-a.e.}$$

The removal of this condition is possible by restricting slightly the class of random fields on which the uniqueness is considered, which requires some extra properties of the stochastic viscosity solutions. But, in order not to over-complicate this already technical section we prefer to work under the above assumption.

The main result of this section is the following *comparison theorem*:

Theorem 7.1 ([8], Theorem 3.1) *Assume A.i) – A.v). Suppose that the function* $v_1 \in C(\mathbf{F}^B, [0, T] \times R^n)$ *is a stochastic viscosity subsolution (resp., supersolution) of* (7.1) *and* $v_2 \in C(\mathbf{F}^B, [0, T] \times R^n)$ *is an ω-wise viscosity supersolution (resp., subsolution) of* (7.1), *such that they are both stochastically bounded. Then it holds that*

$$v_1(t, x) \le \ (\text{resp. } \ge \,) \ v_2(t, x), \qquad \forall (t, x) \in [0, T] \times R^n, \ P\text{-a.e.} \qquad (7.2)$$

The uniqueness results are contained in the following corollary. We recall that in the case of $g = 0$, an ω-wise viscosity solution is necessarily a stochastic viscosity solution, cf. Remark 4.3 (i).

Corollary 7.2 *Under the assumptions A.i) – A.v) we have:*

(i) *If* $v_1 \in C(\mathbf{F}^B, [0, T] \times R^n)$ *is a stochastic viscosity solution and the function* $v_2 \in C(\mathbf{F}^B, [0, T] \times R^n)$ *is an ω-wise viscosity solution of* (7.1) *and both are stochastically bounded, then* $v_1(t, x) \equiv v_2(t, x)$, *for all* $(t, x) \in [0, T] \times R^n$, *P-a.e.*

(ii) *If in addition (A.iv') also holds, then the stochastic viscosity solution to equation* SPDE(f, g) *is unique among the stochastically bounded random fields in the space* $C(\mathbf{F}^B, [0, T] \times R^n)$.

Proof. Statement (i) is obvious. Let us prove (ii). To begin with, let $u_1, u_2 \in C(\mathbf{F}^B, [0, T] \times R^n)$ be two stochastically bounded stochastic viscosity solutions of SPDE(f, g). By the Doss-Sussmann transformation we define

$$v_i(t, x) = \mathcal{E}(t, x, u_i(t, x)), \quad (t, x) \in [0, T] \times R^n, \ i = 1, 2.$$

Then by Proposition 4.4, v_1 and v_2 are both stochastic viscosity solutions to the equation SPDE$(\widetilde{f}, 0)$, with $\widetilde{f}$ as defined in Section 4; moreover, v_1, v_2 are both stochastically bounded, thanks to Proposition 4.7. On the other hand, the proof of Theorem 6.3 has shown that SPDE$(\widetilde{f}, 0)$ admits a stochastically bounded ω-wise viscosity solution $v \in$

$C(\mathbf{F}^B, [0, T] \times R^n)$. This now allows to apply statement (i) separately to v_1, v and to v_2, v. Consequently, $v_1(t, x) = v(t, x) = v_2(t, x)$, for all $(t, x) \in [0, T] \times R^n$, P-a.e. This completes the proof. $\qquad \square$

Since the proof of Theorem 7.1 is quite lengthy, we shall break it into a sequence of lemmata, which will be proved throughout the rest of the section. To begin with, without loss of generality let us assume that v_1 is a stochastic viscosity subsolution and v_2 is an ω-wise viscosity supersolution of (7.1). In analogy to the proof of Theorem 3.6, the deterministic counterpart of the comparison Theorem 7.1, we define

$$\zeta = \sup_{(t,x)\in[0,T]\times R^n} (v_1(t, x) - v_2(t, x)).$$

Clearly, $P(\zeta > 0) = 0$ means that the conclusion of Theorem 7.1 holds, so we shall make the hypothesis that $P(\zeta > 0) > 0$ and then try to find a set $\Gamma^* \in \mathcal{F}_T^B$ with $P(\Gamma^*) > 0$ on which a contradiction can be drawn.

Let us begin by making some reductions and by clarifying necessary notations. Recall that v_1 and v_2 are both uniformly stochastically bounded, that is, there exists some $\mathcal{F}_T^B$-measurable real-valued random variable θ such that $|v_1(t, x)|, |v_2(t, x)| \leq \theta$, for all $(t, x) \in [0, T] \times R^n$, P-a.e. Hence, $|\zeta| \leq 2\theta < \infty$, P-a.e. and $P(\zeta > 0) > 0$ implies that $P(\zeta > 0, \theta \leq \ell) > 0$ for some $\ell > 0$. Consequently,

$$\zeta^\ell = \text{ess} \sup_{\omega\in\{\theta\leq\ell\}} \zeta(\omega) > 0. \tag{7.3}$$

For each natural $N \geq 1$, we define $K_N = [0, T - \frac{1}{N}] \times \{x \in R^n : |x| \leq N\}$ and

$$\zeta_N(\omega) = \sup_{(t,x)\in K_N} (v_1(\omega, t, x) - v_2(\omega, t, x)). \tag{7.4}$$

Since

$$\lim_{N\to\infty} \uparrow \zeta_N = \zeta, \quad P\text{-a.e.,}$$

for each $\varepsilon > 0$ we can introduce the following integer-valued random variable:

$$N(\varepsilon, \omega) = \inf \left\{ N \geq 1 : \zeta_N(\omega) \geq \zeta(\omega) - \frac{\varepsilon}{3} \right\}. \tag{7.5}$$

Furthermore, since $v_2(\omega, ., .)$ is uniformly continuous on each K_N, for a.e. $\omega \in \Omega$ and $N \geq N(\varepsilon, \omega)$ there is some $\delta(N, \omega) > 0$ such that

$$\begin{aligned} \alpha(\omega, \delta, N) &= \sup_{(t,x)\in K_N, \delta'\in(0,\delta)} |v_2(\omega, t, x) - v_2(\omega, t - \delta', x)| \\ &\leq \frac{\varepsilon}{3}, \quad \forall\delta \in (0, \delta(N, \omega)]. \end{aligned} \tag{7.6}$$

Bearing the random variables $\zeta_N, N(\varepsilon, .), \alpha(., \delta, N)$ in mind, for any $\varepsilon \in (0, \zeta^\ell), N \geq 1$ and $\delta > 0$, we set

$$\Gamma_{\varepsilon,N,\delta}=\{\omega \in \Omega \,|\, \zeta(\omega) \geq \zeta^\ell - \varepsilon, \, \theta(\omega) \leq \ell, \, N \geq N(\varepsilon, \omega), \, \alpha(\omega, \delta, N) \leq \frac{\varepsilon}{3}\}. \tag{7.7}$$

We shall prove that the desired set Γ^* can be chosen from the family

$$\{\Gamma_{\varepsilon,N,\delta} \,|\, \varepsilon > 0, \; N \geq 1, \; \delta > 0\}.$$

To complete the necessary notations, we still define an auxiliary (random) function which will take the role played by

$$\psi_{\delta_1,\delta_2}(t, x, t', x') = v_1(t, x) - v_2(t', x')$$
$$- \left(\frac{1}{2\delta_1}\left(|t - t'|^2 + |x - x'|^2\right) + \frac{\delta_2}{2}\left(|x|^2 + |x'|^2 + \frac{1}{T - t} + \frac{1}{T - t'}\right)\right)$$

in the proof of Theorem 3.6. We point out that the main difficulty here in the stochastic case is that the time variable and the spatial variable *cannot* be treated in the same way. In fact, even the two time variables t, t' will have to be treated differently in order to obtain the appropriate measurability (such as "adaptedness"), as we will see in a moment. In particular, the auxiliary function to define cannot be symmetric in t, t'.

Let us first introduce, for $\delta > 0$, a (generalized) convex function $\psi_\delta : R \to R_+ \cup \{+\infty\}$,

$$\psi_\delta(r) = \begin{cases} -\ln\left(1 - \left(\frac{\delta - r}{\delta}\right)^2\right), & r \in (0, 2\delta); \\ +\infty, & r \notin (0, 2\delta). \end{cases} \tag{7.8}$$

Now, with the convention $v_2(t, x) \equiv v_2(0, x)$, for all $t \leq 0$, $\delta_1, \delta_2 > 0$, we define the random field $\Psi_{\delta_1,\delta_2} : \Omega \times [0, T] \times R^n \times (-\infty, T) \times R^n \longrightarrow R$:

$$\Psi_{\delta_1,\delta_2}(\omega, t, x, t', x') = v_1(\omega, t, x) - v_2(\omega, t', x') \tag{7.9}$$
$$- \left(\frac{1}{2\delta_1}|x - x'|^2 + \frac{1}{2}\psi_{\delta_1}(t - t') + \frac{\delta_2}{2}\left(|x|^2 + |x'|^2\right) + \frac{\delta_2}{2}\frac{1}{T - t}\right),$$

and a process

$$\Phi_{\delta_1,\delta_2}(t) = \sup_{\substack{x,x' \in R^n, \\ t' \in (-\infty, T)}} \Psi_{\delta_1,\delta_2}(t, x, t', x'), \qquad t \in [0, T]. \tag{7.10}$$

We observe that, by the definition of ψ_δ, the "sup" in (7.10) is actually taken over $t' \in (t - 2\delta_1, t)$, for any $t \in [0, T]$.

The following lemma relates the set $\Gamma_{\varepsilon,N,\delta}$ with the auxiliary function Ψ_{δ_1,δ_2}:

Lemma 7.3 (i) *For each $\varepsilon \in (0, \zeta^\ell)$, $N \geq 1$ and $\delta > 0$, we have that*

$$\Gamma_{\varepsilon,N,\delta} \subseteq \bigcap_{\substack{\delta_1 \in (0,\delta) \\ \delta_2 \in (0,\varepsilon/(6N^2))}} \left\{ \sup_{\substack{(t,x) \in [0,T] \times R^n \\ (t',x') \in (-\infty,T) \times R^n}} \Psi_{\delta_1,\delta_2}(t, x, t', x') \geq \zeta^\ell - 2\varepsilon \right\}. \tag{7.11}$$

(ii) *For any $\varepsilon \in (0, \zeta^\ell)$, there exist $N^*(\varepsilon) \geq 1$ and $\delta^*(\varepsilon) = \delta(N^*(\varepsilon)) > 0$ such that* $P(\Gamma_{\varepsilon,N^*(\varepsilon),\delta^*(\varepsilon)}) > 0$.

Proof. (i) Let $\omega \in \Gamma_{\varepsilon,N,\delta}$. Then we get from (7.5), (7.6) and (7.7) that for any $\delta_1 \in (0,\delta)$ and $\delta \in (0, \varepsilon/(6N^2))$,

$$\sup_{\substack{(t,x)\in(0,T)\times R^n \\ (t',x')\in(-\infty,T)\times R^n}} \Psi_{\delta_1,\delta_2}(\omega,t,x,t',x')$$

$$\geq \sup_{(t,x)\in K_N} \left\{ v_1(\omega,t,x) - v_2(\omega,t-\delta_1,x) - \delta_2|x|^2 - \delta_2\frac{1}{T-t} \right\}$$

$$\geq \sup_{(t,x)\in K_N} \left\{ v_1(\omega,t,x) - v_2(\omega,t,x) \right\}$$

$$\quad - \sup_{(t,x)\in K_N} \left\{ v_2(\omega,t,x) - v_2(\omega,t-\delta_1,x) \right\} - 2\delta_2 N^2$$

$$\geq \zeta_N(\omega) - \alpha(\omega,\delta,N) - 2\delta_2 N^2$$

$$\geq \left(\zeta(\omega) - \frac{\varepsilon}{3} \right) - \frac{\varepsilon}{3} - \frac{\varepsilon}{3} \geq \zeta(\omega) - \varepsilon \geq \zeta^\ell - 2\varepsilon,$$

proving (i).

(ii) Since

$$\bigcup_{N\geq 1}\bigcup_{\delta>0} \Gamma_{\varepsilon,N,\delta} = \{\zeta \geq \zeta^\ell - \varepsilon, \theta \leq \ell\};$$

and since the definition of ζ^ℓ implies that $P(\zeta \geq \zeta^\ell - \varepsilon, \theta \leq \ell) > 0$ for any $\varepsilon \in (0,\zeta^\ell)$, the conclusion follows easily. $\qquad\square$

We note that, with the notation $\Gamma_\varepsilon^* = \Gamma_{\varepsilon,N^*(\varepsilon),\delta^*(\varepsilon)}$, $\varepsilon \in (0,\zeta^\ell)$, Lemma 7.3(ii) shows that one needs only to find $\varepsilon_0 \in (0,\zeta^\ell)$ so that the contradiction occurs on $\Gamma^* = \Gamma_{\varepsilon_0}^*$.

In order to follow as well as possible the scheme of the proof of the deterministic counterpart, Theorem 3.6, we shall first study the properties of the functions Φ_{δ_1,δ_2}, Ψ_{δ_1,δ_2} and then generalize the Theorem 3.2 of Crandall, Ishii and Lions [10] to our stochastic setting.

Properties of the Functions Φ_{δ_1,δ_2} and Ψ_{δ_1,δ_2}

We have the following two lemmata.

Lemma 7.4 *Let $\varepsilon > 0$ be fixed and let $N^* = N^*(\varepsilon)$, $\delta^* = \delta^*(\varepsilon)$. Then for all $\delta_1 \in (0,\delta^*)$ and $\delta_2 \in (0, \frac{\varepsilon}{6(N^*)^2})$, the process Φ_{δ_1,δ_2} is $\mathbf{F}^B$-adapted and continuous on $[0,T)$. Furthermore, if we define*

$$\hat{t} = \inf\{t \in [0,T) \,|\, \Phi_{\delta_1,\delta_2} \text{ attains a local maximum at } t; \, \Phi_{\delta_1,\delta_2}(t) > \zeta^\ell - 2\varepsilon\}, \quad (7.12)$$

then $\hat{t}$ is an $\mathbf{F}^B$-stopping time; and for any $\delta_2 > 0$ there exists some $\delta^{**} > 0$ such that $\Gamma_\varepsilon^* \subseteq \{0 < \hat{t} < T\}$, whenever $\delta_1 \leq \delta^{**}$.

Proof. The $\mathbf{F}^B$-adaptedness of Φ_{δ_1,δ_2} is a consequence of the choice of our auxiliary function $\psi_{\delta_1}(r)$. To see the continuity, we note that as the supremum of a family of continuous processes, Φ_{δ_1,δ_2} is lower semicontinuous. Thus it suffices to show its upper semicontinuity at any $t \in [0,T)$.

For $\ell > 0$, we define the set $\Omega_\ell = \{\omega \in \Omega \mid |\theta(\omega)| \leq \ell\}$. Since $\Omega_\ell \uparrow \Omega$, modulo a P-null set, it suffices to show that $\Phi_{\delta_1,\delta_2}(\omega,.)$ is upper semicontinuous for $\omega \in \Omega_\ell$. For this let $\ell > 0$, $\omega \in \Omega_\ell$ and $\eta > 0$ be fixed and put

$$\eta_1 = \sqrt{1 - \frac{1}{2}\exp(-8\ell - 2\eta)}; \quad \eta_2 = \sqrt{1 - \exp(-8\ell - 2\eta)}. \tag{7.13}$$

Let $t \in [0, T)$. For any $\bar{t} \in [0, T)$ such that $|\bar{t} - t| \leq \delta_1^* := \delta_1(\eta_1 - \eta_2)$, let $(\bar{x}, \bar{t}', \bar{x}')$ be an $R^n \times [0, T) \times R^n$-valued random vector such that

$$\Phi_{\delta_1,\delta_2}(\bar{t}) \leq \Psi_{\delta_1,\delta_2}(\bar{t}, \bar{x}, \bar{t}', \bar{x}') + \eta. \tag{7.14}$$

Therefore, the definition of Φ_{δ_1,δ_2} and Ψ_{δ_1,δ_2} (see (7.9) and (7.10)) gives that, for P-a.e. $\omega \in \Omega_\ell$,

$$\begin{aligned}
\Phi_{\delta_1,\delta_2}(\bar{t}) &- \Phi_{\delta_1,\delta_2}(t) \tag{7.15} \\
&\leq \Psi_{\delta_1,\delta_2}(\bar{t}, \bar{x}, \bar{t}', \bar{x}') - \Psi_{\delta_1,\delta_2}(t, \bar{x}, \bar{t}', \bar{x}') + \eta \\
&\leq v_1(\bar{t}, \bar{x}) - v_1(t, \bar{x}) - \frac{1}{2}(\psi_{\delta_1}(\bar{t} - \bar{t}') - \psi_{\delta_1}(t - \bar{t}')) - \frac{\delta_2}{2}\left(\frac{1}{T - \bar{t}} - \frac{1}{T - t}\right).
\end{aligned}$$

Since v_1 and v_2 both have continuous paths, it is readily seen from (7.15) that

$$\limsup_{\bar{t} \to t} \Phi_{\delta_1,\delta_2}(\omega, \bar{t}) \leq \Phi_{\delta_1,\delta_2}(\omega, t) - \liminf_{\bar{t} \to t} \frac{1}{2}(\psi_{\delta_1}(\bar{t} - \bar{t}') - \psi_{\delta_1}(t - \bar{t}')).$$

Thus it suffices to prove that

$$\liminf_{\bar{t} \to t} \frac{1}{2}(\psi_{\delta_1}(\bar{t} - \bar{t}') - \psi_{\delta_1}(t - \bar{t}')) \geq 0. \tag{7.16}$$

To see this we note that

$$\Phi_{\delta_1,\delta_2}(\bar{t}) \geq v_1(\bar{t}, 0) - v_2(\bar{t} - \delta_1, 0) - \frac{\delta_2}{2}\frac{1}{T - \bar{t}} \geq -2\ell - \frac{\delta_2}{2}\frac{1}{T - \bar{t}},$$

therefore

$$\begin{aligned}
\frac{\delta_2}{2}(|\bar{x}|^2 + |\bar{x}'|^2) &+ \frac{1}{2}\psi_{\delta_1}(\bar{t} - \bar{t}') \tag{7.17} \\
&= \left\{v_1(\bar{t}, \bar{x}) - v_2(\bar{t}', \bar{x}') - \frac{1}{2\delta_1}|\bar{x} - \bar{x}'|^2 - \frac{\delta_2}{2}\frac{1}{T - \bar{t}}\right\} - \Psi_{\delta_1,\delta_2}(\bar{t}, \bar{x}, \bar{t}', \bar{x}') \\
&\leq \left(2\ell - \frac{\delta_2}{2}\frac{1}{T - \bar{t}}\right) - \Phi_{\delta_1,\delta_2}(\bar{t}) + \eta \leq 4\ell + \eta.
\end{aligned}$$

Consequently, we have $\psi_{\delta_1}(\bar{t} - \bar{t}') \leq 8\ell + 2\eta$.

Then from the definitions of ψ_{δ_1} in (7.8) and η_1, η_2 in (7.13), as well as δ_1^*, we obtain that

$$|\delta_1 - (\bar{t} - \bar{t}')| \leq \delta_1\eta_2, \quad \text{whenever } \bar{t} \in [0, T), |\bar{t} - t| \leq \delta_1^*,$$

and hence

$$|\delta_1 - (t - \bar{t}')| \le |\delta_1 - (\bar{t} - \bar{t}')| + |t - \bar{t}| \le \delta_1\eta_2 + \delta_1^* = \delta_1\eta_1.$$

Since $\eta_2 < \eta_1 < 1$, we have

$$(\bar{t} - \bar{t}'), (t - \bar{t}') \in [\delta_1(1 - \eta_1), \delta_1(1 + \eta_1)] \subset (0, 2\delta_1).$$

Consequently, by the definition of ψ_{δ_1} one checks easily that there exists a constant $C_1 > 0$, depending on δ_1, ℓ and η, such that

$$|\psi_{\delta_1}(\bar{t} - \bar{t}') - \psi_{\delta_1}(t - \bar{t}')| \le C_1|\bar{t} - t|, \quad \forall \bar{t} \in [0, T), |\bar{t} - t| \le \delta_1^*.$$

Thus (7.16) holds, proving the upper semicontinuity of $\Phi_{\delta_1,\delta_2}(\omega, .)$ at t.

To see that the random time $\hat{t}$ defined by (7.12) is an $\mathbf{F}^B$-stopping time, we have only to observe that the adaptedness and continuity of Ψ_{δ_1,δ_2} imply the following relation: for any $t \in [0, T]$,

$$\{\hat{t} < t\} = \bigcup_{\substack{0 \le r < s \le t \\ r,s \in \mathbf{Q}}} \left\{\zeta^\ell - 2\varepsilon < \Phi_{\delta_1,\delta_2}(r); \Phi_{\delta_1,\delta_2}(r) \ge \Phi_{\delta_1,\delta_2}(s)\right\} \in \mathcal{F}_t^B,$$

where $\mathbf{Q}$ denotes the totality of rational numbers.

It remains to prove the assertion that $\Gamma_\varepsilon^* \subseteq \{0 < \hat{t} < T\}$. In light of (7.11) we have only to show that

$$\left\{\sup_{\substack{(t,x,x') \in (0,T) \times (R^n)^2 \\ t' \in (-\infty, T)}} \Psi_{\delta_1,\delta_2}(t, x, t', x') \ge \zeta^\ell - 2\varepsilon\right\} \subseteq \{0 < \hat{t} < T\}, \tag{7.18}$$

whenever $\delta_1 \le \delta^{**}$. However, since Φ_{δ_1,δ_2} is continuous on $[0, T)$ and $\Phi_{\delta_1,\delta_2}(t) \to -\infty$, as $t \to T$, it suffices to show that $\Phi_{\delta_1,\delta_2}(0) \le 0$.

To see this, we define the "modulus of continuity" of the (deterministic) function $v_2(0, .) : R^n \to R$:

$$\omega(r, s) = \sup\left\{|v_2(0, x) - v_2(0, x')| \,\big|\, |x|^2 + |x'|^2 \le s, |x - x'|^2 \le r\right\}, \quad r, s > 0. \tag{7.19}$$

Since $\omega(0, s) = 0$, $\forall s > 0$, for any $\delta_2 > 0$ there exists $\delta^{**} > 0$ such that

$$\omega\left(4\ell\delta_1, \frac{4\ell}{\delta_2}\right) < \frac{\delta_2}{2T}, \quad \text{whenever } \delta_1 \le \delta^{**}. \tag{7.20}$$

Now we recall that, for all $t' \le 0$, we have $v_2(t', x) = v_2(0, x) \ge v_1(0, x)$ and $v_2(0, x) - v_2(0, x') \le 2\ell$. Thus, thanks to (7.19) and (7.20), for $t' \le 0$, $x, x' \in R^n$ and $\delta_1 \in (0, \delta^{**})$, we have

$$\Psi_{\delta_1,\delta_2}(0, x, t', x')$$
$$= v_1(0, x) - v_2(0, x') - \left(\frac{1}{2\delta_1}|x - x'|^2 + \frac{1}{2}\psi_{\delta_1}(-t')\right) - \frac{\delta_2}{2}\left(|x|^2 + |x'|^2\right) - \frac{\delta_2}{2T}$$

$$\leq \ v_2(0, x) - v_2(0, x') - \frac{1}{2\delta_1}|x - x'|^2 - \frac{\delta_2}{2}\left(|x|^2 + |x'|^2\right) - \frac{\delta_2}{2T}$$

$$\leq \ \begin{cases} -\dfrac{\delta_2}{2T} < 0, & |x|^2 + |x'|^2 \geq \dfrac{4\ell}{\delta_2} \ \text{ or } \ |x - x'|^2 \geq 4\ell\delta_1; \\[2mm] \omega\left(4\ell\delta_1, \dfrac{4\ell}{\delta_2}\right) - \dfrac{\delta_2}{2T} < 0, & |x|^2 + |x'|^2 < \dfrac{4\ell}{\delta_2}, \ \ |x - x'|^2 < 4\ell\delta_1. \end{cases}$$

Now, taking the supremum over $t' \in (-\infty, T)$ and $x, x' \in R^n$ we get the inequality $\Phi_{\delta_1,\delta_2}(0) \leq 0$. This proves the lemma. $\qquad\square$

Lemma 7.5 *For fixed $\delta_1 \in (0, \delta^* \wedge \delta^{**})$ and $\delta_2 \in (0, \varepsilon/(6N^{*2}))$, where $\varepsilon \in (0, \zeta^\ell/3)$, there exist $\mathcal{F}_{\widehat{t}}^B$-measurable random variables $\widehat{t} \in (-\infty, T)$ and $\widehat{x}, \widehat{x}' \in R^n$ such that $\widehat{t}(\omega) - 2\delta_1 < \widehat{t}'(\omega) < \widehat{t}(\omega)$, for P-a.e. $\omega \in \Omega$; and $\Phi_{\delta_1,\delta_2}(\widehat{t}) = \Psi_{\delta_1,\delta_2}(\widehat{t}, \widehat{x}, \widehat{t}', \widehat{x}')$. Furthermore, P-a.e. on $\{0 < \widehat{t} < T\}$, it holds that*
 i) Ψ_{δ_1,δ_2} *attains a local maximum at* $(\widehat{t}, \widehat{x}, \widehat{t}', \widehat{x}')$.
 ii) $\Psi_{\delta_1,\delta_2}(t, x, t', x') \leq \Psi_{\delta_1,\delta_2}(\widehat{t}, \widehat{x}, \widehat{t}', \widehat{x}'), \ \forall (t, x, t', x') \in [0, \widehat{t}] \times R^n \times [0, T) \times R^n.$

Proof. Since $\Phi_{\delta_1,\delta_2}(\widehat{t}) \geq \zeta^\ell - 2\varepsilon$, on a set $\widetilde{\Omega}$ of probability 1, for any $\omega \in \widetilde{\Omega}$ we have

$$\Phi_{\delta_1,\delta_2}(\widehat{t}(\omega)) = \sup_{(t', x, x') \in \mathcal{O}(\omega)} \Psi_{\delta_1,\delta_2}(\omega, \widehat{t}(\omega), x, t', x'),$$

where

$$\mathcal{O}(\omega) = \{(t', x, x') \in (-\infty, T) \times R^n \times R^n \mid \Psi_{\delta_1,\delta_2}(\omega, \widehat{t}(\omega), x, t', x') > \zeta^\ell - 3\varepsilon\}.$$

Furthermore, we note that for $\omega \in \Gamma_\varepsilon^*$ and $(t', x, x') \in \mathcal{O}(\omega)$ we have

$$\zeta^\ell - 3\varepsilon < \Psi_{\delta_1,\delta_2}(\omega, \widetilde{t}(\omega), x, t', x') = v_1(\omega, \widetilde{t}(\omega), x) - v_2(\omega, t', x') \qquad (7.21)$$

$$- \left\{ \frac{1}{2\delta_1}|x - x'|^2 + \frac{1}{2}\psi_{\delta_1}(\widehat{t}(\omega) - t') + \frac{\delta_2}{2}(|x|^2 + |x'|^2) + \frac{\delta_2}{2}\frac{1}{T - \widehat{t}(\omega)} \right\}$$

$$\leq 2\ell - \frac{1}{2}\psi_{\delta_1}(\widehat{t}(\omega) - t') - \frac{\delta_2}{2}\frac{1}{T - \widehat{t}(\omega)} - \frac{\delta_2}{2}(|x|^2 + |x'|^2).$$

Therefore $2\ell - (\zeta^\ell - 3\varepsilon) > 0$; and on $\mathcal{O}(\omega)$ we have

$$|x|^2 + |x'|^2 \leq \frac{2}{\delta_2}\left[2\ell - (\zeta^\ell - 3\varepsilon)\right].$$

Now by the definition of ψ_δ in (7.8), one checks that

$$|t' - (\widehat{t}(\omega) - \delta_1)| \leq \delta_1 \sqrt{1 - \exp\{-2[2\ell - (\zeta^\ell - 3\varepsilon)]\}} < \delta_1. \qquad (7.22)$$

Consequently, $\mathcal{O}(\omega)$ is a bounded set. Thus for each fixed $\omega \in \widetilde{\Omega}$ we can find $(\widetilde{t}', \widetilde{x}, \widetilde{x}')$ in the closure of $\mathcal{O}(\omega)$ such that (i) and (ii) hold. Moreover, (7.22) implies that $\widehat{t}(\omega) - 2\delta_1 < \widehat{t}'(\omega) < \widehat{t}(\omega)$ holds. The lemma then follows from the standard measurable selection theorem. $\qquad\square$

In the following we call the quadruple $(\widehat{t}, \widehat{x}, \widehat{t}', \widehat{x}')$ introduced in Lemma 7.4 an $\mathbf{F}^B$-*maximizer* of Ψ_{δ_1,δ_2}. To follow the scheme given by the proof of Theorem 3.6, we shall establish a

Generalization of Theorem 3.2 [10] to the Stochastic Setting

We remind the notions of superjets and subjets in [10] which we have briefly presented in the passage preceding Lemma 3.8. For notational convenience we introduce the auxiliary function

$$\varphi(t, x, t', x') = \frac{1}{2\delta_1}|x - x'|^2 + \frac{1}{2}\psi_{\delta_1}(t - t') + \frac{\delta_2}{2}\left(|x|^2 + |x'|^2 + \frac{1}{T - t}\right), \quad (7.23)$$

for $t, t' \in (0, T)$ with $0 < t - t' < 2\delta_1$ and $x, x' \in R^n$. We have

Theorem 7.6 ([8], Theorem 5.1) *Assume A.i) – A.iv). Let $(\hat{t}, \hat{x}, \hat{t}', \hat{x}')$ be an $\mathbf{F}^B$-maximizer of $\Psi_{\delta_1, \delta_2}$ defined in Lemma 7.5. Then, for any $\varepsilon, \gamma > 0$, there exist two $\mathcal{F}^B_{\hat{t}}$-measurable S^n-valued random variables $\mathcal{X}, \mathcal{Y}$ such that for P-a.e. $\omega \in \Gamma^*_\varepsilon$ it holds that*

i) $((\partial_t, \nabla_x)\varphi(\hat{t}, \hat{x}, \hat{t}', \hat{x}')(\omega), \mathcal{X}(\omega)) \in \overline{\mathcal{P}}^{1,2,+} v_1(\omega, \hat{t}(\omega), \hat{x}(\omega));$

ii) $(-(\partial_{t'}, \nabla_{x'})\varphi(\hat{t}, \hat{x}, \hat{t}', \hat{x}')(\omega), \mathcal{Y}(\omega)) \in \overline{\mathcal{P}}^{1,2,-} v_2(\omega, \hat{t}'(\omega), \hat{x}'(\omega));$

iii) *with*

$$B = \begin{pmatrix} D^2_{xx}\varphi & D^2_{xx'}\varphi \\ D^2_{xx'}\varphi & D^2_{x'x'}\varphi \end{pmatrix}(\hat{t}, \hat{x}, \hat{t}', \hat{x}')(\omega)$$

$$= \frac{1}{\delta_1}\begin{pmatrix} I & -I \\ I & I \end{pmatrix} + \delta_2\begin{pmatrix} I & 0 \\ 0 & I \end{pmatrix},$$

it holds that

$$\begin{pmatrix} \mathcal{X}(\omega) & 0 \\ 0 & -\mathcal{Y}(\omega) \end{pmatrix} \leq B + \gamma\left\{B^2 + \begin{pmatrix} I & 0 \\ 0 & I \end{pmatrix}\right\}. \quad (7.24)$$

Furthermore, there exists a sequence of $\mathbf{F}^B$-stopping times $(\hat{t}_m)_{m\geq 1}$, a sequence of sets $(\Gamma^m_\varepsilon)_{m\geq 1}$ with $P(\Gamma^m_\varepsilon) > 0$ and $\liminf_{m\to\infty} \Gamma^m_\varepsilon = \Gamma^*_\varepsilon$ and a sequence of $\mathcal{F}^B_{\hat{t}_m}$-measurable, $[0, T] \times (R^n)^2 \times (R \times R^n \times S^{n+1})^2$-valued random variables

$$\{(\hat{t}_m, \hat{x}_m, \hat{x}'_m, (\hat{a}^m_1, \hat{p}^m_1, \hat{S}^m_1), (\hat{a}^m_2, \hat{p}^m_2, \hat{S}^m_2))\}$$

which enjoys, P-a.e on Γ^m_ε, the following properties:

i') $(\hat{a}^m_1, \hat{p}^m_1, \hat{S}^m_1) \in \mathcal{P}^{2,+} v_1(\hat{t}_m, \hat{x}_m)$ and $(\hat{a}^m_2, \hat{p}^m_2, \hat{S}^m_2) \in \mathcal{P}^{2,-} v_2(\hat{t}_m, \hat{x}'_m);$

ii') *P-a.e. on Γ^*_ε, it holds that*

$$\lim_{m\to\infty} (\hat{t}_m, \hat{t}'_m, \hat{x}_m, \hat{x}'_m) = (\hat{t}, \hat{t}', \hat{x}, \hat{x}');$$

$$\lim_{m\to\infty} ((\hat{a}^m_1, \hat{p}^m_1), (\hat{a}^m_2, \hat{p}^m_2)) = ((\partial_t, \nabla_x)\varphi(\hat{t}, \hat{x}, \hat{t}', \hat{x}'), -(\partial_{t'}, \nabla_{x'})\varphi(\hat{t}, \hat{x}, \hat{t}', \hat{x}'));$$

iii') *There exists a couple $(\hat{\mathcal{X}}, \hat{\mathcal{Y}})$ of S^{n+1}-valued, $\mathcal{F}^B_{\hat{t}}$-measurable random variables such that, for P-a.e. $\omega \in \Gamma^*_\varepsilon$, $(\hat{\mathcal{X}}(\omega), \hat{\mathcal{Y}}(\omega))$ is a limit point of the sequence $((\hat{S}^m_1(\omega), \hat{S}^m_2(\omega)))_{m\geq 1}$ and that*

$$\hat{\mathcal{X}} = \begin{pmatrix} x_{11} & * \\ * & \mathcal{X} \end{pmatrix}, \quad \hat{\mathcal{Y}} = \begin{pmatrix} x'_{11} & * \\ * & \mathcal{Y} \end{pmatrix}.$$

For the proof of Theorem 7.6, we need a generalization of the *optional section theorem* of Dellacherie–Meyer [11] to space-time random vectors. Let us first recall some notions from [11]. For any *paved* set $(G, \mathcal{G})$ we denote the class of $\mathcal{G}$-analytic sets by $\mathcal{A}(\mathcal{G})$. Next, let $\mathcal{O}$ denote the optional σ-field on $\Omega \times [0, T]$, that is, the σ-field generated by all stochastic intervals

$$[[\eta_1, \eta_2[[= \{(\omega, t) \in \Omega \times [0, T] \mid \eta_1(\omega) \le t < \eta_2(\omega)\}, \quad \eta_1, \eta_2 \in \mathcal{M}^B_{0,T}.$$

Then for any given Borel space $(\mathbf{X}, \mathcal{B}(\mathbf{X}))$ it holds:

Proposition 7.7 ([8], Lemma 5.2) *Let $\Sigma \subseteq \Omega \times [0, T) \times \mathbf{X}$ be an $\mathcal{O} \otimes \mathcal{B}(\mathbf{X})$-measurable subset. Then for any $\varepsilon > 0$, there exists a mapping*

$$(\widehat{t}, \widehat{\mathbf{x}}) : \Omega \to [0, T] \times \mathbf{X},$$

such that

   (i) *$\widehat{t}$ is an $\mathbf{F}^B$-stopping time and $\widehat{\mathbf{x}}$ is an $\mathcal{F}^B_{\widehat{t}}$-measurable random variable;*
   (ii) *for P-a.e. $\omega \in \Omega$ such that $\widehat{t}(\omega) < T$, $(\omega, \widehat{t}(\omega), \widehat{\mathbf{x}}(\omega)) \in \Sigma$;*
   (iii) *$P(\widehat{t} < T) \ge P(\pi(\Sigma)) - \varepsilon$, where $\pi(\Sigma)$ denotes the projection of Σ into Ω.*

The reader interested in the proof is referred to [8]. Here we don't give the proof of this proposition in order to concentrate to that of the stochastic version of the comparison Theorem 7.1. But first we shall still prove Theorem 7.6.

Proof of Theorem 7.6. Let $(\widehat{t}, \widehat{x}, \widehat{t}', \widehat{x}')$ denote an $\mathbf{F}^B$-maximizer of the random field $\Psi_{\delta_1, \delta_2}$. Recall that $\widehat{t}$ is an $\mathbf{F}^B$-stopping time and $(\widehat{x}, \widehat{t}', \widehat{x}')$ is $\mathcal{F}^B_{\widehat{t}}$-measurable. For each fixed $\omega \in \Gamma^*_\varepsilon (\subseteq \{0 < \widehat{t} < T\})$ we can apply Theorem 3.9 ([10], Theorem 3.2) to find matrices

$$\overline{\mathcal{X}}^\omega = (x_{ij}), \quad \overline{\mathcal{Y}}^\omega = (y_{ij}) \in \mathcal{S}^{n+1}$$

such that

$$\left. \begin{array}{l} (\widehat{a}_1(\omega), \widehat{p}_1(\omega), \overline{\mathcal{X}}^\omega) \in \overline{\mathcal{P}}^{2,+} v_1(\omega, \widehat{t}(\omega), \widehat{x}(\omega)) \\ (\widehat{a}_2(\omega), \widehat{p}_2(\omega), \overline{\mathcal{Y}}^\omega) \in \overline{\mathcal{P}}^{2,-} v_2(\omega, \widehat{t}'(\omega), \widehat{x}'(\omega)) \end{array} \right\} \tag{7.25}$$

and that

$$-(\frac{1}{\gamma} + |\overline{B}|) I_{2(n+1)} \le \begin{pmatrix} \overline{\mathcal{X}}^\omega & 0 \\ 0 & -\overline{\mathcal{Y}}^\omega \end{pmatrix} \le \overline{B} + \gamma \overline{B}^2, \tag{7.26}$$

where I_k denotes the $k \times k$ identity matrix and

$$\left. \begin{array}{rl} (\widehat{a}_1, \widehat{p}_1) = & (\partial_t, \nabla_x)\varphi(\widehat{t}, \widehat{x}, \widehat{t}', \widehat{x}'); \\ (\widehat{a}_2, \widehat{p}_2) = & -(\partial_{t'}, \nabla_{x'})\varphi(\widehat{t}, \widehat{x}, \widehat{t}', \widehat{x}'); \\ \overline{B} = & (\partial_t, \nabla_x, \partial_{t'}, \nabla_{x'}) \otimes (\partial_t, \nabla_x, \partial_{t'}, \nabla_{x'})\varphi(\widehat{t}, \widehat{x}, \widehat{t}', \widehat{x}')(\omega). \end{array} \right\} \tag{7.27}$$

Using definition (7.23) one shows easily that, on the set Γ^*_ε, $|\overline{B}|$ is bounded by some constant $C_{\delta_1, \delta_2} > 0$. In order to translate our argument to the matrix B $(= (\nabla_x, \nabla_{x'}) \otimes (\nabla_x, \nabla_{x'}))$, we introduce the following matrices:

$$\Lambda = \begin{pmatrix} 1 & 0 & 0 & 0 \\ 0 & 0 & I_n & 0 \\ 0 & 1 & 0 & 0 \\ 0 & 0 & 0 & I_n \end{pmatrix}; \quad \widetilde{B}_{\delta_1, \delta_2} = \begin{pmatrix} C_{\delta_1, \delta_2} I_2 & 0 \\ 0 & B \end{pmatrix}. \tag{7.28}$$

Since $\Lambda(\partial_t, \nabla_x, \partial_{t'}, \nabla_{x'})^* = (\partial_t, \partial_{t'}, \nabla_x, \nabla_{x'})^*$ and $\Lambda\Lambda^T = I_{2(n+1)}$, it can be checked that

$$\overline{B} + \gamma\overline{B}^2 \leq \Lambda(\widetilde{B}_{\delta_1,\delta_2} + \gamma\widetilde{B}_{\delta_1,\delta_2}^2)\Lambda^*.$$

Thus, from (7.26) we obtain

$$-(\frac{1}{\gamma} + |\overline{B}|)I_{2(n+1)} \leq \begin{pmatrix} \overline{\mathcal{X}}^\omega & 0 \\ 0 & -\overline{\mathcal{Y}}^\omega \end{pmatrix} \leq \Lambda(\widetilde{B}_{\delta_1,\delta_2} + \gamma\widetilde{B}_{\delta_1,\delta_2}^2)\Lambda^*. \tag{7.29}$$

Further, by the definitions of $\overline{\mathcal{P}}^{2,+}$ and $\overline{\mathcal{P}}^{2,-}$ we see that, for each ω, there exists a sequence $\{(t^\kappa, x^\kappa, a_1^\kappa, p_1^\kappa, S_1^\kappa), (t'^\kappa, x'^\kappa, a_2^\kappa, p_2^\kappa, S_2^\kappa)\}_{\kappa \geq 1}$ such that

$$\left.\begin{array}{rcl}
(a_1^\kappa, p_1^\kappa, S_1^\kappa) & \in & \mathcal{P}^{2,+}v_1(\omega, t^\kappa, x^\kappa); \\
(a_2^\kappa, p_2^\kappa, S_2^\kappa) & \in & \mathcal{P}^{2,-}v_1(\omega, t'^\kappa, x'^\kappa); \\
(t^\kappa, x^\kappa, a_1^\kappa, p_1^\kappa, S_1^\kappa) & \longrightarrow & (\widehat{t}(\omega), \widehat{x}(\omega), \widehat{a}_1(\omega), \widehat{p}_1(\omega), \overline{\mathcal{X}}^\omega), \\
(t'^\kappa, x'^\kappa, a_2^\kappa, p_2^\kappa, S_2^\kappa) & \longrightarrow & (\widehat{t}'(\omega), \widehat{x}'(\omega), \widehat{a}_2(\omega), \widehat{p}_2(\omega), \overline{\mathcal{Y}}^\omega), \quad \kappa \to \infty.
\end{array}\right\} \tag{7.30}$$

In light of (7.22) and (7.29) we may assume without loss of generality that $t'^\kappa < t^\kappa$, $\forall \kappa \geq 1$; and there is some integer $K(\omega, \gamma) \geq 1$ such that whenever $\kappa \geq K(\omega, \gamma)$ one has

$$-(\frac{2}{\gamma} + C_{\delta_1,\delta_2})I_{2(n+1)} \leq \begin{pmatrix} S_1^\kappa & 0 \\ 0 & -S_2^\kappa \end{pmatrix} \tag{7.31}$$

$$\leq \Lambda(\widetilde{B}_{\delta_1,\delta_2} + \gamma(\widetilde{B}_{\delta_1,\delta_2}^2 + \frac{1}{2}I_{2(n+1)}))\Lambda^*.$$

For our measurable selection procedure we now show that the matrices S_1^κ and S_2^κ can be chosen as countable-valued. Indeed, for each $M \geq 1$ and $\gamma \in (0, \infty)$ we define the following two subsets of $\mathcal{S}^{n+1}$:

$$\mathcal{S}^{n+1}(\mathbf{Z}/2^M) = \{S = (s_{ij}) \in \mathcal{S}^{n+1} \mid \text{all } s_{ij} \text{ are of the form} \frac{k}{2^M}, \ k \in \mathbf{Z}\};$$

$$\mathcal{H}_\gamma^{n+1} = \left\{H = (h_{ij}) \in \mathcal{S}^{n+1} \mid 0 \leq h_{ii} - \sum_{j \neq i}|h_{ij}| \leq \sum_{j=1}^d |h_{ij}| \leq \frac{1}{2}\gamma, \ \forall i\right\}.$$

By an elementary algebraic calculus one verifies that for any given $\gamma > 0$ and any $S \in \mathcal{S}^{n+1}$, S allows a decomposition of the form $S = \widehat{S} - H$, where $\widehat{S} \in \mathcal{S}^{n+1}(\mathbf{Z}/2^M)$ and $H \in \mathcal{H}_\gamma^{n+1}$, for M large enough. We will write $S \sim (\widehat{S}, H)$ for this decomposition in the sequel.

Now for any $\gamma > 0$, let $M \geq 1$, $\widehat{S}_1^\kappa, \widehat{S}_2^\kappa \in \mathcal{S}^{n+1}(\mathbf{Z}/2^M)$ and $H_1^\kappa, H_2^\kappa \in \mathcal{H}_\gamma^{n+1}$ be such that $S_i^\kappa \sim (\widehat{S}_i^\kappa, H_i^\kappa)$, $i = 1, 2$. It is easily seen that

$$(a_1^\kappa, p_1^\kappa, \widehat{S}_1^\kappa) \in \mathcal{P}^{2,+}v_1(\omega, t^\kappa, x^\kappa) \quad \text{and} \quad (a_2^\kappa, p_2^\kappa, \widehat{S}_2^\kappa) \in \mathcal{P}^{2,-}v_2(\omega, t'^\kappa, x'^\kappa)$$

still hold, whereas (7.31) becomes, with the notation $I = I_{2(n+1)}$,

$$-(\frac{1}{2}\gamma + \frac{2}{\gamma} + C_{\delta_1,\delta_2})I \leq \begin{pmatrix} \widehat{S}_1^\kappa & 0 \\ 0 & -\widehat{S}_2^\kappa \end{pmatrix} \tag{7.32}$$

$$\leq \Lambda(\widetilde{B}_{\delta_1,\delta_2} + \gamma(\widetilde{B}^2_{\delta_1,\delta_2} + I))\Lambda^*, \quad \kappa \geq \kappa^\omega_\gamma.$$

In particular, for some constant $C_{\delta_1,\delta_2}(\gamma)$,

$$|\widehat{S}^\kappa_1|, |\widehat{S}^\kappa_2| \leq C_{\delta_1,\delta_2}(\gamma), \quad \kappa \geq \kappa^\omega_\gamma. \tag{7.33}$$

Our objective is clear: from the sequence

$$(t^\kappa, t'^\kappa, x^\kappa, x'^\kappa, (a^\kappa_1, p^\kappa_1, \widehat{S}^\kappa_1), (a^\kappa_2, p^\kappa_2, \widehat{S}^\kappa_2)), \quad \kappa \geq 0,$$

which depend on ω, we would like to "select" a sequence of random variables that satisfies (7.30) – (7.32) and the t^κ are replaced by $\mathbf{F}^B$-stopping times. For this we first observe that, since the underlying filtration $\mathbf{F}^B$ is Brownian, all $\mathbf{F}^B$-stopping times are predictable. Hence, there is an increasing sequence of $\mathbf{F}^B$-stopping times $(\tau_\kappa)_{\kappa \geq 1}$ that predicts $\widehat{t}$: $\tau_\kappa \uparrow \widehat{t}$ as $\kappa \to \infty$ and $\tau_\kappa < \widehat{t}$ on $\{\widehat{t} > 0\}$, $\kappa \geq 1$. Now, since

$$E\left[\sup_{\tau_\kappa < t < \widehat{t}+1/\kappa} \left|E\left[(\widehat{t}, \widehat{x}, \widehat{x}', (\widehat{a}_1, \widehat{p}_1), (\widehat{a}_2, \widehat{p}_2)) \mid \mathcal{F}^B_t\right]\right.\right.$$
$$\left.\left. - (\widehat{t}, \widehat{x}, \widehat{x}', (\widehat{a}_1, \widehat{p}_1), (\widehat{a}_2, \widehat{p}_2))\right|^2\right] \longrightarrow 0,$$

as $k \to \infty$, we can find a real sequence $(\rho_m)_{m \geq 1}$ with $0 < \rho_m \leq \frac{1}{m}$, $m \geq 1$, such that, for the sets

$$\Gamma^m_\varepsilon = \Gamma^*_\varepsilon \cap \left\{\sup_{\tau_m < t \leq \widehat{t}+\rho_m} \left|E[(\widehat{t}, \widehat{x}, \widehat{x}', (\widehat{a}_1, \widehat{p}_1), (\widehat{a}_2, \widehat{p}_2)) \mid \mathcal{F}^B_t]\right.\right. \tag{7.34}$$
$$\left.\left. - (\widehat{t}, \widehat{x}, \widehat{x}', (\widehat{a}_1, \widehat{p}_1), (\widehat{a}_2, \widehat{p}_2))\right| \leq \frac{1}{2m}\right\},$$

we have $P(\Gamma^*_\varepsilon \backslash \Gamma^m_\varepsilon) \leq 4^{-m} P(\Gamma^*_\varepsilon)$, $m \geq 1$. Next, for each $m \geq 1$, let Σ_m denote the set of all

$$(\omega, t, t', x, x', (a_1, p_1, S_1), (a_2, p_2, S_2))$$
$$\in \Omega \times [0, T]^2 \times (R^n)^2 \times (R \times R^n \times \mathcal{S}^{n+1}(\mathbf{Z}/2^M))^2$$

which satisfy the following conditions:

$$\left.\begin{array}{l}
\tau_m(\omega) < t \leq (\widehat{t}(\omega) + \rho_m) \wedge T; \quad 0 \leq t' < t; \\
\left|(t', x, x', (a_1, p_1), (a_2, p_2)) - E\left[(\widehat{t}, \widehat{x}, \widehat{x}', (\widehat{a}_1, \widehat{p}_1), (\widehat{a}_2, \widehat{p}_2)) \mid \mathcal{F}^B_t\right]\right| \leq \frac{1}{m}; \\
(a_1, p_1, S_1) \in \mathcal{P}^{2,+} v_1(\omega, t, x); \\
(a_2, p_2, S_2) \in \mathcal{P}^{2,-} v_2(\omega, t', x'); \\
|S_1|, |S_2| \leq C_{\delta_1,\delta_2}(\gamma); \\
\begin{pmatrix} S_1 & 0 \\ 0 & -S_2 \end{pmatrix} \leq \Lambda(\widetilde{B}_{\delta_1,\delta_2} + \gamma(\widetilde{B}^2_{\delta_1,\delta_2} + I_{2(n+1)}))\Lambda^*.
\end{array}\right\} \tag{7.35}$$

Let $\mathcal{O}$ denote the optional σ-field on $\Omega \times [0, T]$ and

$$\mathbf{X} = [0, T] \times (R^n)^2 \times (R \times R^n \times \mathcal{S}^{n+1}(\mathbf{Z}/2^M))^2$$

with generic element of $\mathbf{X}$ being

$$\mathbf{x} = (t', x, x', (a_1, p_1, S_1), (a_2, p_2, S_2)).$$

Since $(a_1, p_1, S_1) \in \mathcal{P}^{2,+}v_1(\omega, t, x)$, $(a_2, p_2, S_2) \in \mathcal{P}^{2,-}v_2(\omega, t', x')$ if and only if

$$\limsup_{\substack{s \to 0 \\ y \to 0}} \frac{1}{|s| + |y|^2} \Big\{ v_1(\omega, t + s, x + y) - v_1(\omega, t, x) - (a_1, p_1)(s, y)$$

$$- \frac{1}{2}(S_1(s, y)) \cdot (s, y) \Big\} \leq 0$$

$$\liminf_{\substack{s \to 0 \\ y \to 0}} \frac{1}{|s| + |y|^2} \Big\{ v_2(\omega, t' + s, x' + y) - v_2(\omega, t', x') - (a_2, p_2) \cdot (s, y)$$

$$- \frac{1}{2}(S_2(s, y)) \cdot (s, y) \Big\} \geq 0,$$

from (7.35) we get that $\Sigma_m \in \mathcal{O} \otimes \mathcal{B}(\mathbf{X})$-measurable. Now for each $m \geq 1$ we apply Proposition 7.7 to conclude that for each $m \geq 1$ there exists some mapping $(\hat{t}_m, \hat{\mathbf{x}}_m) : \Omega \to [0, T] \times \mathbf{X}$ such that

(i) $\hat{t}_m$ is an $\mathbf{F}^B$-stopping time and $\hat{\mathbf{x}}_m$ is an $\mathcal{F}^B_{\hat{t}_m}$-measurable random variable;

(ii) for P-a.e. $\omega \in \{\hat{t}_m < T\}$, $(\omega, \hat{t}_m(\omega), \hat{\mathbf{x}}_m(\omega)) \in \Sigma_m$;

(iii) $P(\hat{t}_m < T) \geq P(\Gamma_\varepsilon^m) - \dfrac{1}{m}$. We recall that

$$(\hat{t}_m, \hat{\mathbf{x}}_m) =: (\hat{t}_m, \hat{x}_m, \hat{x}'_m, (\hat{a}_1^m, \hat{p}_1^m, \hat{S}_1^m), (\hat{a}_2^m, \hat{p}_2^m, \hat{S}_2^m)),$$

$$\Gamma_\varepsilon^* = \liminf_{m \to +\infty} \Gamma_\varepsilon^m.$$

From (7.35) we get that, P-a.e. on Γ_ε^*,

$$(\hat{t}_m, \hat{x}_m, \hat{x}'_m, (\hat{a}_1^m, \hat{p}_1^m), (\hat{a}_2^m, \hat{p}_2^m)) \xrightarrow{m \to +\infty} (\hat{t}, \hat{x}, \hat{x}', (\hat{a}_1, \hat{p}_1), (\hat{a}_2, \hat{p}_2)). \qquad (7.36)$$

Furthermore, for any $\gamma > 0$, $M \geq 1$, the set

$$\hat{S}_M^{2(n+1)}(\gamma) = \left\{ (S_1, S_2) \in \left(S^{n+1}(\mathbf{Z}/2^M) \right)^2 \mid |S_1|, |S_2| \leq C_{\delta_1, \delta_2}(\gamma) \right\}$$

is finite. Thus, without specifying its cardinality we can write $\hat{S}_M^{2(n+1)}(\gamma)$ in form of a finite or countably infinite sequence $(H_\alpha)_{\alpha \geq 1}$. For any $\gamma > 0$, we fix $M > 0$ so that the preceding arguments go through. Then, outside a P-null set,

$$\Gamma_\varepsilon^* \subseteq \bigcup_{H_\alpha \in \hat{S}_M^{2(n+1)}(\gamma)} \left\{ (\hat{S}_1^m(\omega), \hat{S}_2^m(\omega)) = H_\alpha \mid \text{for infinitely many } m \geq 1 \right\}.$$

For each $\alpha \geq 1$ we introduce

$$\Omega_\alpha = \{\omega \in \Omega \mid (\hat{S}_1^m, \hat{S}_2^m) = H_\alpha \text{ for infinitely many } m \geq 1\},$$
$$\overline{\Omega}_\alpha = \Omega_\alpha \setminus \cup_{\ell=1}^{\alpha-1} \Omega_\ell;$$

and we define

$$(\widehat{\mathcal{X}}, \widehat{\mathcal{Y}})(\omega) = \sum_{\alpha=1}^{\infty} H_\alpha(\omega) \mathbf{I}_{\overline{\Omega}_\alpha}(\omega).$$

Since all sets $\overline{\Omega}_\alpha$ are in $\mathcal{F}_{\hat{t}}^B$, $(\widehat{\mathcal{X}}, \widehat{\mathcal{Y}})$ is an $\mathcal{S}^{n+1} \times \mathcal{S}^{n+1}$-valued $\mathcal{F}_{\hat{t}}^B$-measurable random variable and, P-a.e., it is a limit point of the sequence $\{(\widehat{S}_1^m, \widehat{S}_2^m)\}_{m \geq 0}$. Furthermore, by its construction, for P-a.e. $\omega \in \Gamma_\varepsilon^*$,

$$\left((\partial_t, \nabla_x) \varphi(\widehat{t}, \widehat{x}, \widehat{t}', \widehat{x}'), \widehat{\mathcal{X}} \right)(\omega) \in \overline{\mathcal{P}}^{2,+} v_1(\omega, \widehat{t}(\omega), \widehat{x}(\omega)),$$

$$\left(-(\partial_{t'}, \nabla_{x'}) \varphi(\widehat{t}, \widehat{x}, \widehat{t}', \widehat{x}'), \widehat{\mathcal{Y}} \right)(\omega) \in \overline{\mathcal{P}}^{2,-} v_1(\omega, \widehat{t}'(\omega), \widehat{x}'(\omega)),$$

and, with $\overline{B}$ defined by (7.27) we have

$$-(\frac{1}{2}\gamma + \frac{2}{\gamma} + |\overline{B}|)I \leq \begin{pmatrix} \widehat{\mathcal{X}}(\omega) & 0 \\ 0 & -\widehat{\mathcal{Y}}(\omega) \end{pmatrix}$$

$$\leq \Lambda(\widetilde{B}_{\delta_1, \delta_2} + \gamma(\widetilde{B}_{\delta_1, \delta_2}^2 + I))\Lambda^*.$$

Finally, we define the $\mathcal{F}_{\hat{t}}^B$-measurable, $\mathcal{S}^n \times \mathcal{S}^n$-valued random variables $(\mathcal{X}, \mathcal{Y})$ such that

$$\widehat{\mathcal{X}}^\omega = \begin{pmatrix} x_{11} & * \\ * & \mathcal{X} \end{pmatrix}, \quad \widehat{\mathcal{Y}} = \begin{pmatrix} y_{11} & * \\ * & \mathcal{Y} \end{pmatrix}.$$

Then, using definitions of superjets (resp., subjets) it is easy to check that

$$(\widehat{a}_1(\omega), \widehat{p}_1(\omega), \mathcal{X}(\omega)) \in \overline{\mathcal{P}}^{1,2,+} v_1(\omega, \widehat{t}(\omega), \widehat{x}(\omega));$$

$$(\widehat{a}_2(\omega), \widehat{p}_2(\omega), \mathcal{Y}(\omega)) \in \overline{\mathcal{P}}^{1,2,-} v_2(\omega, \widehat{t}'(\omega), \widehat{x}'(\omega)).$$

Moreover, for all $(x, x') \in (R^n)^2$, with the notations $\overline{x} = (0, x)$, $\overline{x}' = (0, x')$, we have

$$\left(\left[B + \gamma B^2 - \begin{pmatrix} \mathcal{X}(\omega) & 0 \\ 0 & -\mathcal{Y}(\omega) \end{pmatrix} \right] \begin{pmatrix} x \\ x' \end{pmatrix} \right) \begin{pmatrix} x \\ x' \end{pmatrix}$$

$$= \left(\left[\overline{B} + \gamma \overline{B}^2 - \begin{pmatrix} \widehat{\mathcal{X}}(\omega) & 0 \\ 0 & -\widehat{\mathcal{Y}}(\omega) \end{pmatrix} \right] \begin{pmatrix} \overline{x} \\ \overline{x}' \end{pmatrix} \right) \begin{pmatrix} \overline{x} \\ \overline{x}' \end{pmatrix} \geq 0.$$

Consequently, the inequality (7.26) remains true with $\overline{\mathcal{X}}^\omega$, $\overline{\mathcal{Y}}^\omega$ and $|\overline{B}|$ being replaced by $\mathcal{X}(\omega)$, $\mathcal{Y}(\omega)$ and C_{δ_1, δ_2}, respectively. The proof is now complete. $\qquad \square$

We are now ready to give the

Proof of the Comparison Theorem 7.1

To begin with, we first claim that under the assumption A.v) we can assume without loss of generality that there exists a constant $\mu > 0$ such that for all $h > 0$ it holds that

$$\widetilde{f}(t, x, y + h, z) - \widetilde{f}(t, x, y, z) \leq -\mu h, \quad \forall (t, x, y, z), \quad P\text{-a.e.} \tag{7.37}$$

The argument that justifies (7.37) is essentially the same as that given for (3.11), the corresponding condition in the deterministic setting.

Now we recall the set $\Gamma_\varepsilon^* = \Gamma_{\varepsilon,N^*(\varepsilon),\delta^*(\varepsilon)}$ defined immediately after Lemma 7.3. We shall prove that a contradiction can be drawn for $\varepsilon > 0$ sufficiently small. Theorem 7.6 allows an approach rather similar to that for the proof of Theorem 3.6, modulo some obviously necessary modifications reflecting the stochastic setting of Theorem 7.1.

Now let $\varepsilon \in (0,\zeta^\ell)$, $\delta_2 \in (0,\varepsilon/(6N^*(\varepsilon)^2)]$, and $\delta_1 \in (0,\delta^*(\varepsilon)) \wedge \delta^{**})$ (recall the Lemmata 7.3 and 7.4). Then as in the deterministic case we choose

$$\gamma = \min\left(\frac{\delta_1}{2(1+\delta_1\delta_2)}, \frac{\delta_2}{(1+\delta_2^2)}\right)$$

and, applying Theorem 7.6, for any $\varepsilon > 0$ we can find a sequence

$$(\widehat{t}_m, \widehat{x}_m, \widehat{t}'_m, \widehat{x}'_m, (\widehat{a}_1^m, \widehat{p}_1^m, \widehat{S}_1^m), (\widehat{a}_2^m, \widehat{p}_2^m, \widehat{S}_2^m))$$

and a sequence of sets Γ_ε^m satisfying i') – iii') of Theorem 7.6, i.e., for $\omega \in \Gamma_\varepsilon^m$,

$$\begin{aligned}
v_1(\omega,t,x) \qquad\qquad\qquad\qquad\qquad\qquad\qquad\qquad\qquad\qquad &(7.38)\\
\leq \quad v_1(\omega,\widehat{t}_m(\omega),\widehat{x}_m(\omega)) + \widehat{a}_1^m(\omega)(t - \widehat{t}_m(\omega)) + \widehat{p}_1^m(\omega)\,(x - \widehat{x}_m(\omega))&\\
+ \frac{1}{2}\left(\widehat{S}_1^m(\omega)(t - \widehat{t}_m(\omega), x - \widehat{x}_m(\omega))\right)(t - \widehat{t}_m(\omega), x - \widehat{x}_m(\omega))&\\
+ o(|t - \widehat{t}_m(\omega)|^2 + |x - \widehat{x}_m(\omega)|^2),&
\end{aligned}$$

as $t \to \widehat{t}_m(\omega)$, $x \to \widehat{x}_m(\omega)$. Clearly, the set

$$\widehat{\Gamma}_\varepsilon^m := \{\omega \in \Omega \mid (6.5)\text{ holds}\} \in \mathcal{F}_{\widehat{t}_m}^B.$$

Hence, setting $\widehat{t}_m = T$ on $\left(\widehat{\Gamma}_\varepsilon^m\right)^c$, $\widehat{t}_m$ remains an $\mathbf{F}^B$-stopping time and

$$\Gamma_\varepsilon^m \subseteq \widehat{\Gamma}_\varepsilon^m \subseteq \{0 < \widehat{t}_m < T\}.$$

Now, for fixed $m \geq 1$ and $(\omega,t,x) \in \Omega \times [0,T] \times R^n$, we define

$$\begin{aligned}
&\varphi_m(\omega,t,x)\\
&= v_1(\omega,\widehat{t}_m(\omega),\widehat{x}_m(\omega)) + (\widehat{a}_1^m(\omega),\widehat{p}_1^m(\omega))\left(t - \widehat{t}_m(\omega), x - \widehat{x}_m(\omega)\right)\\
&\quad + \frac{1}{2}\left(\left(\widehat{S}_1^m(\omega) + \frac{1}{m}I_{2(n+1)}\right)\left(t - \widehat{t}_m(\omega), x - \widehat{x}_m(\omega)\right)\right)\left(t - \widehat{t}_m(\omega), x - \widehat{x}_m(\omega)\right).
\end{aligned}$$

Then $\varphi_m \in C^2(\mathcal{F}_{\widehat{t}_m}^B; [0,T] \times R^n)$; and P-a.e. on $\{0 < \widehat{t}_m < T\}$ it holds that

$$\begin{aligned}
&v_1(\omega,t,x) - \varphi_m(\omega,t,x)\\
&\leq \frac{1}{m}(|t - \widehat{t}_m(\omega)|^2 + |x - \widehat{x}_m(\omega)|^2) + o(|t - \widehat{t}_m(\omega)|^2 + |x - \widehat{x}_m(\omega)|^2),
\end{aligned}$$

as $t \to \widehat{t}_m(\omega)$, $x \to \widehat{x}_m(\omega)$. Since $\varphi_m(\omega,\widehat{t}_m(\omega),\widehat{x}_m(\omega)) = v_1(\omega,\widehat{t}_m(\omega),\widehat{x}_m(\omega))$, one has $v_1(\omega,t,x) \geq \varphi_m(\omega,t,x)$, for all (t,x) in a neighbourhood of $(\widehat{t}_m(\omega),\widehat{x}_m(\omega))$, for P-a.e. $\omega \in \{0 < \widehat{t}_m < T\}$. Therefore, by definition of a stochastic viscosity subsolution (with $g \equiv 0$) we have that

$$\mathcal{A}\varphi_m(\widehat{t}_m,\widehat{x}_m) + \widetilde{f}(\widehat{t}_m,\widehat{x}_m,\varphi_m(\widehat{t}_m,\widehat{x}_m),\nabla_x\varphi_m(\widehat{t}_m,\widehat{x}_m)\sigma^*(\widehat{x}_m)) \geq \partial_t\varphi_m(\widehat{t}_m,\widehat{x}_m)$$

on $\{0 < \widehat{t}_m < T\}$. That is, P-a.e. on Γ_ε^m,

$$\widehat{a}_1^m \;\leq\; \frac{1}{2}\mathrm{tr}\left(\sigma\sigma^*(\widehat{x}_m)\widehat{S}_1^m\right) + \frac{1}{2m}|\sigma(\widehat{x}_m)|^2 + \beta(\widehat{x}_m)\widehat{p}_1^m$$
$$+ \widetilde{f}(\widehat{t}_{.,}, \widehat{x}_m, v_1(\widehat{t}_m, \widehat{x}_m), \widehat{p}_1^m\sigma(\widehat{x}_m))$$

Now, thanks to Theorem 7.6, for P-a.e. $\omega \in \Gamma_\varepsilon^* = \liminf_{m\to\infty}\Gamma_\varepsilon^m$, we can take the limit as $m \to \infty$ to obtain that

$$\widehat{a}_1 \leq \frac{1}{2}\mathrm{tr}\left(\sigma\sigma^*(\widehat{x})\widehat{\mathcal{X}}\right) + \beta(\widehat{x})^*\widehat{p}_1 + \widetilde{f}(\widehat{t}, \widehat{x}, v_1(\widehat{t}, \widehat{x}), \widehat{p}_1\sigma(\widehat{x})),$$

in view of the continuity of the function $\widetilde{f}$. With the notation

$$\widetilde{F}(t, x, y, p, S) = \frac{1}{2}\mathrm{tr}\left(\sigma\sigma^*(x)S\right) + \beta(x)^*p + \widetilde{f}(\omega, t, x, y, p\sigma(x)),$$

then from the definition of $(\widehat{a}_1, \widehat{p}_1)$ it follows,

$$\widetilde{F}\left(\widehat{t}, \widehat{x}, v_1(\widehat{t}, \widehat{x}), \frac{1}{\delta_1}(\widehat{x} - \widehat{x}') + \delta_2\widehat{x}, \mathcal{X}\right) \;\geq\; \frac{1}{2\delta_1}\partial_t\psi_{\delta_1}(\widehat{t} - \widehat{t}') + \delta_1\frac{1}{(T - \widehat{t})^2}. \tag{7.39}$$

Similarly, since v_2 is an ω-wise viscosity supersolution, we easily derive that for all $\omega \in \Gamma_\varepsilon^*$

$$\widetilde{F}\left(\widehat{t}', \widehat{x}', v_2(\widehat{t}', \widehat{x}'), \frac{1}{\delta_1}(\widehat{x} - \widehat{x}') - \delta_2\widehat{x}', \mathcal{Y}\right) \leq \frac{1}{2\delta_1}\partial_t\psi_{\delta_1}(\widehat{t} - \widehat{t}'). \tag{7.40}$$

Now, combining (7.38) – (7.40) and noting that $\Gamma_\varepsilon^* \subseteq \{0 < \widehat{t} < T\}$ by Lemma 5.1, we have, P-a.e. on Γ_ε^*,

$$\begin{aligned}
0 \;<\; & \mu\{\zeta^\ell - 2\varepsilon\} \;\leq\; \mu\{v_1(\widehat{t}, \widehat{x}) - v_2(\widehat{t}', \widehat{x}')\} \\
\leq\; & \widetilde{F}\left(\widehat{t}, \widehat{x}, v_2(\widehat{t}', \widehat{x}'), \frac{1}{\delta_1}(\widehat{x} - \widehat{x}') + \delta_2\widehat{x}, \mathcal{X}\right) \\
& - \widetilde{F}\left(\widehat{t}, \widehat{x}, v_1(\widehat{t}, \widehat{x}), \frac{1}{\delta_1}(\widehat{x} - \widehat{x}') + \delta_2\widehat{x}, \mathcal{X}\right) \\
\leq\; & \widetilde{F}\left(\widehat{t}, \widehat{x}, v_2(\widehat{t}', \widehat{x}'), \frac{1}{\delta_1}(\widehat{x} - \widehat{x}') + \delta_2\widehat{x}, \mathcal{X}\right) \\
& - \widetilde{F}\left(\widehat{t}', \widehat{x}', v_2(\widehat{t}', \widehat{x}'), \frac{1}{\delta_1}(\widehat{x} - \widehat{x}') - \delta_2\widehat{x}', \mathcal{Y}\right) \\
=\; & \frac{1}{2}\mathrm{tr}(\sigma(\widehat{x})\sigma(\widehat{x})^*\mathcal{X} - \sigma(\widehat{x}')\sigma(\widehat{x}')^*\mathcal{Y}) \\
& + \left\{\beta(\widehat{x})(\frac{1}{\delta_1}(\widehat{x} - \widehat{x}') + \delta_2\widehat{x}) - \beta(\widehat{x}')(\frac{1}{\delta_1}(\widehat{x} - \widehat{x}') - \delta_2\widehat{x}')\right\} \\
& + \left\{\widetilde{f}(\widehat{t}, \widehat{x}, v_2(\widehat{t}', \widehat{x}'), (\frac{1}{\delta_1}(\widehat{x} - \widehat{x}') + \delta_2\widehat{x})\sigma(\widehat{x}))\right. \\
& \left. - \widetilde{f}(\widehat{t}', \widehat{x}', v_2(\widehat{t}', \widehat{x}'), (\frac{1}{\delta_1}(\widehat{x} - \widehat{x}') - \delta_2\widehat{x}')\sigma(\widehat{x}'))\right\}
\end{aligned}$$

$$= I_1 + I_2 + I_3,$$

where

$$I_1 = \mathrm{tr}\big(\sigma(\widehat{x})\sigma(\widehat{x})^*\mathcal{X} - \sigma(\widehat{x}')\sigma(\widehat{x}')^*\mathcal{Y}\big)$$

$$= \left(\left(\begin{array}{cc} \mathcal{X} & 0 \\ 0 & -\mathcal{Y} \end{array}\right)\left(\begin{array}{c} \sigma(\widehat{x}) \\ \sigma(\widehat{x}') \end{array}\right)\right)\left(\begin{array}{c} \sigma(\widehat{x}) \\ \sigma(\widehat{x}') \end{array}\right)$$

$$\leq \frac{2}{\delta_1}|\sigma(\widehat{x}) - \sigma(\widehat{x}')|^2 + 2\delta_2\left(|\sigma(\widehat{x})|^2 + |\sigma(\widehat{x}')|^2\right)$$

$$\leq \frac{2L^2}{\delta_1}|\widehat{x} - \widehat{x}'|^2 + \delta_2 C_L\left(1 + |\widehat{x}|^2 + |\widehat{x}'|^2\right);$$

$$I_2 = \beta(\widehat{x})^*\left(\frac{1}{\delta_1}(\widehat{x} - \widehat{x}') + \delta_2\widehat{x}\right) - \beta(\widehat{x}')^*\left(\frac{1}{\delta_1}(\widehat{x} - \widehat{x}') - \delta_2\widehat{x}'\right)$$

$$\leq \frac{L}{\delta_1}|\widehat{x} - \widehat{x}'|^2 + \delta_2 c_L\left(1 + |\widehat{x}|^2 + |\widehat{x}'|^2\right);$$

$$I_3 = \widetilde{f}\big(\widehat{t}, \widehat{x}, v_2(\widehat{t}, \widehat{x}'), (\frac{1}{\delta_1}(\widehat{x}\widehat{x}') + \delta_2\widehat{x})\sigma(\widehat{x})\big)$$

$$- \widehat{f}\big(\widehat{t}', \widehat{x}', v_2(\widehat{t}', \widehat{x}'), (\frac{1}{\delta_1}(\widehat{x} - \widehat{x}') - \delta_2\widehat{x}')\sigma(\widehat{x}')\big)$$

$$\leq L\left(|\widehat{t} - \widehat{t}'| + |\widehat{x} - \widehat{x}'| + \frac{1}{\delta_1}|\widehat{x} - \widehat{x}'|^2\right) + \delta_2 C_L\left(1 + |\widehat{x}|^2 + |\widehat{x}'|^2\right).$$

Since, by the definition of ψ_{δ_1} (recall that $\delta_1 \leq 1$),

$$|\widehat{t} - \widehat{t}'| \leq \delta_1 + |(\widehat{t} - \widehat{t}') - \delta_1|$$

$$\leq \delta_1 + \frac{1}{2}\left(\delta_1^2 + (\frac{\delta_1 - (\widehat{t} - \widehat{t}')}{\delta_1})^2\right)$$

$$\leq 2\delta_1 - \frac{1}{2}\ln\left(1 - (\frac{\delta_1 - (\widehat{t} - \widehat{t}')}{\delta_1})^2\right)$$

$$= 2\delta_1 + \psi_{\delta_1}(\widehat{t} - \widehat{t}'),$$

we find a constant $C_L > 0$ depending only on the Lipschitz constant L and $\sigma(0)$ such that, P-a.e on Γ_ε^*,

$$0 < \mu(\zeta^\ell - 2\varepsilon)$$

$$\leq C_L\left(|\widehat{t} - \widehat{t}'| + |\widehat{x} - \widehat{x}'| + \frac{1}{\delta_1}|\widehat{x} - \widehat{x}'|^2 + \delta_2(1 + |\widehat{x}|^2 + |\widehat{x}'|^2)\right)$$

$$\leq C_L\left(3\delta_1 + \psi_{\delta_1}(\widehat{t} - \widehat{t}') + \frac{2}{\delta_1}|\widehat{x} - \widehat{x}'|^2 + \delta_2(1 + |\widehat{x}|^2 + |\widehat{y}|^2)\right).$$

On the other hand, from the definition of Ψ_{δ_1,δ_2} in (7.9), the point $(\widehat{t}, \widehat{x}, \widehat{t}', \widehat{x}')$ as $\mathbf{F}^B$-maximizer of Ψ_{δ_1,δ_2} and from the relations (7.12) and (7.18) we have

$$\frac{1}{2\delta_1}|\widehat{x} - \widehat{x}'|^2 + \frac{1}{2}\psi_{\delta_1}(\widehat{t} - \widehat{t}') + \frac{\delta_2}{2}\left(|\widehat{x}|^2 + |\widehat{x}'|^2 + \frac{1}{T - \widehat{t}}\right) \qquad (7.41)$$

$$= v_1(\widehat{t},\widehat{x}) - v_2(\widehat{t'},\widehat{x'}) - \Psi_{\delta_1,\delta_2}(\widehat{t},\widehat{x},\widehat{t'},\widehat{x'})$$
$$\leq [v_1(\widehat{t},\widehat{x}) - v_2(\widehat{t},\widehat{x})] + [v_2(\widehat{t},\widehat{x}) - v_2(\widehat{t'},\widehat{x'})] - (\zeta^\ell - 2\varepsilon).$$

On Γ_ε^*, it holds that $\theta \leq \ell$. Hence, by definition (7.3) we have

$$v_1(\omega,\widehat{t}(\omega),\widehat{x}(\omega)) - v_2(\widehat{t}(\omega),\widehat{x}(\omega)) \leq \zeta^\ell, \qquad \omega \in \Gamma_\varepsilon^*;$$

and (7.41) yields that

$$\frac{1}{2\delta_1}|\widehat{x}-\widehat{x'}|^2 + \frac{1}{2}\psi_{\delta_1}(\widehat{t}-\widehat{t'}) + \frac{\delta_2}{2}\left(|\widehat{x}|^2 + |\widehat{x'}|^2 + \frac{1}{T-\widehat{t}}\right)$$
$$\leq |v_2(\widehat{t},\widehat{x}) - v_2(\widehat{t'},\widehat{x'})| + 2\varepsilon \leq 2\ell + 2\varepsilon.$$

Therefore, for any fixed $\delta_2 \in (0, \varepsilon/[6N^*(\varepsilon)^2])$ and any $\delta_1 \in (0, \delta^*(\varepsilon) \wedge \delta^{**})$, we must have

$$|\widehat{x}|^2 + |\widehat{x'}|^2 \leq \frac{4}{\delta_2}(\ell+\varepsilon), \quad \widehat{t} \in \left(0, T - \frac{1}{4\delta_2(\ell+\varepsilon)}\right],$$

and

$$|\widehat{x}-\widehat{x'}|^2 + \frac{(\delta_1 - (\widehat{t}-\widehat{t'}))^2}{\delta_1} \leq |\widehat{x}-\widehat{x'}|^2 + \delta_1\psi_{\delta_1}(\widehat{t}-\widehat{t'}) \leq 4\delta_1(\ell+\varepsilon).$$

Using the ω-wise continuity of v_2 we then conclude that

$$v_2(\widehat{t},\widehat{x}) - v_2(\widehat{t'},\widehat{x'}) \longrightarrow 0 \text{ as } \delta_1 \to 0.$$

Consequently, for $\delta_2 \in (0, \varepsilon/(6N^*(\varepsilon)^2))$, from the estimates made above we obtain P-a.e. on Γ_ε^*

$$0 < \mu(\zeta^\ell - 2\varepsilon)$$
$$\leq C_L\left(3\delta_1 + \psi_{\delta_1}(\widehat{t}-\widehat{t'}) + \frac{2}{\delta_1}|\widehat{x}-\widehat{x'}|^2 + \delta_2(1 + |\widehat{x}|^2 + |\widehat{y}|^2)\right)$$
$$\leq 4C_L\Big\{\delta_1 + \delta_2$$
$$+ \left(\frac{1}{2\delta_1}|\widehat{x}-\widehat{x'}|^2 + \frac{1}{2}\psi_{\delta_1}(\widehat{t}-\widehat{t'}) + \frac{\delta_2}{2}(|\widehat{x}|^2 + |\widehat{x'}|^2 + \frac{1}{T-\widehat{t}})\right)\Big\}$$
$$\leq 4C_L\left(\delta_1 + \delta_2 + 2\varepsilon + |v_2(\widehat{t},\widehat{x}) - v_2(\widehat{t'},\widehat{x'})|\right) \longrightarrow 4C_L(\delta_2 + 2\varepsilon), \delta_1 \to 0.$$

Since $P(\Gamma_\varepsilon^*) > 0$, for all $\varepsilon \in (0, \zeta^\ell)$ and some $\delta \leq \delta(N,\varepsilon)$, $N \geq N^*(\varepsilon)$ and for all $\delta_2 < \varepsilon/(6N^2)$, we deduce from the above estimate that

$$0 < \mu(\zeta^\ell - 2\varepsilon) \leq 8C_L\varepsilon, \ \forall \varepsilon \in (0, \tfrac{1}{2}\zeta^\ell).$$

Letting $\varepsilon \to 0$ we obtain a contradiction to the assumption $P(\zeta > 0) > 0$. Therefore we must have $P(\zeta \leq 0) = 1$, that is, $v_1(t,x) \leq v_2(t,x)$ for all $(t,x) \in [0,T] \times R^d$ P-a.e. The proof of the Comparison Theorem 7.1 is now complete. $\square$

REFERENCES

[1] V. Arkin and M. Saksonov, Necessary optimality conditions for stochastic differential equations, *Soviet Math. Dokl.* **20** (1979), 1–5.

[2] M. Bardi, M.G. Crandall, L.C. Evans, H.M. Soner and P.E. Souganidis, *Viscosity solutions and applications*, Lecture Notes in Math. **1660**, Springer-Verlag, 1997.

[3] G. Barles, R. Buckdahn and E. Pardoux, Backward stochastic differential equations and integral-partial differential equations, *Stochastics Stochastics Rep.* **60** (1997), 57–83.

[4] A. Bensoussan, *Lectures on Stochastic Control, Nonlinear Filtering and Control Theory*, Proceedings of the 3rd 1981 Session, CIME, Lecture Notes in Math. **972**, Springer-Verlag, 1981.

[5] J.M. Bismut, Conjugate convex functions in optimal stochastic control, *J. Math. Anal. Appl.* **44** (1973), 384–404.

[6] J.M. Bismut, Contrôle des systèmes linéaires quadratiques: applications de l'intégrale stochastique, *Sémininaire de Probab.* XII, Lecture Notes in Math. **649**, Springer-Verlag, 1978.

[7] R. Buckdahn and J. Ma, Stochastic viscosity solution for nonlinear partial differential equations (I), To appear in *Stochastic Processes Appl.*

[8] R. Buckdahn and J. Ma, Stochastic viscosity solution for nonlinear partial differential equations (II), To appear in *Stochastic Processes Appl.*

[9] M. G. Crandall and P. L. Lions, Viscosity solutions of Hamilton–Jacobi equations, *Trans. Amer. Math. Soc.* **277** (1983), 1–42.

[10] M. G. Crandall, H. Ishii and P. L. Lions, User's guide to viscosity solutions of second order partial differential equations, *Bull. Amer. Math. Soc. (NS)* **27** (1992), 1–67.

[11] C. Dellacherie and P. Meyer, *Probabilities and Potential*, North Holland, 1978.

[12] H. Doss, Lien entre équations différentielles stochastiques et ordinaires, *Ann. Inst. Henri Poincaré* **13** (1977), 99–125.

[13] D. Duffie and L. Epstein, Stochastic differential utility, *Econometrica* **60** (1992), 353–394.

[14] D. Duffie and L. Epstein, Asset pricing with stochastic differential utility, *The Review of Financial Studies* **5** (1992), 411–436.

[15] D. Duffie and P.L. Lions, PDE solutions of stochastic differential utility, *J. Math. Econom.* **21** (1992), 577–606.

[16] D. Duffie, J. Ma and J. Yong, Black's consol rate conjecture, *Ann. Appl. Probab.* **5** (1994), 356–382.

[17] N. El Karoui and L. Mazliak (eds.), *Backward stochastic differential equations*, Pitman Research Notes in Math. **364**, Addison Wesley Longman Limited, 1997.

[18] N. El Karoui, S. Peng and M. C. Quenez, Backward stochastic differential equations in finance, *Math. Finance* **7** (1997), 1–71.

[19] N. El Karoui and M.C. Quenez, Nonlinear pricing theory and backward stochastic differential equations, in: *B. Biais et al. (eds.), Financial Mathematics*, Lectures given at the 3rd Session of the Centro Internazionale Matematico Estivo (CIME), held in Bressanone, Italy, July 8–13, 1996, Lecture Notes in Math. **1656**, 191–246, Springer-Verlag, Berlin, 1997.

[20] W. H. Fleming and H. M. Soner, *Controlled Markov processes and viscosity solutions*, Springer-Verlag, New York, 1992.

[21] S. Hamadène, Equations différentielles stochastiques rétrogrades: le cas localement lipschitzien, *Ann. Inst. Henri Poincaré* **32** (1996), 645–660.

[22] S. Hamadène, Multidimensional backward SDEs with uniformly continuous coefficients, *Preprint*, 2000.

[23] S. Hamadène and J.P. Lepeltier, Backward equations, stochastic control and zero-sum stochastic differential games, *Stochastics Stochastics Rep.* **54** (1995), 221–231.

[24] S. Hamadène, J.P. Lepeltier and S. Peng, BSDEs with continuous coefficients and application to Markovian nonzero sum stochastic differential games, to appear.

[25] Y. Hu and S. Peng, Probabilistic interpretation for systems of quasilinear elliptic partial differential equations, *Stochastics Stochastics Rep.* **37** (1991), 61–74.

[26] Y. Kabanov, On the Pontryagin Maximum Principle for the the linear stochastic differential equations, *Probabilistic Models and Control of Economical Processes*, CEMI, 1978.

[27] M. Kobylansky, Résultats d'existence et d'unicité pour des équations différentielles stochastiques rétrogrades avec des générateurs à croissance quadratique, *C. R. Acad. Sci. Paris Sér. I Math.* **324** (1) (1997), 81–86.

[28] H. Kunita, *Stochastic Flows and Stochastic Differential Equations*, Cambridge Studies in Advanced Math. **24**, Cambridge University Press, 1990.

[29] J.P. Lepeltier and J. San Martin, Backward stochastic differential equations with continuous generator, *Stat. Probab. Lett.*, to appear.

[30] P.-L. Lions and P.E. Souganidis, Fully nonlinear stochastic partial differential equations, *C.R. Acad. Sci. Paris* **326**, Série 1 (1998), 1085–1092.

[31] P.-L. Lions and P.E. Souganidis, Fully nonlinear stochastic partial differential equations: non-smooth equations and applications, *C.R. Acad. Sci. Paris* **327**, Série 1 (1998), 735–741.

[32] D. Nualart and E. Pardoux, Stochastic calculus with anticipating integrands, *Probab. Theory Relat. Fields* **78** (1988), 535–581.

[33] D. Ocone and E. Pardoux, A generalized Itô–Ventzell formula. Application to a class of anticipating stochastic differential equations, *Ann. Inst. Henri Poincaré* **25**, 1 (1989), 39–71.

[34] E. Pardoux, Stochastic partial differential equations and filtering of diffusion processes, *Stochastics* **3** (1979), 127–167.

[35] E. Pardoux, Quelques méthodes probabilistes pour les équations aux dérivées partielles, *Actes du 30ème Congrès d'Analyse Numérique*: CANum '98 (Arles, 1998), 91–109, ESAM Proc. **6**, Soc. Math. Appl. Indust., Paris, 1999.

[36] E. Pardoux and S. Peng, Adapted solution of a backward stochastic differential equation, *Syst. Control Lett.* **14** (1990), 55–61.

[37] E. Pardoux and S. Peng, Backward stochastic differential equations and quasilinear parabolic partial differential equations, *Lecture Notes in Control and Information Sciences* **176**, 200–217, Springer-Verlag, Berlin, 1992.

[38] E. Pardoux and S. Peng, Backward doubly stochastic differential equations and systems of quasilinear SPDEs, *Probab. Theory Relat. Fields* **98** (1994), 209–227.

[39] S. Peng, Probabilistic interpretation for systems of quasilinear parabolic partial differential equations, *Stochastics Stochastics Rep.* **37** (1991), 61–74.

[40] S. Peng, A generalized dynamic programming principle and Hamilton-Jacobi-Bellman equation, *Stochastics Stochastics Rep.* **38** (1992), 119–134.

[41] S. Peng, A nonlinear Feynman–Kac formula and applications, *Proceedings of Symposium of System Sciences and Control Theory*, Chen and Yong (eds.), 173–184, World Scientific, Singapore, 1992.

[42] S.Peng, Backward stochastic differential equation and its application in optimal control, *Appl. Math. Optim.* **27** (1993), 125–144.

[43] H. Sussmann, On the gap between deterministic and stochastic differential equations, *Ann. Probab.* **6** (1978), 19–41.

ISOLATED SINGULAR POINTS
OF STOCHASTIC DIFFERENTIAL
EQUATIONS*

A.S. CHERNY[1] and H.-J. ENGELBERT[2]

[1] Moscow State University
Department of Probability Theory
119899 Moscow, Russia

[2] Friedrich-Schiller-Universität Jena
Institut für Stochastik
D-07743 Jena, Germany

Abstract We consider a one-dimensional homogeneous stochastic differential equation of the form

$$dX_t = b(X_t)dt + \sigma(X_t)dB_t, \quad X_0 = x,$$

where b and σ are supposed to be measurable functions and $\sigma \neq 0$. No assumptions of boundedness (or boundedness away from zero) are imposed. We introduce a class of points called *isolated singular points* and investigate the weak existence as well as the uniqueness in law of the solution in the neighbourhood of such a point. A complete qualitative classification of these points is presented. There are 63 different types. The constructed classification allows us to find out whether a solution can reach an isolated singular point, whether it can leave this point, and so on. It has been found that, for 59 types, there exists a unique solution in the neighbourhood of the corresponding isolated singular point. Moreover, the solution is a strong Markov process. The remaining 4 types of isolated singular points (we call them *branch types*) disturb the uniqueness. One can construct various "bad" solutions in the neighbourhood of a branch point. In particular, there exist non-Markov solutions. As an application of the obtained results, we consider equations of the form

$$dX_t = \mu|X_t|^\alpha dt + \nu|X_t|^\beta dB_t, \quad X_0 = x,$$

and present the classification for this case.

*Research supported by Grant INTAS 97-30204.

Key Words Stochastic differential equations, singular coefficients, solutions up to a random time, continuous strong Markov processes, local characteristics of a diffusion, speed measure, scale function.

1. INTRODUCTION

1.1. The basis of the theory of *diffusion processes* was formed by Kolmogorov in [18] (the Chapman–Kolmogorov equation, forward and backward partial differential equations). This theory was further developed in a series of papers by Feller (see, for example, [9], [10]). In particular, Feller described the *boundary behaviour* of a diffusion process.

Itô [12], [13] proposed an alternative approach to constructing diffusions. He introduced the notion of a *stochastic differential equation* (abbreviated below as *SDE*). Stroock and Varadhan [24] introduced the concept of a *martingale problem* which is closely connected with the notion of a SDE.

Itô, McKean [14] and Dynkin [4] proposed another approach to the diffusion processes. They proved that a one-dimensional continuous *strong Markov* process that satisfies an additional *regularity* condition can be obtained from a Brownian motion by the following three operations: *random time-change, transformation of the phase space* and *killing at a random time*.

The relationship between continuous strong Markov processes and martingale or semimartingale solutions of SDEs is still an interesting problem to be studied. Engelbert and Schmidt proved in [8] that any continuous strong Markov local martingale can be obtained from a solution of a SDE without drift through a special form of *time delay*. Çinlar, Jacod, Protter and Sharpe presented in [3] conditions for a regular continuous strong Markov process to be a semimartingale. Schmidt [23] gave a criterion for a regular continuous strong Markov process to be a solution of a SDE. Similar problems for continuous strong Markov processes with no regularity assumptions were treated by Assing and Schmidt in [1].

1.2. In this paper, we investigate one-dimensional homogeneous SDEs of the form

$$dX_t = b(X_t)\,dt + \sigma(X_t)\,dB_t, \quad X_0 = x, \tag{1.1}$$

where $(B_t)_{t\geq 0}$ is a standard linear Brownian motion and x is a real number.

We will study the following main problems:

I. *Does there exist a solution of* (1.1) *and is it unique?*

II. *Does it have the strong Markov property?*

III. *What is the qualitative behaviour of the solution?*

Let us first describe the known results related to the existence and uniqueness of solutions of such equations. Most of these results are connected with more general multidimensional inhomogeneous SDEs, i.e., equations of the form

$$dX_t^i = b^i(t, X_t)\,dt + \sum_{j=1}^{n} \sigma^{ij}(t, X_t)\,dB_t^j, \quad X_0^i = x^i \quad (i = 1, \dots, n). \tag{1.2}$$

The first sufficient condition for the existence and the uniqueness of a solution of (1.2) was obtained by Itô [13]. This condition requires that the coefficients b and σ are locally Lipschitzian.

Stroock and Varadhan [24] proved that there exists a unique solution of (1.2) under the assumption that b is measurable and bounded, while σ is continuous and strictly elliptic. They also proved the following statement. If the coefficients b and σ do not depend on t and, for any x, there exists a unique solution of (1.2), then this solution is a strong Markov process.

Krylov [19], [20] considered multidimensional *homogeneous* SDEs and proved the existence and the uniqueness for the case where b is measurable and bounded, while σ is measurable and strictly elliptic (no continuity assumption on σ was imposed). For the case $n > 2$, an additional assumption was made to guarantee the uniqueness.

The conditions imposed on b and σ in the papers of Stroock, Varadhan, Krylov were much weaker than Itô's condition. Ershov and Shiryaev introduced the notions of *weak* and *strong* solutions (the definitions can be found, for example, in the book [21] by Liptser and Shiryaev). According to this terminology, the solution constructed by Itô is a *strong* solution while the solutions constructed in the later papers (under much weaker assumptions) are *weak* solutions. The relationship between weak and strong solutions was investigated in the paper [25] by Zvonkin and Krylov.

For the special case of one-dimensional homogeneous SDEs (i.e., SDEs of the form (1.1)), there exist much better sufficient conditions for the weak existence and the uniqueness of a solution. This was shown by Engelbert and Schmidt in [5] – [8] (for the case $b = 0$, they gave necessary and sufficient conditions). Engelbert and Schmidt proved that if $\sigma(x) \neq 0$ for any $x \in \mathbb{R}$ and

$$\frac{1 + |b|}{\sigma^2} \in L^1_{\text{loc}}(\mathbb{R}), \tag{1.3}$$

then, for any $x \in \mathbb{R}$, there exists a unique (weak) solution of (1.1).

1.3. The main goal of this paper is to investigate one-dimensional homogeneous SDEs for which condition (1.3) is violated, i.e., SDEs with *singular coefficients*. The only assumption we make from the outset is that $\sigma(x) \neq 0$ for any $x \in \mathbb{R}$. We will only deal with weak solutions and study the above mentioned Problems I, II, III.

The importance of the stochastic differential equations with singular coefficients both for the theory and for the practical applications can be shown by the following arguments.

There are many examples of SDEs that arise naturally in the stochastic analysis and do not satisfy condition (1.3). Such are, for example, the equations for *Bessel processes* and for the *squares of Bessel processes*.

SDEs with singular coefficients are essential for various applications of the stochastic analysis. Indeed, suppose that we model some process as a solution of SDE (1.1). Assume that this process is positive by its nature (for example, it is the price of a stock on the securities market or the size of a population). Then the SDE used to model such a process should have singular coefficients. The reason is as follows. If condition (1.3) is satisfied, then, for any $x > 0$ and any $a < 0$, the solution started at x reaches the level a with positive probability (see Theorem 7.1 in Section 7).

In order to investigate SDEs with singular coefficients, we introduce the following definition. A point $d \in \mathbb{R}$ is called a *singular* point for SDE (1.1) if

$$\frac{1 + |b|}{\sigma^2} \notin L^1_{\text{loc}}(d)$$

(see Definition 4.1 for the notation $L^1_{\text{loc}}(d)$). According to this definition, any point $d \in \mathbb{R}$ is either a singular point or a regular one. It turns out that there exists a *qualitative* difference between these two classes of points. This difference is expressed in terms of the behaviour of a solution in the neighbourhood of the corresponding point (see Section 4).

Using the above terminology, we can say that a SDE has singular coefficients if and only if the set of its singular points is nonempty. It is worth noting that in practice one often comes across SDEs that have only one singular point (usually, it is zero). Thus, the most important class of singular points is formed by the *isolated singular points*. (We call $d \in \mathbb{R}$ an isolated singular point if d is singular and there exists a deleted neighbourhood of d that consists of regular points).

In this paper, we present a complete qualitative classification of isolated singular points. This classification allows us to make conclusions about the qualitative behaviour of a solution in the neighbourhood of the corresponding point. In particular, the classification allows us to find out whether a solution can reach an isolated singular point and whether it can leave this point. This is done through the coefficients b and σ.

In order to perform this classification, we investigate the behaviour of a solution first in the right-hand neighbourhood of an isolated singular point and then in the left-hand neighbourhood. It has been found that there exist 8 qualitative types of the behaviour of a solution in the one-sided neighbourhood of a point. Therefore, there exist $63(= 8^2 - 1)$ qualitative types of isolated singular points (see Section 7).

1.4. This paper is arranged as follows. Section 2 contains some definitions related to SDEs.

In Section 3, we cite some definitions and statements related to regular continuous strong Markov processes. The behaviour of such a process in the right-hand (left-hand) neighbourhood of a point d may be described by 4 parameters $e_1, \ldots, e_4$. These parameters were introduced by Feller [9], Itô and McKean [14]. The parameters show whether the process may leave the point d in the right (left) direction, whether it may reach d from the right (left) side, and so on.

In Section 4, we give the definition of a *singular point*. We prove several statements which show that there exists a qualitative difference between the singular points and the regular ones. This confirms that the given definition of a singular point is reasonable.

In Section 5, we present two examples of a SDE with a singular point. These examples show how a solution may behave in the neighbourhood of such a point. In particular, we investigate the existence and the uniqueness of a solution for SDEs governing Bessel processes.

In Section 6, we define a *solution up to a random time*. This notion is necessary for several reasons (in particular, for treating the explosions). Solutions up to a random time were also considered in [7], [8], [17; Ch. 5, (5.1)].

The most important part of this paper is Section 7. In this section, we investigate the behaviour of a solution in the right-hand neighbourhood of an isolated singular point. We prove that, for any x out of this neighbourhood, there exists a solution defined up to a random time of a special form. Moreover, this solution is a strong Markov process. The local characteristics $e_1, \ldots, e_4$ of this process as well as its *speed measure* and *scale function* are expressed by b and σ. This leads to the qualitative classification of *right types* of isolated singular points. It has been found that there are 8 different types. Furthermore, we show that an isolated singular point can have one of 63 possible types. The one-sided classification of isolated singular points is illustrated diagrammatically in Figure 1.

In Section 8, the above classification is applied to the power equations, i.e., equations of the form

$$dX_t = \mu|X_t|^\alpha dt + \nu|X_t|^\beta dB_t, \quad X_0 = x.$$

The right types of zero for this SDE can easily be expressed by μ, ν, α and β (see Figure 2).

2. STOCHASTIC DIFFERENTIAL EQUATIONS

Definition 2.1 A *solution* of SDE (1.1) is a *pair* (Y, B) of adapted processes on a filtered probability space $(\Omega, \mathcal{G}, (\mathcal{G}_t), Q)$ such that

i) B is a $(\mathcal{G}_t, Q)$-Brownian motion (i.e., it is a Brownian motion and a $(\mathcal{G}_t, Q)$-martingale);

ii) for any $t \geq 0$,

$$\int_0^t \left(|b(Y_s)| + \sigma^2(Y_s)\right) ds < \infty \quad \text{Q-a.s.};$$

iii) for any $t \geq 0$,

$$Y_t = x + \int_0^t b(Y_s)\, ds + \int_0^t \sigma(Y_s)\, dB_s \quad \text{Q-a.s.}$$

Definition 2.2 A solution (Y, B) is called a *strong* solution if the process Y is adapted to the filtration $(\overline{\mathcal{F}}_t^B)$, i.e., the completed natural filtration of B.

A solution in the sense of Definition 2.1 is called a *weak* solution.

Definition 2.3 There is *uniqueness in law* for (1.1) if whenever (Y, B) and $(\widetilde{Y}, \widetilde{B})$ are two solutions (which may be defined on different probability spaces) with the same starting point, then the laws of Y and $\widetilde{Y}$ are equal.

Definition 2.4 There is *pathwise uniqueness* for (1.1) if whenever (Y, B) and $(\widetilde{Y}, B)$ are two solutions on the same filtered probability space with the same starting point, then Y and $\widetilde{Y}$ are indistinguishable.

From here on, it will be more convenient for us to use another definition of a solution (which is equivalent to Definition 2.1). In order to give this definition, we need some notation.

Let $C(\mathbb{R}_+)$ be the space of continuous functions $\mathbb{R}_+ \to \mathbb{R}$, where $\mathbb{R}_+ = [0, \infty)$. Let $X = (X_t)_{t \geq 0}$ denote the *coordinate process* on $C(\mathbb{R}_+)$, i.e., X is defined by

$$X_t : C(\mathbb{R}_+) \ni \omega \longmapsto \omega(t) \in \mathbb{R}. \tag{2.1}$$

Let $(\mathcal{F}_t)$ be the *canonical filtration* on $C(\mathbb{R}_+)$, i.e., $\mathcal{F}_t = \sigma(X_s; \ s \leq t)$, and $\mathcal{F}$ be the Borel σ-field on $C(\mathbb{R}_+)$. Note that $\mathcal{F} = \bigvee_{t \geq 0} \mathcal{F}_t$.

Definition 2.5 A *solution* of SDE (1.1) is a *measure* P on $\mathcal{F}$ such that
  i) $P\{X_0 = x\} = 1$;
  ii) for any $t \geq 0$,

$$\int_0^t \left(|b(X_s)| + \sigma^2(X_s) \right) ds < \infty \quad \text{P-a.s.};$$

  iii) the process

$$M_t = X_t - \int_0^t b(X_s)\, ds \tag{2.2}$$

is a $(\mathcal{F}_t, P)$-local martingale;
  iv) the process

$$M_t^2 - \int_0^t \sigma^2(X_s)\, ds \tag{2.3}$$

is a $(\mathcal{F}_t, P)$-local martingale.
  In the following, we will call P a *solution started at* x.

Remarks (i) If one accepts Definition 2.5, then the *uniqueness* of a solution does not need a special definition.

(ii) Definitions 2.1 and 2.5 do not cover the case of *exploding* solutions. In Section 6, we give the definition of a *solution up to a random time*. This makes it possible to consider explosions. □

The following statement relates Definition 2.1 and Definition 2.5.

Theorem 2.6 *Suppose that* $\sigma(x) \neq 0$ *for all* $x \in \mathbb{R}$.
  (i) *Let* (Y, B) *be a solution of* (1.1) *in the sense of Definition 2.1. Then the measure* $P = \text{Law}(Y_t; \ t \geq 0)$ *is a solution of* (1.1) *in the sense of Definition 2.5.*
  (ii) *Let* P *be a solution of* (1.1) *in the sense of Definition 2.5. Then the pair* (Y, B) *defined by*

$$Y_t = X_t, \quad B_t = \int_0^t \frac{1}{\sigma(X_s)}\, dX_s - \int_0^t \frac{b(X_s)}{\sigma(X_s)}\, ds \tag{2.4}$$

is a solution of (1.1) *on the filtered probability space* $(C(\mathbb{R}_+), \mathcal{F}, (\mathcal{F}_t), P)$ *in the sense of Definition 2.1.*

  The proof is straightforward.

3. CONTINUOUS STRONG MARKOV PROCESSES

We will add an isolated point $\{\Delta\}$ to the real line and consider the functions taking values in $\mathbb{R} \cup \{\Delta\}$.

Definition 3.1 The space $\overline{C}(\mathbb{R}_+)$ consists of the functions $f : \mathbb{R}_+ \to \mathbb{R} \cup \{\Delta\}$ with the following property: there exists a time $\xi(f) \in [0, \infty]$ such that f is continuous on $[0, \xi(f))$ and $f = \Delta$ on $[\xi(f), \infty)$. The time $\xi(f)$ is called the *killing time* of f.

Throughout this section, $X = (X_t)_{t \geq 0}$ denotes the coordinate process on $\overline{C}(\mathbb{R}_+)$, i.e.,

$$X_t : \overline{C}(\mathbb{R}_+) \ni \omega \mapsto \omega(t) \in \mathbb{R} \cup \{\Delta\}; \tag{3.1}$$

$(\mathcal{F}_t)$ will be the canonical filtration on $\overline{C}(\mathbb{R}_+)$, i.e., $\mathcal{F}_t = \sigma(X_s; s \leq t)$; $\mathcal{F}$ will stand for the σ-field $\bigvee_{t \geq 0} \mathcal{F}_t = \sigma(X_s; s \geq 0)$. Note that $(\mathcal{F}_t)$ is *not* right-continuous. We therefore introduce the filtration $\mathcal{F}_t^+ = \bigcap_{\varepsilon > 0} \mathcal{F}_{t+\varepsilon}$ $(t \geq 0)$.

Remark The space $\overline{C}(\mathbb{R}_+)$ may be endowed with a metric that turns it into a *Polish* space. Moreover, the corresponding Borel σ-field coincides with $\mathcal{F}$. The space $C(\mathbb{R}_+)$ is a closed subset of $\overline{C}(\mathbb{R}_+)$ in this metric. □

In the following reasoning, we will use the notations:

$$T_a = \inf\{t \geq 0 : X_t = a\}, \tag{3.2}$$
$$\overline{T}_a = \sup_n \inf\{t \geq 0 : |X_t - a| \leq 1/n\}, \tag{3.3}$$
$$T_{a,b} = T_a \wedge T_b, \tag{3.4}$$
$$\overline{T}_{a,b} = \overline{T}_a \wedge \overline{T}_b. \tag{3.5}$$

Here, $a, b \in \mathbb{R}$. On the set $\{X_0 = \Delta\}$, we have $T_a = \overline{T}_a = \infty$. Note that $T_a \neq \overline{T}_a$ because the process X may be killed just before it reaches a.

Definition 3.2 Let $I \subseteq \mathbb{R}$ be an interval which may be closed, open or semi-open. A *continuous strong Markov process on* I is a family $(\mathsf{P}_x)_{x \in I}$ of probability measures on $\mathcal{F}$ such that
i) for any $x \in I$,

$$\mathsf{P}_x\{X_0 = x\} = 1, \quad \mathsf{P}_x\{\forall t \geq 0, X_t \in I \cup \{\Delta\}\} = 1;$$

ii) for any $A \in \mathcal{F}$, the map $x \mapsto \mathsf{P}_x(A)$ is Borel-measurable;
iii) for any $(\mathcal{F}_t^+)$-stopping time T, any $\mathcal{F}$-measurable nonnegative function Ψ and any $x \in I$,

$$\mathsf{E}_{\mathsf{P}_x}[\Psi \circ \Theta_T | \mathcal{F}_T^+] = \mathsf{E}_{\mathsf{P}_{X_T}}[\Psi] \quad \mathsf{P}_x\text{-a.s.}$$

on the set $\{X_T \neq \Delta\}$. Here, Θ_T is the shift on $\overline{C}(\mathbb{R}_+)$ defined as follows:

$$(\Theta_T \circ X)_t = \begin{cases} X_{t+T} & \text{if } T < \infty, \\ \Delta & \text{if } T = \infty. \end{cases} \tag{3.6}$$

(Obviously, Θ_T is $\mathcal{F}|\mathcal{F}$-measurable).

Definition 3.3 A *regular* continuous strong Markov process on I is a family $(P_x)_{x \in I}$ that satisfies properties i) – iii) of Definition 3.2 as well as the following conditions:

iv) for any $x \in I$, we have on the set $\{\xi < \infty\}$: $\lim_{t \uparrow \xi} X_t$ exists and does not belong to I P_x-a.s. (here, $\xi = \inf\{t \geq 0 : X_t = \Delta\}$). In other words, X can be killed only at the endpoints of I that do not belong to I;

v) for any $x \in \overset{\circ}{I}$ and any $y \in I$, we have $P_x\{\exists t \geq 0 : X_t = y\} > 0$, where $\overset{\circ}{I}$ denotes the interior of I.

From here on, we will call regular continuous strong Markov processes simply *regular* processes.

Proposition 3.4 *Suppose that* $(P_x)_{x \in I}$ *is a regular process. There exists a continuous strictly increasing function* $s : \overset{\circ}{I} \to \mathbb{R}$ *such that* $s(X^{T_{a,b}})$ *is a* P_x*-local martingale for any* $a \leq x \leq b$ *in* $\overset{\circ}{I}$ *(here,* $X^{T_{a,b}}$ *is the process* X *stopped at* $T_{a,b}$, *where* $T_{a,b}$ *is defined in (3.4)). Furthermore, the function* s *is determined uniquely up to an affine transformation, and it satisfies the following property: for any* $a \leq x \leq b$ *in* I,

$$P_x\{T_b < T_a\} = \frac{s(x) - s(a)}{s(b) - s(a)}.$$

For the proof, see [22; Ch. VII, (3.2)] or [16; (20.7)].

Definition 3.5 A function s with the properties stated in Proposition 3.4 is called the *scale function* of the process $(P_x)_{x \in I}$.

We now turn to another characteristic of a regular process. For $a \leq b$ in I, set

$$G_{a,b}(x,y) = \frac{\big(s(x) \wedge s(y) - s(a)\big)\big(s(b) - s(x) \vee s(y)\big)}{s(b) - s(a)}, \quad x,y \in [a,b],$$

where s is a variant of the scale function.

Proposition 3.6 *For a regular process* $(P_x)_{x \in I}$, *there exists a unique measure* m *on* $\overset{\circ}{I}$ *such that, for any nonnegative function* f *and any* $a \leq x \leq b$ *in* I,

$$E_{P_x}\left[\int_0^{T_{a,b}} f(X_s)\, ds\right] = 2 \int_a^b G_{a,b}(x,y)\, f(y)\, m(dy). \tag{3.7}$$

For the proof, see [22; Ch. VII, (3.6)] or [16; (20.10)].

Definition 3.7 The measure m given by Proposition 3.6 is called the *speed measure* of the process $(P_x)_{x \in I}$.

Remarks (i) Some authors use the term *speed measure* for $2m$ instead of m.

(ii) The measure m is unique for a fixed choice of the scale function. If another variant of the scale function is taken, then one gets a different G and, as a result, a new m.   □

Let $(P_x)_{x \in I}$ be a continuous strong Markov process and $d \in I \setminus \{r\}$, where r denotes the right endpoint of I. The behaviour of the process (P_x) in the right-hand neighbourhood of d may be described by the following parameters:

$$
\begin{aligned}
e_1 &= \lim_{x \downarrow d} P_d\{T_x < \theta\}, \\
e_2 &= \lim_{y \downarrow d} \lim_{x \downarrow d} P_x\{T_y < \theta\}, \\
e_3 &= \lim_{x \downarrow d} P_x\{T_d < \theta\}, \\
e_4 &= \lim_{x \downarrow d} P_x\{\overline{T}_d < \theta\},
\end{aligned}
\tag{3.8}
$$

where T_d, $\overline{T}_d$ are defined in (3.2), (3.3) and $\theta > 0$ is an arbitrary constant.

Proposition 3.8 *The values $e_1, \ldots, e_4$ do not depend on the choice of $\theta > 0$. Moreover, they can form only the following combinations:*

e_1	e_2	e_3	e_4
1	1	1	1
1	1	0	0
0	0	p	1
0	0	0	0
0	1	0	0

Here, p may take any value from $[0, 1]$.

For the proof, see [14; §3.3].

Remark The value $p \in (0, 1)$ corresponds to the case where the process is killed with probability $1 - p$ just before it reaches d. $\qquad\square$

4.   ISOLATED SINGULAR POINTS: THE REASONING

Throughout this section, we assume that $\sigma(x) \neq 0$ for all $x \in \mathbb{R}$. By X we denote the coordinate process on $C(\mathbb{R}_+)$ (see (2.1)).

Definition 4.1 A measurable function $f : \mathbb{R} \to \mathbb{R}$ is *locally integrable at a point $d \in \mathbb{R}$* if there exists $\delta > 0$ such that

$$
\int_{d-\delta}^{d+\delta} |f(x)| \, dx < \infty.
$$

We will use the notation: $f \in L^1_{\text{loc}}(d)$.

Definition 4.2 A measurable function f is *locally integrable on a set $D \subseteq \mathbb{R}$* if f is locally integrable at each point $d \in D$. Notation: $f \in L^1_{\text{loc}}(D)$.

Proposition 4.3 *Suppose that, for SDE* (1.1),

$$\frac{1 + |b|}{\sigma^2} \in L^1_{loc}(\mathbb{R}).$$

Then, for any $x \in \mathbb{R}$, there exists a unique solution of (1.1).

For the proof, see [7] or [8].

Remark The solution constructed in Proposition 4.3 may explode at a finite time. □

Section 7 contains the following local analog of Proposition 4.3 (see Theorem 7.1). *If the function $(1 + |b|)/\sigma^2$ is locally integrable at a point d, then there exists a unique solution of* (1.1) *"in the neighbourhood of d"*. Therefore, such a point should be called "regular".

Definition 4.4 A point $d \in \mathbb{R}$ is called a *singular point* for SDE (1.1) if

$$\frac{1 + |b|}{\sigma^2} \notin L^1_{loc}(d).$$

A point that is not singular will be called *regular*.

Definition 4.5 A point $d \in \mathbb{R}$ is called an *isolated singular point* for (1.1) if d is singular and there exists a deleted neighbourhood of d that consists of regular points.

The next four statements are intended to show that the singular points in the sense of Definition 4.4 are indeed "singular".

Proposition 4.6 *Suppose that $|b|/\sigma^2 \in L^1_{loc}(\mathbb{R})$ and $1/\sigma^2 \notin L^1_{loc}(d)$. Then there exists no solution of* (1.1) *started at d.*

For the proof, see [7] or [8].

Theorem 4.7 *Let I be an open interval. Suppose that $|b|/\sigma^2 \notin L^1_{loc}(x)$ for any $x \in I$. Then, for any $x \in I$, there exists no solution of* (1.1) *started at x.*

Proof. Suppose that P is a solution started at $x \in I$. By the occupation times formula (see [22; Ch. VI, (1.6)]) and by the definition of a solution, we have

$$\int_0^t |b(X_s)|\, ds = \int_0^t \frac{|b(X_s)|}{\sigma^2(X_s)}\, d\langle X \rangle_s = \int_{\mathbb{R}} \frac{|b(y)|}{\sigma^2(y)} L^y_t(X)\, dy < \infty \quad \text{P-a.s.} \qquad (4.1)$$

Here, $L^y_t(X)$ denotes the local time of X spent in y up to t. As $L^y_t(X)$ is right-continuous in y (see [22; Ch. VI, (1.7)]), we deduce that

$$P\{\forall t \ge 0,\ \forall y \in I,\ L^y_t(X) = 0\} = 1.$$

Therefore, for the stopping time $T = 1 \wedge \inf\{t \ge 0 : X_t \notin I\}$, one has

$$T = \int_0^T 1\, ds = \int_0^T \sigma^{-2}(X_s)\, d\langle X \rangle_s$$

$$= \int_{\mathbb{R}} \sigma^{-2}(y) L^y_T(X)\, dy = \int_I \sigma^{-2}(y) L^y_T(X)\, dy = 0.$$

(We used the fact that $L^y_T(X) = 0$ for $y \notin I$; see [22; Ch. VI, (1.3)]). This leads to a contradiction since $T > 0$.   □

Theorem 4.8 *Suppose that d is a singular point for* (1.1) *and* P *is a solution started at a point x. Then*

$$L_t^d(X) = L_t^{d-}(X) = 0 \quad \text{P-a.s.}$$

for all $t \geq 0$.

Proof. Since d is a singular point, we have

$$\forall \varepsilon > 0, \quad \int_d^{d+\varepsilon} \frac{1 + |b(x)|}{\sigma^2(x)} \, dx = \infty \tag{4.2}$$

or

$$\forall \varepsilon > 0, \quad \int_{d-\varepsilon}^d \frac{1 + |b(x)|}{\sigma^2(x)} \, dx = \infty. \tag{4.3}$$

If (4.2) is satisfied, then (4.1), together with the right-continuity of $L_t^y(X)$ in y, guarantees that $\forall t \geq 0$, $L_t^d(X) = 0$ P-a.s. If (4.3) is satisfied, then $\forall t \geq 0$, $L_t^{d-}(X) = 0$ P-a.s.

Let B be the process defined in (2.4). Then

$$\int_0^t I(X_s = d) \, dX_s = \int_0^t I(X_s = d) \, b(X_s) \, ds + \int_0^t I(X_s = d) \, \sigma(X_s) \, dB_s$$

$$= \int_0^t I(X_s = d) \, b(X_s) \, ds + M_t,$$

where M is a $(\mathcal{F}_t, P)$-local martingale (here, $(\mathcal{F}_t)$ is the canonical filtration on $C(\mathbb{R}_+)$). By the occupation times formula (see [22; Ch. VI, (1.6)]),

$$\int_0^t I(X_s = d) \, b(X_s) \, ds = \int_0^t \frac{I(X_s = d) \, b(X_s)}{\sigma^2(X_s)} \, d\langle X \rangle_s$$

$$= \int_{\mathbb{R}} \frac{I(x = d) \, b(x)}{\sigma^2(x)} \, L_t^x(X) \, dx = 0 \quad \text{P-a.s.}$$

Similarly,

$$\langle M \rangle_t = \int_0^t I(X_s = d) \, \sigma^2(X_s) \, ds = 0 \quad \text{P-a.s.}$$

Therefore,

$$\int_0^t I(X_s = d) \, dX_s = 0.$$

This equality, combined with the properties of the local times (see [22; Ch. VI, (1.7)]), guarantees that

$$\forall t \geq 0, \quad L_t^d(X) = L_t^{d-}(X) \quad \text{P-a.s.} \tag{4.4}$$

We have already proved that $L_t^d(X) = 0$ or $L_t^{d-}(X) = 0$. This, together with (4.4), leads to the desired statement. $\qquad\square$

Theorem 4.9 *Suppose that d is a regular point for SDE* (1.1) *and* P *is a solution started at a point x. Suppose moreover that* $P\{T_d < \infty\} > 0$ *(*T_d *is defined in* (3.2)*). Then, on the set* $\{t > T_d\}$, *we have*

$$L_t^d(X) > 0, \quad L_t^{d-}(X) > 0 \quad \text{P-a.s.}$$

This theorem can be derived from Theorem 7.1 in Section 7.

Theorem 4.10 *Suppose that*

$$\frac{1 + |b|}{\sigma^2} \in L^1_{\text{loc}}(\mathbb{R} \setminus \{0\}), \quad \frac{1 + |b|}{\sigma^2} \notin L^1_{\text{loc}}(0).$$

Then there are only 4 possibilities:
1. *There is no solution started at zero.*
2. *There exists a unique solution started at zero, and it is nonnegative (i.e., $P\{\forall t \geq 0, X_t \geq 0\} = 1$).*
3. *There exists a unique solution started at zero, and it is nonpositive.*
4. *There exist a nonnegative solution as well as a nonpositive solution started at zero. In this case, alternating solutions may also exist.*

This theorem follows from the results of Section 7.

Proposition 4.3 and Theorem 4.10 illustrate the qualitative difference between the singular points and the regular ones. If the conditions of Proposition 4.3 are satisfied (in this case, zero is a regular point), then there exists a unique solution P started at zero. Moreover, this solution has alternating signs, i.e.,

$$P\{\forall \varepsilon > 0 \ \exists t < \varepsilon : X_t > 0\} = 1, \quad P\{\forall \varepsilon > 0 \ \exists t < \varepsilon : X_t < 0\} = 1.$$

These properties follow from the construction of the solution (see [7], [8]). On the other hand, if the conditions of Theorem 4.10 are satisfied (in this case, zero is an isolated singular point), then the above situation is impossible.

5. ISOLATED SINGULAR POINTS: EXAMPLES

Throughout this section, X denotes the coordinate process on $C(\mathbb{R}_+)$.

Example 5.1 For the SDE

$$dX_t = -\frac{1}{2X_t} I(X_t \neq 0)\, dt + dB_t, \quad X_0 = x, \tag{5.1}$$

there exists no solution started at zero.

Proof. Suppose that P is a solution started at zero. Let B be the process defined in (2.4). By Itô's formula,

$$X_t^2 = -\int_0^t I(X_s \neq 0)\, ds + 2 \int_0^t X_s\, dB_s + t$$
$$= \int_0^t I(X_s = 0)\, ds + 2 \int_0^t X_s\, dB_s.$$

By the occupation times formula (see [22; Ch. VI, (1.6)]),

$$\int_0^t I(X_s = 0)\, ds = \int_0^t I(X_s = 0)\, d\langle X \rangle_s = \int_{\mathbb{R}} I(x = 0)\, L_t^x(X)\, dx = 0.$$

Thus, X^2 is a local martingale with $X_0^2 = 0$. Consequently, $X_t^2 = 0$ P-a.s. On the other hand, the measure concentrated on $X \equiv 0$ is not a solution. □

Remark If $x \neq 0$, then (5.1) possesses no solution in the sense of Definition 2.5. However, (5.1) has a solution defined up to the moment $T_0 = \inf\{t \geq 0 : X_t = 0\}$ in the sense of Definition 6.1. Moreover, this solution is unique. □

In the following example, we investigate SDEs for *Bessel processes*.

Example 5.2 Let us consider the SDE

$$dX_t = \frac{\delta - 1}{2X_t} I(X_t \neq 0)\, dt + dB_t, \quad X_0 = x, \qquad (5.2)$$

with $\delta > 1, x \in \mathbb{R}$.

   (i) If $x \neq 0$ and $\delta \geq 2$, then (5.2) has a unique solution.

   (ii) If $x = 0$ or $1 < \delta < 2$, then (5.2) possesses different solutions.

Proof. **(i)** With no loss of generality, we may assume that $x > 0$. Let P be the distribution of a δ-dimensional Bessel process started at x. It is well known (see, for example, [22; Ch. XI, §1]) that P is a solution of (5.2) (in the sense of Definition 2.5). Let P′ be another solution. Set

$$Q = \mathrm{Law}(X_t^2; t \geq 0 \,|\, P), \quad Q' = \mathrm{Law}(X_t^2; t \geq 0 \,|\, P').$$

By Itô's formula, both Q and Q′ are solutions of SDE

$$dX_t = \delta dt + 2\sqrt{|X_t|}\, dB_t, \quad X_0 = x^2. \qquad (5.3)$$

For this equation, the drift b is constant and the diffusion coefficient σ is Hölder continuous of order $1/2$. Therefore, there is even strong existence and strong uniqueness for (5.3) (see [22; Ch. IX, (3.5)]). By the theorem of Yamada and Watanabe (see [22; Ch. IX, (1.7)]), there is weak uniqueness for (5.3), i.e., Q′ = Q. Hence,

$$\mathrm{Law}(|X_t|; t \geq 0 \,|\, P) = \mathrm{Law}(|X_t|; t \geq 0 \,|\, P'). \qquad (5.4)$$

Furthermore, the properties of the Bessel processes guarantee that, for $\delta \geq 2$, $P\{\forall t \geq 0,\ X_t > 0\} = 1$ (see [22; Ch. XI, §1]). This, together with (5.4), implies that $P'\{\forall t \geq 0,\ X_t \neq 0\} = 1$. Since the paths of X are continuous and $P'\{X_0 = x > 0\} = 1$, we get $P'\{\forall t \geq 0,\ X_t > 0\} = 1$. Using (5.4) once again, we obtain P = P′.

   (ii) We will first suppose that $x = 0$. Let P be defined as above and P′ be the image of P under the map

$$C(\mathbb{R}_+) \in \omega \longmapsto -\omega \in C(\mathbb{R}_+).$$

It is easy to verify that P′ is also a solution of (5.2) started at zero. The solutions P and P′ are different since

$$P\{\forall t \geq 0,\ X_t \geq 0\} = 1, \quad P'\{\forall t \geq 0,\ X_t \leq 0\} = 1.$$

Moreover, for any $\alpha \in (0, 1)$, the measure $P^\alpha = \alpha P + (1 - \alpha)P'$ is also a solution.

Suppose now that $x > 0$. Let P denote the distribution of a δ-dimensional Bessel process started at x. Since $1 < \delta < 2$, we have: $P\{\exists t > 0 : X_t = 0\} = 1$ (see [22; Ch. XI, §1]). Let P' be the image of P under the map

$$C(\mathbb{R}_+) \ni \omega \longmapsto \omega' \in C(\mathbb{R}_+),$$

$$\omega'(t) = \begin{cases} \omega(t) & \text{if } t \le T_0(\omega), \\ -\omega(t) & \text{if } t > T_0(\omega), \end{cases}$$

where T_0 is defined in (3.2). Then P' is also a solution of (5.2). $\square$

Remark If $x = 0$ or $1 < \delta < 2$, then SDE (5.2) possesses different strong solutions as well as solutions that are not strong (see [2]). However, strong solutions are not investigated in this paper. $\square$

6. SOLUTIONS UP TO A RANDOM TIME

Throughout this section, X denotes the coordinate process on $\overline{C}(\mathbb{R}_+)$ (see (3.1)) and $(\mathcal{F}_t)$ denotes the canonical filtration.

In the following, we will need two different definitions: a solution up to T and a solution up to $T-$.

Definition 6.1 Let T be a stopping time on $\overline{C}(\mathbb{R}_+)$. A *solution of* (1.1) *up to* T (or a solution *defined* up to T) is a measure P on $\mathcal{F}_T$ such that
  i) $P\{X_0 = x\} = 1$;
  ii) $\int_0^T \left(|b(X_s)| + \sigma^2(X_s)\right) ds < \infty$ P-a.s.;
  iii) $T < \infty$ P-a.s.;
  iv) the process

$$M_t = X_{t \wedge T} - \int_0^{t \wedge T} b(X_s) \, ds$$

is a $(\mathcal{F}_t, P)$-local martingale;
  v) the process

$$M_t^2 - \int_0^{t \wedge T} \sigma^2(X_s) \, ds$$

is a $(\mathcal{F}_t, P)$-local martingale.
In the following, we will often say that (P, T) is a solution of (1.1) started at x.

Remarks (i) Let $\xi = \inf\{t \ge 0 : X_t = \Delta\}$. The properties i) – v) imply that $T < \xi$ P-a.s..

(ii) The measure P is defined on $\mathcal{F}_T$ and not on $\mathcal{F}$ since otherwise it would not be unique. $\square$

We remind that T is called a *predictable* stopping time if there exists an increasing sequence $(T_n)_{n=1}^\infty$ of stopping times such that $T_n < T$, $\lim_n T_n = T$. Such a sequence is called a *predicting sequence* for T.

Definition 6.2 Let T be a predictable stopping time on $\overline{C}(\mathbb{R}_+)$ with a predicting sequence (T_n). A *solution of* (1.1) *up to* $T-$ (or a solution *defined* up to $T-$) is a measure P on $\mathcal{F}_{T-}$ such that, for any n, the restriction of P to $\mathcal{F}_{T_n}$ is a solution up to T_n.

In the following, we will often say that $(P, T-)$ is a solution of (1.1) started at x.

Remarks (i) Obviously, this definition does not depend on the choice of a predicting sequence for T.

(ii) Definition 6.2 implies that $T \leq \xi$ P-a.s. $\qquad\square$

Let us now clarify the relationship between the definitions of a solution up to a random time and the standard Definition 2.5.

Theorem 6.3 (i) *Suppose that* $(P, T-)$ *is a solution of* (1.1) *in the sense of Definition 6.2 and* $T = \infty$ *P-a.s. Then* P *admits a unique extension* $\widetilde{P}$ *to* $\mathcal{F}$. *Let* Q *be the measure on* $C(\mathbb{R}_+)$ *defined as the restriction of* $\widetilde{P}$ *to the set* $\{\xi = \infty\}$. *(We use here the obvious property* $\overline{C}(\mathbb{R}_+) \cap \{\xi = \infty\} = C(\mathbb{R}_+)$). *Then* Q *is a solution of* (1.1) *in the sense of Definition 2.5.*

(ii) *Let* Q *be a solution of* (1.1) *in the sense of Definition 2.5. Let* P *be the measure on* $\overline{C}(\mathbb{R}_+)$ *defined as* $P(A) = Q(A \cap \{\xi = \infty\})$. *Then* $(P, \infty-)$ *is a solution of* (1.1) *in the sense of Definition 6.2.*

Proof. (i) The existence and the uniqueness of $\widetilde{P}$ follow from the equality

$$\mathcal{F}|\{T = \infty\} = \mathcal{F}_{T-}|\{T = \infty\}.$$

The latter part of the statement as well as (ii) are obvious. $\qquad\square$

7. THE CLASSIFICATION OF ISOLATED SINGULAR POINTS

We will first investigate the behaviour of a solution of (1.1) in the right-hand neighbourhood of an isolated singular point which is supposed to be equal to zero. A complete qualitative classification in terms of the parameters $e_1, \ldots, e_4$ defined in (3.8) is presented.

Throughout this section, we suppose that $\sigma(x) \neq 0$ for all $x \in \mathbb{R}$.

As zero is an isolated singular point, there exists $a > 0$ such that

$$\frac{1 + |b|}{\sigma^2} \in L^1_{\text{loc}}((0, a]). \tag{7.1}$$

We note that the integral

$$\int_0^a \frac{1 + |b(x)|}{\sigma^2(x)} \, dx$$

may converge if zero is an isolated singular point. In this case, the corresponding integral diverges in the left-hand neighbourhood of zero.

A solution P defined up to T will be called *nonnegative* if

$$P\{\forall t \leq T, \, X_t \geq 0\} = 1.$$

We will use the functions

$$\rho(x) = \exp\left\{ \int_x^a \frac{2b(y)}{\sigma^2(y)}\, dy \right\}, \quad x \in (0, a], \tag{7.2}$$

$$s(x) = \begin{cases} \int_0^x \rho(y)\, dy & \text{if } \int_0^a \rho(y)\, dy < \infty, \\ -\int_x^a \rho(y)\, dy & \text{if } \int_0^a \rho(y)\, dy = \infty \end{cases} \tag{7.3}$$

and the measure

$$m(dx) = \frac{I(0 < x < a)}{\rho(x)\,\sigma^2(x)}\, dx. \tag{7.4}$$

For a stopping time T, we will consider the map

$$\Phi_T : \overline{C}(\mathbb{R}_+) \ni \omega \longmapsto \omega^T \in \overline{C}(\mathbb{R}_+) \tag{7.5}$$

defined as $\omega^T(t) = \omega(t \wedge T(\omega))$ and the map

$$\overline{\Phi}_T : \overline{C}(\mathbb{R}_+) \ni \omega \longmapsto \overline{\omega}^T \in \overline{C}(\mathbb{R}_+) \tag{7.6}$$

defined as

$$\overline{\omega}^T(t) = \begin{cases} \omega(t) & \text{if } t < T(\omega), \\ \Delta & \text{if } t \geq T(\omega). \end{cases}$$

Throughout this section, $e_1, \ldots, e_4$ mean the values defined in (3.8).

Theorem 7.1 *Suppose that*

$$\int_0^a \frac{1 + |b(x)|}{\sigma^2(x)}\, dx < \infty.$$

(i) *For any* $x \in [0, a]$, *there exists a unique solution* P_x *of (1.1) defined up to* $T_{0,a}$ *(cf. (3.4)).*

(ii) *Set* $\widetilde{\mathsf{P}}_x = \mathsf{P}_x \circ \Phi_{T_{0,a}}^{-1}$ *(cf. (7.5)). Then* $(\widetilde{\mathsf{P}}_x)_{x\in[0,a]}$ *is a regular process whose scale function and speed measure are given by (7.3) and (7.4). Moreover, for this process,*

$$e_1 = 0, \quad e_2 = 0, \quad e_3 = 1, \quad e_4 = 1.$$

If the conditions of Theorem 7.1 are satisfied, we will say that zero has *right type 0*.

Remark Condition iii) of Definition 6.1 guarantees that $T_{0,a}$ is P_x-a.s. finite. Moreover, it follows from (3.7) that $\mathsf{E}_{\mathsf{P}_x} T_{0,a} < \infty$. □

Theorem 7.2 *Suppose that*

$$\int_0^a \rho(x)\, dx < \infty, \quad \int_0^a \frac{1 + |b(x)|}{\rho(x)\,\sigma^2(x)}\, dx < \infty, \quad \int_0^a \frac{|b(x)|}{\sigma^2(x)}\, dx = \infty.$$

(i) *For any* $x \in [0, a]$, *there exists a nonnegative solution* P_x *defined up to* T_a. *Moreover, it is unique in the class of nonnegative solutions.*

(ii) *Set* $\widetilde{\mathsf{P}}_x = \mathsf{P}_x \circ \Phi_{T_a}^{-1}$. *Then* $(\widetilde{\mathsf{P}}_x)_{x\in[0,a]}$ *is a regular process whose scale function and speed measure are given by (7.3) and (7.4). Moreover, for this process,*

$$e_1 = 1, \quad e_2 = 1, \quad e_3 = 1, \quad e_4 = 1.$$

If the conditions of Theorem 7.2 are satisfied, we will say that zero has *right type 2*.

Remark Under the conditions of Theorem 7.2, we have $E_{P_x} T_a < \infty$. This can be derived from a formula similar to (3.7) that is related to regular processes with a reflecting point (see [22; Ch. VII, (3.10)]). $\square$

Theorem 7.3 *Suppose that*

$$\int_0^a \rho(x)\, dx < \infty, \quad \int_0^a \frac{1 + |b(x)|}{\rho(x)\, \sigma^2(x)}\, dx = \infty, \quad \int_0^a \frac{1 + |b(x)|}{\rho(x)\, \sigma^2(x)}\, s(x)\, dx < \infty.$$

(i) *For any solution* (P, T), *we have* $P\{\forall t \in [T_0, T], \ X_t \leq 0\} = 1$.
(ii) *For any* $x \in [0, a]$, *there exists a unique solution* P_x *defined up to* $T_{0,a}$.
(iii) *Set* $\tilde{P}_x = P_x \circ \Phi_{T_{0,a}}^{-1}$. *Then* $(\tilde{P}_x)_{x \in [0,a]}$ *is a regular process whose scale function and speed measure are given by (7.3) and (7.4). Moreover, for this process,*

$$e_1 = 0, \quad e_2 = 0, \quad e_3 = 1, \quad e_4 = 1.$$

If the conditions of Theorem 7.3 are satisfied, we will say that zero has *right type 1*.

Remark Statement (i) implies that any solution (P, T) started at $x \leq 0$ is nonpositive. $\square$

Theorem 7.4 *Suppose that*

$$\int_0^a \rho(x)\, dx < \infty, \quad \int_0^a \frac{|b(x)|\, s(x)}{\rho(x)\, \sigma^2(x)}\, dx = \infty, \quad \int_0^a \frac{s(x)}{\rho(x)\, \sigma^2(x)}\, dx < \infty.$$

(i) *If* (P, T) *is a solution started at* $x > 0$, *then* $T < T_0$ *P-a.s.*
(ii) *If* (P, T) *is a solution started at* $x \leq 0$, *then* $P\{\forall t \leq T, \ X_t \leq 0\} = 1$.
(iii) *For any* $x \in (0, a)$, *there exists a unique solution* P_x *defined up to* $\overline{T}_{0,a}-$ *(cf. (3.5))*.
(iv) *Set* $\overline{P}_x = P_x \circ \overline{\Phi}_{T_{0,a}}^{-1}$ *(cf. (7.6)). Then* $(\overline{P}_x)_{x \in (0,a)}$ *is a regular process whose scale function and speed measure are given by (7.3) and (7.4). Let* $\overline{P}_0$ *be the measure concentrated on* $X \equiv 0$. *Then* $(\overline{P}_x)_{x \in [0,a)}$ *is a continuous strong Markov process with*

$$e_1 = 0, \quad e_2 = 0, \quad e_3 = 0, \quad e_4 = 1.$$

If the conditions of Theorem 7.4 are satisfied, we will say that zero has *right type 6*.

Theorem 7.5 *Suppose that*

$$\int_0^a \rho(x)\, dx < \infty, \quad \int_0^a \frac{s(x)}{\rho(x)\, \sigma^2(x)}\, dx = \infty.$$

(i) *If* (P, T) *is a solution started at* $x > 0$, *then* $T < T_0$ *P-a.s.*
(ii) *If* (P, T) *is a solution started at* $x \leq 0$, *then* $P\{\forall t \leq T, \ X_t \leq 0\} = 1$.
(iii) *For any* $x \in (0, a)$, *there exists a unique solution* P_x *defined up to* $\overline{T}_a-$. *Moreover,* $P_x\{\overline{T}_a = \infty \text{ and } X_t \xrightarrow[t \to \infty]{} 0\} > 0$.
(iv) *Set* $\overline{P}_x = P_x \circ \overline{\Phi}_{T_a}^{-1}$. *Then* $(\overline{P}_x)_{x \in (0,a)}$ *is a regular process whose scale function and speed measure are given by (7.3) and (7.4). Let* $\overline{P}_0$ *be the measure concentrated on* $X \equiv 0$. *Then* $(\overline{P}_x)_{x \in [0,a)}$ *is a continuous strong Markov process with*

$$e_1 = 0, \quad e_2 = 0, \quad e_3 = 0, \quad e_4 = 0.$$

If the conditions of Theorem 7.5 are satisfied, we will say that zero has *right type 4*.

Theorem 7.6 *Suppose that*

$$\int_0^a \rho(x)\, dx = \infty, \qquad \int_0^a \frac{1 + |b(x)|}{\rho(x)\, \sigma^2(x)} |s(x)|\, dx < \infty.$$

(i) *If* (P, T) *is a solution started at* $x > 0$, *then* $T < T_0$ P-*a.s.*

(ii) *For any* $x \in (0, a]$, *there exists a unique solution* P_x *defined up to* T_a. *For* $x = 0$, *there exists a nonnegative solution* P_0 *defined up to* T_a, *and it is unique in the class of nonnegative solutions.*

(iii) *Set* $\widetilde{P}_x = P_x \circ \Phi_{T_a}^{-1}$. *Then* $(\widetilde{P}_x)_{x\in(0,a]}$ *is a regular process whose scale function and speed measure are given by* (7.3) *and* (7.4). *Moreover,* $(\widetilde{P}_x)_{x\in[0,a]}$ *is a continuous strong Markov process with*

$$e_1 = 1, \quad e_2 = 1, \quad e_3 = 0, \quad e_4 = 0.$$

If the conditions of Theorem 7.6 are satisfied, we will say that zero has *right type 3*.

Theorem 7.7 *Suppose that*

$$\int_0^a \rho(x)\, dx = \infty, \qquad \int_0^a \frac{|b(x)\, s(x)|}{\rho(x)\, \sigma^2(x)}\, dx = \infty, \qquad \int_0^a \frac{|s(x)|}{\rho(x)\, \sigma^2(x)}\, dx < \infty.$$

(i) *If* (P, T) *is a solution started at* $x > 0$, *then* $T < T_0$ P-*a.s.*

(ii) *If* (P, T) *is a solution started at* $x \le 0$, *then* $P\{\forall t \le T, \ X_t \le 0\} = 1$.

(iii) *For any* $x \in (0, a]$, *there exists a unique solution* P_x *defined up to* T_a.

(iv) *Set* $\widetilde{P}_x = P_x \circ \Phi_{T_a}^{-1}$. *Then* $(\widetilde{P}_x)_{x\in(0,a]}$ *is a regular process whose scale function and speed measure are given by* (7.3) *and* (7.4). *Let* $\widetilde{P}_0$ *be the measure concentrated on* $X \equiv 0$. *Then* $(\widetilde{P}_x)_{x\in[0,a]}$ *is a continuous strong Markov process with*

$$e_1 = 0, \quad e_2 = 1, \quad e_3 = 0, \quad e_4 = 0.$$

If the conditions of Theorem 7.7 are satisfied, we will say that zero has *right type 7*.

Theorem 7.8 *Suppose that*

$$\int_0^a \rho(x)\, dx = \infty, \qquad \int_0^a \frac{|s(x)|}{\rho(x)\, \sigma^2(x)}\, dx = \infty.$$

(i) *If* (P, T) *is a solution started at* $x > 0$, *then* $T < T_0$ P-*a.s.*

(ii) *If* (P, T) *is a solution started at* $x \le 0$, *then* $P\{\forall t \le T, \ X_t \le 0\} = 1$.

(iii) *For any* $x \in (0, a]$, *there exists a unique solution* P_x *defined up to* T_a.

(iv) *Set* $\widetilde{P}_x = P_x \circ \Phi_{T_a}^{-1}$. *Then* $(\widetilde{P}_x)_{x\in(0,a]}$ *is a regular process whose scale function and speed measure are given by* (7.3) *and* (7.4). *Let* $\widetilde{P}_0$ *be the measure concentrated on* $X \equiv 0$. *Then* $(\widetilde{P}_x)_{x\in[0,a]}$ *is a continuous strong Markov process with*

$$e_1 = 0, \quad e_2 = 0, \quad e_3 = 0, \quad e_4 = 0.$$

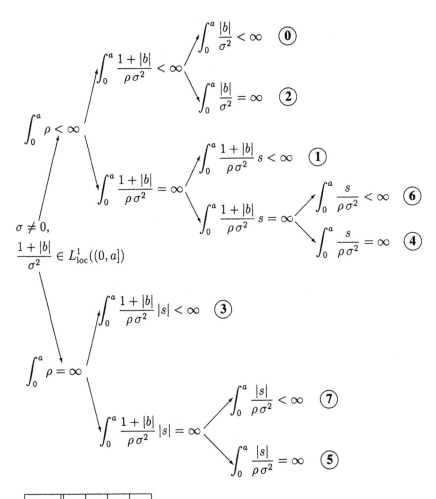

Type	e_1	e_2	e_3	e_4
0	0	0	1	1
1	0	0	1	1
2	1	1	1	1
3	1	1	0	0
4	0	0	0	0
5	0	0	0	0
6	0	0	0	1
7	0	1	0	0

Types	Exit	Non-exit
Entrance	2	3
Non-entrance	0 1	4 5 6 7

FIGURE 1. One-sided classification of isolated singular points

If the conditions of Theorem 7.8 are satisfied, we will say that zero has *right type 5*.

For the sake of brevity, we present here only the statements of the results and give no proofs. The proofs can be found in the forthcoming paper by the same authors.

Figure 1 represents the one-sided classification of isolated singular points. We note that the integrability conditions given on Figure 1 do not have the same form as those given by Theorems 7.1 – 7.8. Nevertheless, they are equivalent. For example,

$$\int_0^a \frac{1 + |b(x)|}{\sigma^2(x)} \, dx < \infty$$

if and only if

$$\int_0^a \rho(x) \, dx < \infty, \quad \int_0^a \frac{1 + |b(x)|}{\rho(x) \, \sigma^2(x)} \, dx < \infty, \quad \int_0^a \frac{|b(x)|}{\sigma^2(x)} \, dx < \infty.$$

(In this case zero has right type 0).

If zero has right type 2 or 3, then there exist positive solutions started at zero. Thus, types 2 and 3 may be called *entrance* types. On the other hand, types 1, 4, 5, 6, 7 are *non-entrance* ones: any solution started at zero is nonpositive.

If zero has right type 0, 1 or 2, then, for $x \in (0, a)$, there exists a solution started at x that reaches zero with positive probability. Thus, types 0, 1, 2 may be called *exit* types. If the right type of zero is one of $3, \ldots, 7$, then any solution with $x > 0$ does not reach zero. So, these types are *non-exit* ones.

Let us now informally describe how a solution behaves in the right-hand neighbourhood of an isolated singular point for each of the types $0, \ldots, 7$.

If zero has right type **0**, then, for any $x \in [0, a]$, there exists a unique solution defined up to $T_{0,a}$. This solution reaches zero with positive probability. An example of a SDE for which zero has right type 0 is provided by the equation

$$dX_t = dB_t, \quad X_0 = x.$$

If zero has right type **1**, then, for any $x \in [0, a]$, there exists a unique solution defined up to $T_{0,a}$. This solution reaches zero with positive probability. Any solution started at zero (it may be defined up to another stopping time) is nonpositive. In other words, a solution may leave zero only in the negative direction. The SDE for the square of a 0-dimensional Bessel process

$$dX_t = 2\sqrt{|X_t|} \, dB_t, \quad X_0 = x$$

provides an example of a SDE for which zero has right type 1.

If zero has right type **2**, then, for any $x \in [0, a]$, there exists a unique *nonnegative* solution defined up to T_a. This solution reaches zero with positive probability and is reflected at this point. There may exist other solutions up to T_a (these solutions may take negative values). For the SDE

$$dX_t = \frac{\delta - 1}{2|X_t|} I(X_t \neq 0) \, dt + dB_t, \quad X_0 = x$$

with $1 < \delta < 2$ (this is the SDE for a δ-dimensional Bessel process), zero has right type 2.

If zero has right type **3**, then, for any $x \in (0, a]$, there exists a unique solution defined up to T_a. This solution never reaches zero. There exists a unique *nonnegative* solution started at $x = 0$ and defined up to T_a (for $x = 0$, there may exist other solutions that take negative values and are defined up to T_a). For the SDE

$$dX_t = \frac{\delta - 1}{2|X_t|} I(X_t \neq 0)\, dt + dB_t, \quad X_0 = x$$

with $\delta \geq 2$, zero has right type 3.

If zero has right type **4**, then, for any $x \in (0, a)$, there exists a unique solution defined up to $\overline{T}_a-$. This solution never reaches zero. There exists no solution up to T_a for the following reason. The above mentioned solution tends to zero with positive probability as $t \to \infty$. So, this solution never reaches the point a with positive probability. On the other hand, if (P, T_a) is a solution, then T_a should be P-a.s. finite. For type 4 as well as for types 5, 6, 7 below, any solution started at zero is nonpositive. An example of a SDE for which zero has right type 4 is provided by the equation

$$dX_t = \frac{1}{3}|X_t|\, dt + |X_t|\, dB_t, \quad X_0 = x.$$

If zero has right type **5**, then, for any $x \in (0, a]$, there exists a unique solution defined up to T_a. This solution never reaches zero. As opposed to the previous case, the solution reaches the point a a.s. For the SDE

$$dX_t = \frac{1}{2}|X_t|\, dt + |X_t|\, dB_t, \quad X_0 = x,$$

zero has right type 5.

If zero has right type **6**, then, for any $x \in (0, a)$, there exists a unique solution P_x defined up to $\overline{T}_{0,a}-$. Moreover, $\overline{T}_{0,a}$ is finite P_x-a.s. However, there exists no solution up to $T_{0,a}$ because the integral $\int_0^{T_{0,a}} |b(X_s)|\, ds$ equals infinity with positive probability.

If zero has right type **7**, then the qualitative behaviour of a solution is almost the same as for right type 5. The only difference is in the value e_2. We do not give the examples of SDEs for which zero has right type 6 or 7 because these examples are rather complicated.

So far, we have investigated the behaviour of a solution in the right-hand neighbourhood of zero (recall that zero is assumed to be an isolated singular point). According to the classification given above, zero has one of 8 possible right types. If zero has right type **0**, then

$$\exists a > 0 : \int_0^a \frac{1 + |b(x)|}{\sigma^2(x)}\, dx < \infty. \tag{7.7}$$

If the right type of zero is one of $1, \ldots, 7$, then

$$\forall a > 0, \int_0^a \frac{1 + |b(x)|}{\sigma^2(x)}\, dx = \infty$$

(this can easily be seen from Figure 1).

In a similar way, one can define *left types* of zero (there are 8 left types). We will say that zero *has type* $(i-j)$ if it has left type i and right type j. Thus, there are $64(=8 \times 8)$ possibilities. If zero has type $(0-0)$, then, in view of (7.7), $(1+|b|)/\sigma^2 \in L^1_{\text{loc}}(0)$ and so, zero is not a singular point. For the other 63 possibilities, this function is not locally integrable at zero. As a result, an isolated singular point can have one of 63 possible types.

It is easy to see that only 4 of these 63 types can disturb the uniqueness of a solution. These are types $(2-2)$, $(2-3)$, $(3-2)$ and $(3-3)$. Indeed, if zero has one of these types, then there exist both nonnegative and nonpositive solutions started at zero. Therefore, we call these 4 types the *branch types* while corresponding isolated singular points are called the *branch points*. If SDE (1.1) has a branch point, then one can easily construct non-Markov solutions. We will illustrate this by the following example.

Example 7.9 Let us consider the SDE

$$dX_t = \frac{\delta - 1}{2|X_t|} I(X_t \neq 0)\, dt + dB_t, \quad X_0 = x \qquad (7.8)$$

with $1 < \delta < 2$. Take $x > 0$ and let P be the nonnegative solution started at x (this is the distribution of the δ-dimensional Bessel process started at x). Let us consider the map

$$C(\mathbb{R}_+) \ni \omega \longmapsto \omega' \in C(\mathbb{R}_+)$$

defined as

$$\omega'(t) = \begin{cases} \omega(t) & \text{if } t \le T_0(\omega), \\ \omega(t) & \text{if } t > T_0(\omega) \text{ and } \omega(T_0(\omega)/2) > 1, \\ -\omega(t) & \text{if } t > T_0(\omega) \text{ and } \omega(T_0(\omega)/2) \le 1. \end{cases}$$

Then the image P′ of P under this map is a non-Markov solution of (7.8).

Let us mention another way to construct non-Markov solutions of one-dimensional homogeneous SDEs. Consider the equation

$$dX_t = |X_t|^\lambda dB_t, \quad X_0 = x \qquad (7.9)$$

with $0 < \lambda < 1/2$. This example was first considered by Girsanov in [11] (also see the discussion by Jacod and Yor [15; p. 122]). SDE (7.9) possesses different solutions started at zero. A trivial solution is concentrated on $\omega \equiv 0$ but there also exists a non-trivial solution (note that, for this equation, $1/\sigma^2 \in L^1_{\text{loc}}(\mathbb{R})$). In order to construct a non-Markov solution of (7.9), we start a solution at $x \neq 0$. When it first reaches zero, we hold it at zero for a time period that depends on the past of the path and then restart it from zero in a non-trivial way. This way of constructing non-Markov solutions is well known (see [7] or [14; p. 79]). Actually, the same trick can be performed with the SDE

$$dX_t = I(X_t \neq 0)\, dB_t, \quad X_0 = 0.$$

It is important for both examples that σ vanishes at zero. On the other hand, in Example 7.9, $\sigma \equiv 1$.

8. THE APPLICATION TO POWER EQUATIONS

Let us consider the SDE

$$dX_t = \mu |X_t|^\alpha I(X_t \neq 0) \, dt + \left(\nu |X_t|^\beta I(X_t \neq 0) + \eta I(X_t = 0) \right) dB_t. \qquad (8.1)$$

Here, $\nu \neq 0$, $\eta \neq 0$. Obviously, all the properties of (8.1) are the same for any $\eta \neq 0$. We add the term $\eta I(X_t \neq 0)$ to guarantee that $\sigma \neq 0$ at each point.

Theorem 8.1 *Set* $\lambda = \mu/\nu^2$, $\gamma = \alpha - 2\beta$. *Then right types of zero for* (8.1) *are those given in Figure 2.*

The proof of this statement easily follows from the classification of the right types given above.

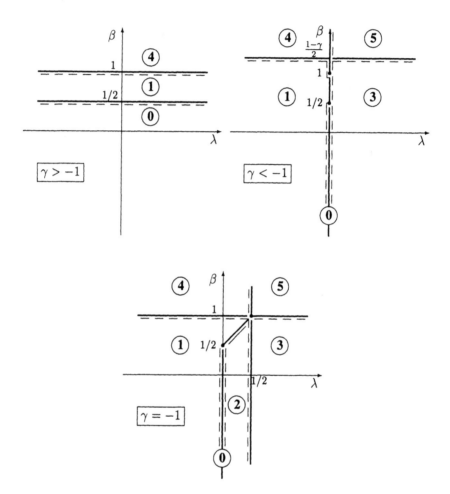

FIGURE 2. One-sided classification for the power equations

Here, $\lambda = \mu/\nu^2$, $\gamma = \alpha - 2\beta$, where α, β, μ and ν are given by (8.1). One should first calculate γ and select the corresponding graph (out of the three graphs shown). Then one should plot the point (λ, β) on this graph and find the part of the graph the point lies in. The number ⓘ marked in this part indicates that, for equation (8.1), zero has right type i. For example, if $\gamma < -1$, $\lambda > 0$ and $\beta \geq (1 - \gamma)/2$, then zero has right type 5.

REFERENCES

[1] S. Assing, W. Schmidt, *Continuous strong Markov processes in dimension one*, Lecture Notes in Mathematics **1688**, Springer-Verlag, Berlin, 1998.

[2] A.S. Cherny, On the strong and weak solutions of stochastic differential equations governing Bessel processes, *Stochastics Stochastics Rep.* **70** (2000), No. 3–4, 213–219.

[3] E. Çinlar, J. Jacod, P. Protter, M.J. Sharpe, Semimartingales and Markov processes, *Z. Wahrscheinlichkeitstheorie verw. Gebiete* **54** (1980), 161–219.

[4] E.B. Dynkin, *Markov Processes.* Springer-Verlag, 1965.

[5] H.-J. Engelbert, W. Schmidt, On the behaviour of certain functionals of the Wiener process and applications to stochastic differential equations, *Stochastic Differential Systems*, Lecture Notes in Control and Information Sciences **36**, 47–55, Springer-Verlag, Berlin, 1981.

[6] H.-J. Engelbert, W. Schmidt, On one-dimensional stochastic differential equations with generalized drift, *Stochastic Differential Systems*, Lecture Notes in Control and Information Sciences **69**, 143–155, Springer-Verlag, Berlin, 1985.

[7] H.-J. Engelbert, W. Schmidt, On solutions of one-dimensional stochastic differential equations without drift, *Z. Wahrscheinlichkeitstheorie verw. Gebiete* **68** (1985), 287–314.

[8] H.-J. Engelbert, W. Schmidt, Strong Markov continuous local martingales and solutions of one-dimensional stochastic differential equations, I, II, III, *Math. Nachr* **143** (1989), 167–184; **144** (1989), 241–281; **151** (1991), 149–197.

[9] W. Feller. The parabolic differential equations and the associated semi-groups of transformations, *Ann. Math.* **55** (1952), 468–519.

[10] W. Feller, Diffusion processes in one dimension, *Trans. Amer. Math. Soc.* **77** (1954), 1–31.

[11] I.V. Girsanov, An example of the non-uniqueness of a solution of a stochastic equation of K. Itô, *Theory Probab. Appl.* **7** (1962), No. 3, 336–342.

[12] K. Itô, On a stochastic integral equation, *Proc. Jap. Acad.* **22** (1946), 32–35.

[13] K. Itô, On stochastic differential equations, *Mem. Amer. Math. Soc.* **4** (1951), 1–51.

[14] K. Itô, H.P. McKean, *Diffusion processes and their sample paths*, Springer-Verlag, 1965.

[15] J. Jacod, M. Yor, Étude des solutions extrémales et représentation intégrale des solutions pour certains problèmes de martingales, *Z. Wahrscheinlichkeitstheorie verw. Gebiete* **38** (1977), 83–125.

[16] O. Kallenberg, *Foundations of modern probability*, Springer-Verlag, 1997.

[17] I. Karatzas, S.E. Shreve, *Brownian motion and stochastic calculus*, Springer-Verlag, 1988.

[18] A.N. Kolmogoroff, Über die analytischen Methoden in der Wahrscheinlichkeitsrechnung, *Math. Annalen* **104** (1931), 415–458.

[19] N.V. Krylov, On quasidiffusion processes, *Theory Probab. Appl.* **11** (1966), No. 3, 424–442.

[20] N.V. Krylov, On Itô's stochastic differential equations, *Theory Probab. Appl.* **14** (1969), No. 2, 340–348.

[21] R.S. Liptser, A.N. Shiryaev, *Statistics of random processes*, Springer-Verlag, 2001.

[22] D. Revuz, M. Yor, *Continuous martingales and Brownian motion*, Springer-Verlag, 1994.

[23] W. Schmidt, On semimartingale diffusions and stochastic differential equations, *Stochastics Stochastics Rep.* **29** (1990), 407–424.

[24] D.W. Stroock, S.R.S. Varadhan, Diffusion processes with continuous coefficients, I, II, *Communications in Pure and Applied Mathematics* **22** (1969), 345–400; 479–530.

[25] A.K. Zvonkin, N.V. Krylov, On the strong solutions of stochastic differential equations, In: *Proceedings of the School-Seminar on Random Processes* (Druskininkai, 1974), part 2, 8–99, Vilnius, 1975.

ON ONE-DIMENSIONAL STOCHASTIC EQUATIONS DRIVEN BY SYMMETRIC STABLE PROCESSES*

H.-J. ENGELBERT[1] and V.P. KURENOK[2]

[1] Friedrich-Schiller-Universität Jena
Institut für Stochastik
D-07743 Jena, Germany

[2] Belorussian State University
Department of Mathematics and Mechanics
F. Skoriny Av. 4, 220050 Minsk, Belarus

Abstract We study stochastic equations $X_t = x_0 + \int_0^t b(u, X_{u-}) \, dZ_u$, where Z is an one-dimensional symmetric stable process of index α with $0 < \alpha \le 2$, $b : [0, \infty) \times \mathbb{R} \to \mathbb{R}$ is a measurable diffusion coefficient, and $x_0 \in \mathbb{R}$ is the initial value. We give sufficient conditions for the existence of weak solutions. Our main results generalize results of P. A. Zanzotto [20] who dealt with homogeneous diffusion coefficients b. In the nonhomogeneous case we present new sufficient conditons for the existence of (nonexploding) solutions even if Z is a Brownian motion. Using the property that appropriate time changes of stochastic integrals with respect to stable processes are again stable processes with the same index, we present a new proof of the main result which simplifies the approach given by P. A. Zanzotto [20].

Key Words One-dimensional stochastic equations, measurable coefficients, Wiener process, Cauchy process, symmetric stable processes, time change, purely discontinuous processes.

MR Subject Classification 60H10, 60J60, 60J65, 60G44.

1. INTRODUCTION

We consider the stochastic equation

$$X_t = x_0 + \int_0^t b(u, X_{u-}) \, dZ_u, \quad t < \tau_\Delta(X), \tag{1.1}$$

*Research supported by Grant INTAS 97-30204. First version: March 17, 1999.

where $\tau_\Delta(X)$ is the explosion time of X, $x_0 \in \mathbb{R}$, $b : [0, \infty) \times \mathbb{R} \to \mathbb{R}$ is a measurable function, and $(Z, \mathbb{F})$ is an one-dimensional symmetric stable process with $Z_0 = 0$. This is a process with homogeneous and $\mathcal{F}_s$-independent increments $Z_t - Z_s$ for $0 \le s \le t$ and right continuous left hand limited (càdlàg) trajectories such that the characteristic function of Z_t has the form

$$\mathbb{E} \exp(i\lambda Z_t) = \exp(-t|\lambda|^\alpha), \quad \lambda \in \mathbb{R}, \quad t \ge 0.$$

Here the parameter α satisfies $0 < \alpha \le 2$ and is called the index of the symmetric stable process. For $\alpha = 2$, Z is a Brownian motion, the only symmetric stable process with continuous trajectories. For $\alpha = 1$ we obtain the Cauchy process, a càdlàg process Z with homogeneous and independent increments such that Z_t is Cauchy-distributed with parameter t.

It will be useful to have in mind the following equivalent description of symmetric α-stable processes $(Z, \mathbb{F})$: These are càdlàg processes such that, for every $\lambda \in \mathbb{R}$, $\exp(i\lambda Z_t + |\lambda|^\alpha t)$, $t \ge 0$, is a (complex-valued) $\mathbb{F}$-martingale. This can be seen as a substitute for P. Lévy's martingale characterization of Brownian motion.

There are several approaches to the definition of stochastic integrals for symmetric stable processes Z with $0 < \alpha < 2$. Since every symmetric stable process is a semi-martingale, the stochastic integral with respect to Z can be understood as a stochastic integral with respect to a semimartingale ([11], Section 4.7). The difficulty is that these processes are purely discontinuous with unbounded jumps and, for $\alpha \le 1$, even the first moment does not exist. Therefore, in this approach stochastic measures defined by the jumps of the process and their compensators are strongly involved.

Another approach to stochastic integrals with respect to symmetric stable processes is due to J. Rosiński and W. Woyczyński [14] which generalizes the classical concept of K. Itô for Brownian motion to symmetric stable processes.

We shall make use of the semimartingale integral and the integral introduced by J. Rosiński and W. Woyczyński [14] as well. Of course, if one of the integrals exists then there also exists the other integral and both integrals coincide. Whereas the first concept yields general results, in particular, on the time change of such integrals, the advantage of the second concept consists in giving better conditions on the integrands for the existence of the integral.

In this paper we are mainly interested in the existence of weak solutions of Eq. (1.1). The homogeneous Eq. (1.1), i.e., if $b(t, x) = b(x)$, $(t, x) \in [0, \infty) \times \mathbb{R}$, driven by Brownian motion was investigated by the first author and W. Schmidt [5], [6], [7]. There were given necessary and sufficient conditions for existence as well as for uniqueness in law of solutions of Eq. (1.1) for all initial values $x_0 \in \mathbb{R}$. In particular, it was shown that the local integrability of b^{-2} is necessary and sufficient for the existence of nontrivial solutions for all initial values $x_0 \in \mathbb{R}$. Eq. (1.1) with time-dependent coefficients b and $\alpha = 2$ was investigated in several papers ([10], [17], [18], [15]), in which various generalizations of the results mentioned above were obtained. For symmetric stable processes of index $0 < \alpha \le 2$ Eq. (1.1) with homogeneous coefficients was considered by P. A. Zanzotto [20]. In particular, for the case $1 < \alpha \le 2$ he proved that there exists a nontrivial solution of Eq. (1.1) for all initial values $x_0 \in \mathbb{R}$ if and only if the function $|b|^{-\alpha}$ is locally integrable. This result is surprising in view of its complete analogy to that obtained in [5] for $\alpha = 2$. For $\alpha \leq 1$ the problem is a little more delicate: For $\alpha < 1$ P.A. Zanzotto gave sufficient conditions, and it seems that necessary conditions can hardly be established. The case $\alpha = 1$ was left open in [20].

The main objective of the present paper is to investigate Eq. (1.1) for symmetric stable processes Z with arbitrary index $\alpha \in (0, 2]$ and for general, nonhomogeneous diffusion coefficients b and to give sufficient conditions for the existence of (weak) solutions. We generalize results obtained by P. A. Zanzotto [20] for homogeneous diffusion coefficients b and for indexes α with $0 < \alpha < 2$ and $\alpha \neq 1$. We also include the case if Z is a Cauchy process which corresponds to the case $\alpha = 1$. In particular, we improve the conditions found in [20] in the homogeneous case for indexes α with $0 < \alpha < 1$. Furthermore, even if Z is a Brownian motion, which corresponds to the case $\alpha = 2$, we present new sufficient conditons for the existence of nonexploding solutions for nonhomogeneous diffusion coefficients that are different from those obtained by A. Rozkosz and L. Słomiński [15] and by T. Senf [17], [18]. The basic tool of the paper is time change of symmetric stable processes. Contrary to P. A. Zanzotto [20] we need not examine the jump measures and their compensators associated with symmetric stable processes, their stochastic integrals and time changes. Using the property that appropriate time changes of stochastic integrals with respect to symmetric stable processes are again symmetric stable processes with the same index, we present a new proof of the main result which simplifies the approach given in [20] and unifies the treatment of the continuous case ($\alpha = 2$) on the one side and the purely discontinuous case ($0 < \alpha < 2$) on the other side resulting in a complete analogy between these quite different cases.

In Section 2, we collect some properties on the behaviour of integral functionals of symmetric stable processes which are the key for the construction of the time change processes given in Section 3. These results seem to be of interst in its own right. In Section 4, we briefly review the results of J. Rosiński and W. Woyczyński [14] on stochastic integrals with respect to symmetric stable processes and their time changes giving them a slightly more general form. In Section 5, we then state sufficient conditions for the existence of (weak) solutions of Eq. (1.1). Finally, in Section 6 we deal with the homogeneous case and, in particular, with the uniqueness in law of solutions. The paper is concluded by an example which contains a complete description of the case if b is an arbitrary power function.

After finishing the first version of the paper, the authors became aquainted with the papers of P.A. Zanzotto [21] and [22], the first being a technical report of the University of Pisa and the second being part of mini-proceedings of the Conference on Lévy processes in Aarhus 1999 appeared in April 1999. It turns out that there is a certain overlap with the present paper concerning the homogeneous case. In particular, there is also considered the case of a Cauchy process ($\alpha = 1$). Detailed comments are postponed to Section 6 giving a survey on the present state of the homogeneous case.

2. INTEGRAL FUNCTIONALS OF STABLE PROCESSES

Let $(S, \mathbb{F})$ be a symmetric stable process with index $\alpha \in (0, 2]$ defined on some probability space $(\Omega, \mathcal{F}, \mathbb{P})$ such that $S_0 = 0$. For any nonnegative Borel (or only Lebesgue) measurable function f, we consider the integral functional

$$T_t = \int_0^{t+} f(S_u)\, du, \quad t \geq 0. \tag{2.1}$$

By definition, the process $T = (T_t)_{t \geq 0}$ is increasing and right-continuous taking values in $[0, +\infty]$. First we investigate the case $1 < \alpha \leq 2$. For this we introduce the set

$$E := E(f) = \{x \in \mathbb{R} : \int_{x-}^{x+} f(y)\, dy = +\infty\} \tag{2.2}$$

and note that E is a closed subset of $\mathbb{R}$. We also define the first entry time D_E of S into E by $D_E = \inf\{t \geq 0 : S_t \in E\}$. In the sequel, the Lebesgue measure on $\mathbb{R}$ will always be denoted by λ. The following proposition generalizes Lemma 1 of [6] to symmetric stable processes with index $1 < \alpha \leq 2$.

Proposition 2.1 *Suppose that $1 < \alpha \leq 2$ and $\lambda(\{f > 0\}) > 0$. We then have:*

(i) $T_t < +\infty$ *for all $t < D_E$* $\mathbb{P}$*-a.s.*

(ii) $T_{D_E} = +\infty$ $\mathbb{P}$*-a.s.*

Proof. Since $1 < \alpha \leq 2$, as for Brownian motion, S has a local time $L^S(t, a)$ jointly continuous in (t, a) and such that $L^S(t, 0) > 0$ $\mathbb{P}$-a.s. for every $t > 0$ (cf., e.g., H. Kesten [9], E. S. Boylan [3], Ch. Stone [19]). Moreover, for every $x \in \mathbb{R}$, the first hitting time $\tau_x = \inf\{t > 0 : S_t = x\}$ is $\mathbb{P}$-a.s. finite (cf. S. C. Port [12]). The proof now follows similar lines as in [6], Lemma 1, and [4], Theorem 1. For (i), we notice that the local time $L^S(t, \cdot, \omega)$ has compact support $K(t, \omega) = \{S_{u-}(\omega), S_u(\omega) : 0 \leq u \leq t\}$ contained in E^c if $t < D_E(\omega)$, as in the case $\alpha = 2$. The function f being integrable on $K(t, \omega)$ and the local time being bounded there, the assertion immediately follows from the occupation time formula. For (ii), if $E = \emptyset$ one has to show $T_\infty = +\infty$ $\mathbb{P}$-a.s. which is also a consequence of Proposition 2.6 below. If $E \neq \emptyset$ then $D_E < +\infty$ $\mathbb{P}$-a.s. Using the strong Markov property, $\tilde{S}_t = S_{D_E + t} - S_{D_E}$, $t \geq 0$, is again a symmetric α-stable process which is independent of S_{D_E}. In this way, the statement reduces to $T_0 = +\infty$ $\mathbb{P}$-a.s. if $0 \in E$. This is established precisely as in [4] (Lemma). $\qquad\square$

The following 0–1-law is an immediate consequence of Proposition 2.1 and was already stated in [4] for $\alpha = 2$ and by P. A. Zanzotto [20] for $1 < \alpha \leq 2$.

Corollary 2.2 *Suppose that $1 < \alpha \leq 2$. The following conditions are equivalent:*

(i) $\mathbb{P}(\{T_t < +\infty$ *for every* $t \geq 0\}) > 0$.

(ii) $\mathbb{P}(\{T_t < +\infty$ *for every* $t \geq 0\}) = 1$.

(iii) $E = \emptyset$, *i.e., f is locally integrable.*

Corollary 2.3 *Let $1 < \alpha \leq 2$ and $\lambda(\{f > 0\}) > 0$. Then $T_\infty = +\infty$ $\mathbb{P}$-a.s.*

Before we investigate the case $\alpha = 1$, let us introduce the following definition.

Definition 2.4 A measurable set $B \subseteq \mathbb{R}$ is called x_0-polar if $\mathbb{P}(\{T(B-x_0) = +\infty\}) = 1$. The set B is called polar if B is x_0-polar for every $x_0 \in \mathbb{R}$.

Here $T(B - x_0)$ is just the first hitting time of $B - x_0$ by S or, equivalently, of B by $x_0 + S$. (In the definition of the first entry time, $\geq$ should be replaced by $>$.) For $0 < \alpha \leq 1$ it is well-known that the singletons $B = \{x\}$, $x \in \mathbb{R}$, and hence all denumerable sets B consisting of isolated points, are polar (cf. [2], Chapter II, (3.15)). In the following, if $0 < \alpha \leq 1$, we frequently use the hypothesis that the set $E = E(f)$ is 0-polar. We emphasize that, in particular, this is satisfied if f is integrable over a sufficiently small neighbourhood of every $x \in \mathbb{R}$, except for, possibly, a denumerable set of isolated points.

Proposition 2.5 *Suppose that $\alpha = 1$ and that the following condition is satisfied:*

$$E \quad \text{is 0-polar and} \quad \int_{0-}^{0+} |\ln|y|| f(y) \, dy < +\infty. \tag{2.3}$$

Then we have $\quad T_t < +\infty \quad$ *for every* $\quad t \geq 0 \quad$ $\mathbb{P}$*-a.s.*

Proof. Let G_N be an increasing sequence of open sets with compact closure $\bar{G}_N \subseteq E^c$ such that $E^c = \bigcup_{N=1}^{\infty} G_N$ and σ_N defined by

$$\sigma_N = \inf\{t \geq 0 : S_t \in G_N^c\}. \tag{2.4}$$

We note that f is integrable over G_N. Using the quasi-left continuity of S it can easily be seen that σ_N increases to D_E and, because E is 0-polar and $T(E) = D_E$ since $0 \notin E$, to infinity as $N \to \infty$. Hence it is sufficient to verify that, for all $t \geq 0$ and $N \geq 1$, $\mathbb{E}T_{t \wedge \sigma_N} < +\infty$. Using the theorem of Fubini and that S_u is Cauchy-distributed with parameter u we obtain

$$\mathbb{E} \int_0^{t \wedge \sigma_N} f(S_u) \, du \leq \int_0^t \mathbb{E}(1_{G_N}(S_u) f(S_u)) \, du = \int_0^t \int_{G_N} f(y) \frac{u}{\pi(u^2 + y^2)} \, dy \, du$$

$$= \pi^{-1} \int_{G_N} f(y) \int_0^t \frac{u}{u^2 + y^2} \, du \, dy = (2\pi)^{-1} \int_{G_N} f(y) \ln \frac{t^2 + y^2}{y^2} \, dy.$$

Now $\ln \dfrac{t^2 + y^2}{y^2}$ is continuous at every $y \neq 0$ and behaves like $|\ln|y||$ as $y \to 0$. Hence the right hand side is finite proving the assertion. $\qquad \square$

The next proposition for $\alpha = 1$ is a consequence of the Harris recurrence of the Cauchy process and is known (cf. D. Revuz and M. Yor [13], Proposition X.3.11). As we shall see later, it ensures that for solutions to Eq. (1.1) for $\alpha = 1$ explosion does not occur. As the proof is the same, we state the result for general indexes α with $1 \leq \alpha \leq 2$.

Proposition 2.6 *Suppose that S is a symmetric α-stable process with $1 \le \alpha \le 2$ and that $\lambda(\{f > 0\}) > 0$. Then*

$$T_\infty = +\infty \quad \text{P-a.s.}$$

Proof. For the convenience of the reader, we include an alternative and simple proof of this fact which will basically be used in the following. To this end, we regard the symmetric α-stable process S as a Markov process defined on a family $(\Omega, \mathcal{F}, \mathbb{F}, \mathbb{P}_z, z \in \mathbb{R})$ of filtered probability spaces such that $\mathbb{P}_z(\{S_0 = z\}) = 1$ for every $z \in \mathbb{R}$. For $\varepsilon \ge 0$ we define

$$\varphi_\varepsilon(z) = \mathbb{P}_z(\{\int_0^\infty f(S_u)\,du > \varepsilon\}), \quad z \in \mathbb{R}.$$

Since $\lambda(\{f > 0\}) > 0$, it is clear that $\mathbb{E}_z \int_0^\infty f(S_u)\,du > 0$ and consequently $\varphi_0(z) > 0$. For some fixed $z \in \mathbb{R}$ and sufficiently small $\varepsilon > 0$, it now follows that $\varphi_\varepsilon(z) > 0$. On the other side, the function φ_ε is excessive and the recurrence properties of S imply that $\varphi_\varepsilon \equiv a$ is constant (cf. R. M. Blumenthal and R. K. Getoor [2], Chapter II, (4.19)). From the Markov property, for all $z \in \mathbb{R}$ we get

$$a = \lim_{N \to \infty} \varphi_\varepsilon(S_N) = \lim_{N \to \infty} \mathbb{P}_{S_N}(\{\int_0^\infty f(S_u)\,du > \varepsilon\})$$

$$= \lim_{N \to \infty} \mathbb{P}_z(\{\int_N^\infty f(S_u)\,du > \varepsilon\} \mid \mathcal{F}_N) = 1_{\bigcap_{N=1}^\infty \{\int_N^\infty f(S_u)\,du > \varepsilon\}} \quad \mathbb{P}_z\text{-a.s.},$$

where the last equality follows from the theorem of Lebesgue-Lévy on the convergence of conditional expectations. But $a > 0$ and, consequently,

$$\mathbb{P}_z(\bigcap_{N=1}^\infty \{\int_N^\infty f(S_u)\,du > \varepsilon\}) = 1, \quad z \in \mathbb{R}.$$

In view of the theorem of Lebesque on majorized convergence

$$\int_0^\infty f(S_u)\,du = +\infty \quad \mathbb{P}_z\text{-a.s.}, \quad z \in \mathbb{R},$$

must hold. Setting $z = 0$ this yields the assertion.   □

We now consider the case of indexes α with $0 < \alpha < 1$.

Proposition 2.7 *Suppose that $0 < \alpha < 1$ and that the following condition is satisfied:*

$$E \quad \text{is 0-polar and} \quad \int_{0-}^{0+} |y|^{\alpha-1} f(y)\,dy < +\infty. \tag{2.5}$$

We then have $\quad T_t < +\infty \quad$ *for all* $\quad t \ge 0 \quad$ *P-a.s.*

Proof. Let G_N and σ_N be defined by (2.4). As in the proof of Proposition 2.5 it is sufficient to verify that $\mathbb{E}\,T_{t \wedge \sigma_N} < +\infty$. Using (1.7) in Chapter II of [2] we estimate

$$\mathbb{E}\int_0^{t \wedge \sigma_N} f(S_u)\,du \le \mathbb{E}\int_0^\infty 1_{G_N}(S_u)\,f(S_u)\,du = c_\alpha \int_{G_N} |y|^{\alpha-1} f(y)\,dy.$$

The right member is finite in view of the definition of G_N and assumption (2.5). $\square$

The next lemma is of interest only for $0 < \alpha \leq 1$: For $1 < \alpha \leq 2$ it already follows from Proposition 2.1.

Lemma 2.8 *Suppose that* $0 < \alpha \leq 2$. *Let* I *be a subset of* $[0, +\infty)$ *containing* $[0, t)$ *or* $[t, +\infty)$ *for some* $t > 0$. *Then*

$$\int_I |S_u|^{-\alpha} \, du = +\infty \quad \mathbb{P}\text{-a.s.}$$

Proof. Let $N \geq 1$ be fixed and set $A_N = \{\int_I |S_u|^{-\alpha} \, du \leq N\}$. Using that $u^{-\frac{1}{\alpha}} S_u$ has the same distribution as S_1 we now estimate

$$N \geq \mathbb{E}\left(1_{A_N} \int_I |S_u|^{-\alpha} \, du\right) = \int_I \mathbb{E}\left(1_{A_N} (u^{-\frac{1}{\alpha}}|S_u|)^{-\alpha}\right) u^{-1} \, du$$

$$= \int_I \int_0^\infty \mathbb{P}\left(A_N \cap \{(u^{-\frac{1}{\alpha}}|S_u|)^{-\alpha} \geq v\}\right) dv \, u^{-1} \, du$$

$$\geq \int_I \int_0^\infty \left(\mathbb{P}(A_N) - \mathbb{P}(\{(u^{-\frac{1}{\alpha}}|S_u|)^{-\alpha} < v\})\right)^+ dv \, u^{-1} \, du$$

$$= \left(\int_I u^{-1} \, du\right) \cdot \left(\int_0^\infty \left(\mathbb{P}(A_N) - \mathbb{P}(\{|S_1|^{-\alpha} < v\})\right)^+ dv\right).$$

The first integral on the right hand side is $+\infty$, hence the second integral has to be zero which in view of $\mathbb{P}(\{|S_1|^{-\alpha} > 0\}) = 1$ is only possible if $\mathbb{P}(A_N) = 0$. $\square$

Example 2.9 We have the following properties:

(i) If $\beta \geq \alpha \wedge 1$ then $\int_0^t |S_u|^{-\beta} \, du = +\infty$ for all $t \geq 0$ $\mathbb{P}$-a.s.

(ii) If $\beta < \alpha \wedge 1$ then $\int_0^t |S_u|^{-\beta} \, du < +\infty$ for all $t \geq 0$ $\mathbb{P}$-a.s.

Indeed, for $1 < \alpha \leq 2$ and $\beta \geq 1$, (i) follows from Proposition 2.1 and for $0 < \alpha \leq 1$ and $\beta \geq \alpha$ from Lemma 2.8 for $I = [0, t]$. Similarly, if $1 < \alpha \leq 2$ and $\beta < 1$, (ii) is a consequence of Proposition 2.1 and, if $0 < \alpha \leq 1$ and $\beta < \alpha$, of Propositions 2.5 and 2.7. This example shows that $\alpha \wedge 1$ is the critical exponent for convergence or divergence and that, at least for power functions, the conditions (2.3) and (2.5) of Propositions 2.5 and 2.7 cannot be improved. $\square$

Finally, we deal with sufficient conditions for $T_\infty = +\infty$ $\mathbb{P}$-a.s. for $0 < \alpha < 1$. Then S is transient in the sense that $\lim_{t \to \infty} |S_t| = +\infty$ $\mathbb{P}$-a.s. Hence the behaviour of f near $-\infty$ and $+\infty$ will be substantial. We introduce the measures μ_α by

$$\mu_\alpha(A) = \int_A |y|^{\alpha-1} \, dy, \quad A \in \mathcal{B}(\mathbb{R}), \quad 0 \leq \alpha < 1. \tag{2.6}$$

Lemma 2.10 *Suppose that* $0 < \alpha < 1$. *We then have:*

(i) *For every measurable subset* A *of* $\mathbb{R}$ *such that* $\mu_\alpha(A) < +\infty$

$$\int_0^\infty 1_A(S_u) \, du < +\infty \quad \mathbb{P}\text{-a.s.}$$

In particular, this is true if $\lambda(A) < +\infty$.

(ii) $\lim_{t\to\infty} |S_t| = +\infty$ *P-a.s.*

(iii) $\int_t^\infty |S_u|^{-\alpha} du = +\infty$ *for every* $t \geq 0$ *P-a.s.*

(iv) *For every measurable subset A of* $\mathbb{R}$ *such that* $\mu_0(A) < +\infty$

$$\int_0^\infty 1_A(S_u) |S_u|^{-\alpha} du < +\infty \quad \text{P-a.s.}$$

In particular, this is true if $A \subseteq \mathbb{R} \setminus (-1, 1)$ *and* $\lambda(A) < +\infty$.

(v) *For every* $\beta > \alpha$

$$\int_0^\infty 1_{\{|S_u| \geq 1\}}) |S_u|^{-\beta} du < +\infty \quad \text{P-a.s.}$$

Proof. First we notice that in view of [2] (Chapter II, (1.7))

$$\mathbb{E} \int_0^\infty g(S_u) du = c_\alpha \int_{-\infty}^{+\infty} |y|^{\alpha-1} g(y) dy \tag{2.7}$$

for all nonnegative measurable g. Setting $g = 1_A$, from (2.7) we obtain

$$\mathbb{E} \int_0^\infty 1_A(S_u) du = c_\alpha \mu_\alpha(A) < +\infty$$

and hence the occupation time of S in A is finite P-a.s. This verifies (i). Part (ii) can be found in the book of K. Sato [16]. For the convenience of the reader, we suggest the following short proof. For this purpose, we set $A = (-N, N)$. Part (i) yields $\limsup_{t\to\infty} |S_t| \geq N$ P-a.s. for all $N \geq 1$ and hence $\limsup_{t\to\infty} |S_t| = +\infty$ P-a.s. On the other side, the function φ_x defined by $\varphi_x(y) = |x-y|^{\alpha-1}$, $y \in \mathbb{R}$, is excessive for the symmetric stable semigroup with parameter $0 < \alpha < 1$ (cf. [2], Chapter II, (3.15)). Thus the process $|x - S|^{\alpha-1}$ is a supermartingale if $x \neq 0$ and $\lim_{t\to\infty} |x - S_t|^{\alpha-1}$ exists P-a.s. Consequently, $\lim_{t\to\infty} |S_t| = +\infty$ P-a.s.. This proves (ii). Part (iii) follows from Lemma 2.8 for $I = [t, +\infty)$. For establishing (iv) we use formula (2.7):

$$\mathbb{E} \int_0^\infty 1_A(S_u) |S_u|^{-\alpha} du = c_\alpha \int_A |y|^{\alpha-1} |y|^{-\alpha} dy = c_\alpha \int_A |y|^{-1} dy = c_\alpha \mu_0(A)$$

which, obviously, yields (iv). Analogously, for proving (v) we observe

$$\mathbb{E} \int_0^\infty 1_{\{|S_u| \geq 1\}}) |S_u|^{-\beta} du = 2c_\alpha \int_1^\infty |y|^{\alpha-1} |y|^{-\beta} dy = 2c_\alpha \int_1^\infty |y|^{\alpha-\beta-1} dy.$$

The right hand side being finite, this completes the proof of the lemma. □

Proposition 2.11 *Suppose that* $0 < \alpha < 1$. *Let f be a nonnegative measurable function such that the following condition is satisfied:*

$$\exists c > 0 : \quad \mu_0(\{x \in \mathbb{R} : |x| \geq 1, f(x) |x|^\alpha < c\}) < +\infty. \tag{2.8}$$

We then have $T_\infty = +\infty$ *P-a.s.*

Proof. We set $A = \{x \in \mathbb{R} : |x| \geq 1,\ f(x)\,|x|^{\alpha} < c\}$ and assume $\mu_0(A) < +\infty$ where μ_0 is defined by (2.6). Using Lemma 2.10 (ii) we choose $v = v(\omega)$ so large that $|S_u(\omega)| \geq 1$ for all $u \geq v$ $\mathbb{P}$-a.s. For any $t \geq v$ we now get

$$
\begin{aligned}
T_t &\geq \int_v^t f(S_u)\,du \geq c \int_v^t \mathbf{1}_{A^c \cap (-1,1)^c}(S_u)\,|S_u|^{-\alpha}\,du \\
&= c \int_v^t \mathbf{1}_{\{|S_u| \geq 1\}}\,|S_u|^{-\alpha}\,du - c \int_v^t \mathbf{1}_A(S_u)\,|S_u|^{-\alpha}\,du \\
&= c \int_v^t |S_u|^{-\alpha}\,du - c \int_v^t \mathbf{1}_A(S_u)\,|S_u|^{-\alpha}\,du,
\end{aligned}
$$

the right hand side converging to $+\infty$ as $t \to \infty$ $\mathbb{P}$-a.s. by Lemma 2.10 (iii), (iv). $\square$

Remark 2.12 Let $0 < \alpha < 1$. Each of the following conditions implies (2.8) and is therefore sufficient for the property $T_\infty = +\infty$ $\mathbb{P}$-a.s.:

$$\exists c > 0: \qquad \lambda(\{x \in \mathbb{R} : f(x)\,|x|^{\alpha} < c\}) < +\infty. \tag{2.9}$$

$$\exists c > 0: \qquad \lambda(\{x \in \mathbb{R} : f(x) < c\}) < +\infty. \tag{2.10}$$

$$|x|^{-\alpha} = O(f(x)), \quad |x| \to \infty, \quad \lambda\text{-a.e.} \tag{2.11}$$

The second condition was employed by P. A. Zanzotto (cf. [20], Lemma 2.7).

The following proposition shows, in particular, that the condition (2.8) of Proposition 2.11 (also see (2.9), (2.10), (2.11)) on the asymptotic behaviour of f, in a certain sense, cannot be improved.

Proposition 2.13 *Let $0 < \alpha < 1$ and suppose that $T_t < +\infty$ $\mathbb{P}$-a.s. for all $t \geq 0$ (cf. Proposition 2.7 for sufficient conditions). If there exists $\beta > \alpha$ such that the condition*

$$f(x) = O(|x|^{-\beta}), \quad |x| \to \infty, \quad \lambda\text{-a.e.} \tag{2.12}$$

holds then $\quad T_\infty < +\infty \quad \mathbb{P}$-a.s.

Proof. If (2.12) is satisfied then there exist $N \geq 1$ and $c > 0$ such that

$$f(x) \leq c\,|x|^{-\beta} \quad \text{for all} \quad |x| \geq N \quad \lambda\text{-a.e.}$$

In view of Lemma 2.10 (ii) we can choose $t = t(\omega)$ such that $|S_u(\omega)| \geq N$ for all $u \geq t$. The occupation time in a set of Lebesgue measure zero being zero we obtain

$$\int_0^\infty f(S_u)\,du \leq T_t + c \int_t^\infty \mathbf{1}_{\{|S_u| \geq N\}}|S_u|^{-\beta}\,du.$$

The right hand side is finite $\mathbb{P}$-a.s. because of the assumption and Lemma 2.10 (v). $\square$

Remarks 2.14 There remain open several problems concerning the behaviour of integral functionals of symmetric stable processes:

(i) Let S be the Cauchy process. Is then the condition (2.3) of Proposition 2.5 also necessary for $T_t < +\infty$, $t \geq 0$, $\mathbb{P}$-a.s.? Is there valid a 0–1-law as for $1 < \alpha \leq 2$?

(ii) In the case $0 < \alpha < 1$, is the condition (2.5) of Proposition 2.7 also necessary for the convergence of the integral functionals T_t for every $t \geq 0$? Because of the transience of S (cf. Lemma 2.10 (ii)) a 0–1-law seems to be not true.

3. CONSTRUCTION OF TIME CHANGE PROCESSES

Let $(S, \mathbb{G})$ be a symmetric α-stable process defined on $(\Omega, \mathcal{F}, \mathbb{P})$ such that $S_0 = 0$. In this section, we consider the following integral equation:

$$T_t = \int_0^{t+} h(T_u, S_u)\, du, \quad t \geq 0, \tag{3.1}$$

where $h : [0, +\infty) \times \mathbb{R} \to [0, +\infty]$ is a nonnegative measurable function of (t, x). A solution T_t is allowed to take the value $+\infty$ for some $t \geq 0$. For this reason we make the convention $h(+\infty, x) = +\infty$ for every $x \in \mathbb{R}$. We will look for conditions on h guaranteeing that there exist solutions T of Eq. (3.1) which are, moreover, adapted to the filtration $\mathbb{G}^S$ generated by S and completed by all $\mathbb{P}$-null sets. The right inverse A of the increasing process T, defined by

$$A_t = \inf\{s \geq 0 : T_s > t\}, \quad t \geq 0, \tag{3.2}$$

can then serve as a time change for S. This will be carried out in Section 5 for constructing solutions $(X, \mathbb{F})$ of Eq. (1.1).

For every $N \geq 1$, let the functions $\underline{h}_N$ and $\overline{h}_N$ be defined by

$$\underline{h}_N(x) = \inf_{0 \leq t \leq N} h(t, x), \quad \overline{h}_N(x) = \sup_{0 \leq t \leq N} h(t, x), \quad x \in \mathbb{R}.$$

We suppose that the function h satisfies the following two assumptions:

(A) The function $t \to h(t, x)$ is finite and continuous for λ-a.a. $x \in \mathbb{R}$.

(B) For every $N \geq 1$,

    (a) if $1 < \alpha \leq 2$, $\overline{h}_N$ is locally integrable,

    (b) if $\alpha = 1$, $f = \overline{h}_N$ satisfies (2.3),

    (c) if $0 < \alpha < 1$, $f = \overline{h}_N$ satisfies (2.5).

Theorem 3.1 *Suppose that h satisfies the conditions **(A)** and **(B)**. Then there exists a $\mathbb{G}^S$-adapted solution T of Eq. (3.1) such that $A_t < A_\infty$ on $\{A_\infty < +\infty\}$ for all $t \geq 0$ $\mathbb{P}$-a.s. where A is defined by (3.2) and $A_\infty = \sup_{t \geq 0} A_t$.*

Proof. First we notice that, instead of **(A)**, we may assume that h is finite and continuous in t *for all* $x \in \mathbb{R}$. Otherwise we find a measurable set $A \subseteq \mathbb{R}$ such that $\lambda(A) = 0$ and $t \to h(t, x)$ is finite and continuous for all $x \in A^c$. The function $\tilde{h}$ defined by $\tilde{h}(t, x) = h(t, x)$ if $x \in A^c$ and 0 otherwise for all $t \geq 0$ satisfies the new hypothesis and if we know that there is a solution T of Eq. (3.1) for $\tilde{h}$ then T is also a solution of Eq. (3.1) for h:

$$\left| \int_0^t h(T_u, S_u)\, du - \int_0^t \tilde{h}(T_u, S_u)\, du \right| = \int_0^t h(T_u, S_u)\, \mathbf{1}_A(S_u)\, du = 0 \quad \mathbb{P}\text{-a.s.}$$

since the occupation time of S in A is zero $\mathbb{P}$-a.s.

In the second step, we set $\overline{h}(x) = \sup_{t \geq 0} h(t, x)$, $x \in \mathbb{R}$, and assume that, instead of $\overline{h}_N$, $\overline{h}$ satisfies the properties stated in assumption **(B)**. For arbitrary $n \in \mathbb{N}$ we define approximating functions $h_n : [0, \infty) \times \mathbb{R} \to [0, \infty)$ by

$$h_n(t, x) = \inf_{s \in [m2^{-n}, (m+1)2^{-n})} h(s, x), \quad \text{if} \quad t \in [m2^{-n}, (m+1)2^{-n}), \quad x \in \mathbb{R}.$$

Clearly, h_n is increasing and converges to h pointwise as $n \to \infty$, and $h_n \leq \overline{h}$. In view of the continuity of h in t for every x, we now obtain that $h_n(\cdot, x)$ converges to $h(\cdot, x)$ as $n \to \infty$ *uniformly* on every compact set of $[0, +\infty)$. Because of $h_n \leq \overline{h}$, Corollary 2.2, Proposition 2.5 and Proposition 2.7 we have

$$\int_0^t h_n(m2^{-n}, S_u)\, du \leq \int_0^t \overline{h}(S_u)\, du < +\infty, \quad t \geq 0, \quad \mathbb{P}\text{-a.s.} \tag{3.3}$$

Inductively over m, we now define a sequence $(\tau_m^n)_{m \geq 0}$ of $\mathbb{G}^S$-stopping times:

$$\tau_0^n = 0, \quad \tau_{m+1}^n = \inf\{t \geq \tau_m^n : \int_{\tau_m^n}^t h_n(m2^{-n}, S_u)\, du \geq 2^{-n}\}.$$

Also we set

$$\rho_m^n = \inf\{t \geq 0 : \int_0^t \overline{h}(S_u)\, du \geq m2^{-n}\}, \quad m \geq 0.$$

From (3.3), by induction we observe that $\rho_m^n \leq \tau_m^n$. But $\lim_{m \to \infty} \rho_m^n = +\infty$ $\mathbb{P}$-a.s. which implies $\lim_{m \to \infty} \tau_m^n = +\infty$ $\mathbb{P}$-a.s. for all $n \geq 1$. We now define

$$T_t^n = m2^{-n} + \int_{\tau_m^n}^t h_n(m2^{-n}, S_u)\, du \quad \text{if} \quad \tau_m^n \leq t < \tau_{m+1}^n, \quad m \geq 0.$$

Obviously, T^n is $\mathbb{G}^S$-adapted. It is now elementary to verify that T^n satisfies

$$T_t^n = \int_0^t h_n(T_u^n, S_u)\, du, \quad t \geq 0,$$

and, moreover, $T_t^n \leq T_t^{n+1}$ for all $t \geq 0$. Consequently, there exists the limit T,

$$T_t = \lim_{n \to \infty} T_t^n \leq \int_0^t \overline{h}(S_u)\, du, \quad t \geq 0,$$

and T is $\mathbb{G}^S$-adapted. We show that T is a solution to Eq. (3.1) for the original function h. Since $h_n(\cdot, x)$ uniformly converges to $h(\cdot, x)$ on every compact interval and T_u^n is bounded uniformly in $u \leq t$ and n $\mathbb{P}$-a.s., we observe

$$\lim_{n \to \infty} h_n(T_u^n, S_u) = h(T_u, S_u), \quad u \leq t, \quad \mathbb{P}\text{-a.s.}$$

Moreover, $h_n(T_u^n, S_u)$ is majorized by $\overline{h}(S_u)$ which is integrable over $[0, t]$ $\mathbb{P}$-a.s. Using the theorem of Lebesgue on majorized convergence we obtain

$$T_t = \lim_{n \to \infty} T_t^n = \lim_{n \to \infty} \int_0^t h_n(T_u^n, S_u)\, du = \int_0^t h(T_u, S_u)\, du, \quad t \geq 0, \quad \mathbb{P}\text{-a.s.}$$

In the third step, we get rid from the additional assumption of the second step by a localization procedure. For this we define

$$h_N(t, x) = h(t \wedge N, x), \quad (t, x) \in [0, +\infty) \times \mathbb{R},$$

for every $N \geq 1$. The function h_N then satisfies the assumptions of the second step and we consider the $\mathbb{G}^S$-adapted and $\mathbb{P}$-a.s. finite solution T^N of Eq. (3.1) for h_N constructed there. Now we set $A_N = \inf\{t \geq 0 : T_t^N > N\}$. By the construction of T^N it is clear that $T_t^N = T_t^{N+1}$ for every $t \leq A_N$ $\mathbb{P}$-a.s. From this it also follows $A_N < A_{N+1}$ on $\{A_N < +\infty\}$ $\mathbb{P}$-a.s. Hence the limit $A_\infty = \lim_{N \to \infty} A_N$ exists and $A_N < A_\infty$ on $\{A_\infty < +\infty\}$ $\mathbb{P}$-a.s. Finally, for every $t \geq 0$, we put $T_t = T_t^N$ if $t \leq A_N$ for some $N \geq 1$ and $+\infty$ if $t \geq A_\infty$. Then T is the required solution of Eq. (3.1). $\square$

In general, the solution T of Eq. (3.1) constructed in the proof of Theorem 3.1 can be bounded or even equal to zero. Roughly speaking, the associated time change A then "explodes" in finite time with positive probability. The following theorem gives conditions that explosions do not occur.

Theorem 3.2 *Let h be a nonnegative measurable function defined on $[0, +\infty) \times \mathbb{R}$. Suppose that h satisfies the following condition:*

(C) *For every $N \geq 1$,*

(a) *if $1 \leq \alpha \leq 2$, then $\lambda(\{\underline{h}_N > 0\}) > 0$.*

(b) *if $0 < \alpha < 1$, then $f = \underline{h}_N$ satisfies (2.8).*

Let T be any supersolution of Eq. (3.1) for h, i.e., $T_t \geq \int_0^t h(T_u, S_u)\, du$, $t \geq 0$. Then we have $T_\infty = +\infty$ $\mathbb{P}$-a.s.

Proof. For every $N \geq 1$, we define $C_N = \{T_\infty \leq N\}$. On C_N we obtain

$$N \geq T_t \geq \int_0^t h(T_u, S_u)\, du \geq \int_0^t \underline{h}_N(S_u)\, du.$$

But in view of Corollary 2.3, Proposition 2.6 and Proposition 2.11 the right hand side converges to $+\infty$ $\mathbb{P}$-a.s. as $t \to +\infty$. This yields $\mathbb{P}(C_N) = 0$. $\square$

4.   STOCHASTIC INTEGRALS WITH RESPECT TO STABLE PROCESSES

For the proof of the main theorem of the next section, we first adapt to our needs some basic properties of stochastic integrals with respect to symmetric stable processes due to J. Rosiński and W. Woyczyński [14]. To begin with, we introduce the notion of a symmetric α-stable process stopped at a random time T.

Definition 4.1 *Let $(Z, \mathbb{F})$ be a càdlàg process and T an $\mathbb{F}$-stopping time. Then $(Z, \mathbb{F})$ is said to be a symmetric α-stable process stopped at T if for every $\lambda \in \mathbb{R}$ the process $\exp(i\lambda Z_t + |\lambda|^\alpha(t \wedge T))$, $t \geq 0$, is a (complex-valued) $\mathbb{F}$-martingale.*

Lemma 4.2 *Every symmetric α-stable process $(Z, \mathbb{G})$ stopped at T can be extended to a symmetric α-stable process $(\tilde{Z}, \tilde{\mathbb{G}})$ on $(\tilde{\Omega}, \tilde{\mathcal{F}}, \tilde{\mathbb{P}})$ such that $Z_t = \tilde{Z}_{t \wedge T}$, $t \geq 0$, $\mathbb{P}$-a.s.*

Proof. Let $(\bar{Z}, \bar{\mathbb{F}})$ be a symmetric α-stable process given on the space $(\bar{\Omega}, \bar{\mathcal{F}}, \bar{\mathbb{P}})$ and introduce $(\tilde{\Omega}, \tilde{\mathcal{F}}, \tilde{\mathbb{P}}) = (\Omega, \mathcal{F}, \mathbb{P}) \times (\bar{\Omega}, \bar{\mathcal{F}}, \bar{\mathbb{P}})$, $\tilde{\mathbb{F}} = \mathbb{F} \times \bar{\mathbb{F}}$ and

$$\tilde{Z}_t(\omega, \bar{\omega}) = Z_{t \wedge T(\omega)}(\omega) + (\bar{Z}_t(\bar{\omega}) - \bar{Z}_{t \wedge T(\omega)}(\bar{\omega})), \quad \tilde{\omega} = (\omega, \bar{\omega}) \in \tilde{\Omega}, \quad t \geq 0.$$

One easily checks that then $(\tilde{Z}, \tilde{\mathbb{F}})$ is a symmetric α-stable process. Completing $\tilde{\mathbb{F}}$ appropriately, the resulting filtration will be right-continuous. Hence, without loss of generality, we may assume that $\tilde{\mathbb{F}}$ satisfies the usual conditions. $\quad\square$

Proposition 4.3 *Let $(S, \mathbb{G})$ be a symmetric α-stable process. Suppose that H is a $\mathbb{G}$-previsible process (with values in $\mathbb{R}$) and set*

$$T_t = \int_0^{t+} |H_u|^\alpha \, du, \quad t \geq 0.$$

Let $A = (A_t)_{t \geq 0}$ be the right inverse of $T = (T_t)_{t \geq 0}$ (see (3.2)). We then have:

(i) *The stochastic integral $\int_0^{A_t} H_u \, dS_u$ is well-defined for all $t \geq 0$.*

(ii) *The process $(Z, \mathbb{F})$ defined by $Z_t = \int_0^{A_t} H_u \, dS_u$, $\mathcal{F}_t = \mathcal{G}_{A_t}$, $t \geq 0$, is a symmetric α-stable process stopped at $U_\infty := T_{A_\infty -}$.*

(iii) *The stochastic integral $\int_0^t H_u \, dS_u$ is well-defined on $\{t < A_\infty\}$ for all $t \geq 0$.*

(iv) *If $\mathbb{E} \int_0^t |H_u|^\alpha \, du = 0$ then $\int_0^t H_u \, dS_u = 0$ $\quad \mathbb{P}$-a.s.*

Proof. We only give a brief sketch referring to [14] for more information on stochastic integrals for symmetric α-stable processes. For the previsible process $1_{[0,A_t]} H$ we have

$$\int_0^\infty |1_{[0,A_t]}(u) H_u|^\alpha \, du = \int_0^{A_t} |H_u|^\alpha \, du = T_{A_t -} = t \wedge T_{A_\infty -} = t \wedge U_\infty \leq t.$$

In the same way as in [14] it can be seen that

$$\int_0^{A_t} H_u \, dS_u := \int_0^\infty 1_{[0,A_t]}(u) H_u \, dS_u$$

exists which is realized by using the estimate

$$\sup_{\lambda > 0} \lambda^\alpha \mathbb{P}(\{ \sup_{0 \leq t < \infty} |\int_0^t F_u \, dS_u| > \lambda\}) \leq c_\alpha \mathbb{E} \int_0^\infty |F_u|^\alpha \, du =: c_\alpha \|F\|_\alpha^\alpha \quad (4.1)$$

for every previsible step process F. The inequality (4.1) then extends to every previsible F such that the right hand side is finite. From this, assertion (iv) follows immediately. For proving (ii), we notice that

$$\exp(i\lambda \int_0^t F_u \, dS_u + |\lambda|^\alpha \int_0^t |F_u|^\alpha \, du), \quad t \geq 0,$$

is a bounded $\mathbb{G}$-martingale if F is previsible and such that $\int_0^\infty |F_u|^\alpha \, du$ is bounded. For previsible step processes F this is an easy consequence of the defining property of symmetric α-stable processes, for general F the result follows approximating F by previsible step processes in the norm $\|\cdot\|_\alpha$. From this follows that

$$\exp(i\lambda \int_0^{t \wedge A_N} H_u \, dS_u + |\lambda|^\alpha \int_0^{t \wedge A_N} |H_u|^\alpha \, du), \quad t \ge 0,$$

is a *bounded* $\mathbb{G}$-martingale and, by Doob's optional sampling theorem,

$$\exp(i\lambda \int_0^{A_t \wedge A_N} H_u \, dS_u + |\lambda|^\alpha \int_0^{A_t \wedge A_N} |H_u|^\alpha \, du)$$

$$= \exp(i\lambda \int_0^{A_t \wedge N} H_u \, dS_u + |\lambda|^\alpha (t \wedge N \wedge U_\infty)), \quad t \ge 0,$$

is a bounded $\mathbb{F}$-martingale and hence (ii) is proven. Finally, for showing (iii), we set

$$\int_0^t H_u \, dS_u := \int_0^{t \wedge A_N} H_u \, dS_u \quad \text{if} \quad t \le A_N$$

for all $t < A_\infty = \lim_{N \to \infty} A_N$ which correctly defines the desired integral on the set $\{t < A_\infty\}$. $\square$

5.   EXISTENCE OF SOLUTIONS

We now come back to Eq. (1.1) and ask for sufficient conditions for the existence of solutions $(X, \mathbb{F})$. To begin with, we give the precise definition of a, possibly, exploding solution $(X, \mathbb{F})$ of Eq. (1.1). First we introduce the path space $D^e = D^e_{\mathbb{R}}([0. + \infty))$. Let Δ be a point not belonging to $\mathbb{R}$ and set $\mathbb{R}_\Delta = \mathbb{R} \cup \{\Delta\}$; Δ is adjoint to $\mathbb{R}$ as an isolated point. If $z : [0, +\infty) \longrightarrow \mathbb{R}_\Delta$, and $n \ge 1$, we define

$$\tau_\Delta(z) = \inf\{t \ge 0 : z(t) = \Delta\}, \qquad \tau_n = \inf\{t \ge 0 : |z(t)| \ge n\}. \tag{5.1}$$

Now let D^e be the set of all right-continuous functions $z : [0, +\infty) \longrightarrow \mathbb{R}_\Delta$ with left hand limits in $\mathbb{R}$ on $(0, \tau_\Delta(z))$ such that $z(t) = \Delta$ whenever $t \ge \tau_\Delta(z)$ and, for every $n \ge 1$, $\tau_n(z) < \tau_\Delta(z)$. For any path $z \in D^e$, $\tau_\Delta(z)$ is called the explosion time of z. It can easily be seen that $\lim_{n \to \infty} \tau_n(z) = \tau_\Delta(z)$ for all $z \in D^e$. By $\mathcal{D}^e$ we define the σ-algebra generated by the coordinate mappings $z \longrightarrow z(t)$, for all $t \ge 0$. Finally, for any $z \in D^e$, by z^- the function $z^-(t) = z(t-)$ if $t < \tau_\Delta(z)$ and $z^-(t) = \Delta$ otherwise is denoted where $z(0-) = z(0)$.

A stochastic process $(X, \mathbb{F})$, defined on a probability space $(\Omega, \mathcal{F}, \mathbb{P})$ with filtration $\mathbb{F}$ (always satisfying the usual conditions), is called a solution of Eq. (1.1) with initial state $x_0 \in \mathbb{R}$, if X has paths in the space D^e and if there exists a symmetric α-stable process Z with respect to $\mathbb{F}$ such that

$$X_t = x_0 + \int_0^t b(u, X_u^-) \, dZ_u \quad \text{on} \quad \{t < \tau_\Delta(X)\} \quad \mathbb{P}\text{-a.s.}$$

for all $t \geq 0$ holds where $\tau_\Delta(X)$ is the composition of τ_Δ (defined by (5.1)) and X and is called the explosion time of X. Of course, $X^-(\omega) = X(\omega)^-, \omega \in \Omega$.

The definition of a solution $(X, \mathbb{F})$ includes the existence of the stochastic integral above on $\{t < \tau_\Delta(X)\}$, which is equivalent to

$$\int_0^t |b|^\alpha(u, X_u^-) \, du < +\infty \quad \text{on} \quad \{t < \tau_\Delta(X)\} \quad \mathbb{P}\text{-a.s.}$$

(cf. Section 4). In other words, the stochastic integral should exist on the stochastic interval $[0, \tau_\Delta(X))$. Note that $\tau_\Delta(X)$ is a previsible $\mathbb{F}$-stopping time.

On the other side, it is not difficult to verify that

$$\int_0^{\tau_\Delta(X)} |b|^\alpha(u, X_u^-) \, du = +\infty \quad \mathbb{P}\text{-a.s.};$$

otherwise there would be a finite limit

$$\lim_{n \to \infty} X_{\tau_n(X)} = x_0 + \lim_{n \to \infty} \int_0^{\tau_n(X)} b(u, X_u^-) \, dZ_u.$$

with positive probability (cf. Section 4) contradicting the property that $\tau_\Delta(X)$ is the explosion time of X.

Before we state the main result we give some preparations. Let b be a measurable function defined on $[0, +\infty) \times \mathbb{R}$. It will be convenient for us to extend b on $[0, +\infty) \times \mathbb{R}_\Delta$ by setting $b(t, \Delta) = 0$ for all $t \geq 0$. If $(X, \mathbb{F})$ is an, in general, exploding solution of Eq. (1.1) we set

$$\tilde{A}_t = \int_0^{t+} |b|^\alpha(u, X_u) \, du \quad \left(= \int_0^{t+} |b|^\alpha(u, X_u^-) \, du \right), \quad t \geq 0, \quad (5.2)$$

and define the right inverse $\tilde{T}$ of the increasing process $\tilde{A}$:

$$\tilde{T}_t = \inf\{s \geq 0 : \tilde{A}_s > t\}, \quad t \geq 0. \quad (5.3)$$

We notice that $\tilde{T}_\infty$ is just $\mathbb{P}$-a.s. equal to the explosion time of the increasing process $\tilde{A}$ and also $\mathbb{P}$-a.s. equal to the explosion time $\tau_\Delta(X)$ of X; see the above explanation. This implies $\tilde{A}_{\tilde{T}_\infty-} = +\infty$ on $\{\tilde{T}_\infty < +\infty\}$ $\mathbb{P}$-a.s. or, equivalently,

$$\tilde{T}_t < \tilde{T}_\infty \quad \text{on} \quad \{\tilde{T}_\infty < +\infty\} \quad \mathbb{P}\text{-a.s.} \quad (5.4)$$

We also introduce the notation

$$\tilde{U}_\infty = \inf\{s \geq 0 : \tilde{A}_s = \tilde{A}_\infty\}. \quad (5.5)$$

Of course, $\tilde{U}_\infty = \tilde{T}_{\tilde{A}_\infty-} \leq \tilde{T}_\infty$ and $\tilde{U}_\infty$ is the smallest stopping time such that $\tilde{A}_t = \tilde{A}_{t \wedge \tilde{U}_\infty}$ or, equivalently, $X_t = X_{t \wedge \tilde{U}_\infty}$. Using Proposition 4.3 (i), (ii) (by changing the roles of A and T) we observe that the time changed process $(S, \mathbb{G})$ with

$$S_t = X_{\tilde{T}_t} - x_0 = \int_0^{\tilde{T}_t} b(u, X_u^-) \, dZ_u, \quad \mathbb{G}_t = \mathcal{F}_{\tilde{T}_t}, \quad t \geq 0, \quad (5.6)$$

is a symmetric α-stable process stopped at $\tilde{A}_{\tilde{T}_\infty-} = \tilde{A}_\infty$. Enlarging the probability space, without loss of generality we can, and we always do, assume that $(S, \mathbb{G})$ is extended to a full symmetric α-stable process. Because X is constant $\mathbb{P}$-a.s. on $(t, \tilde{T}_{\tilde{A}_t}]$ which is an interval of constancy for $\tilde{A}$ (cf. (4.1)) we obtain the important representation

$$X_t = x_0 + S_{\tilde{A}_t}, \quad t < \tilde{T}_\infty, \quad \mathbb{P}\text{-a.s.} \tag{5.7}$$

Definition 5.1 (i) A solution $(X, \mathbb{F})$ is called *basic* if

$$\int_0^{\tilde{U}_\infty} \mathbf{1}_{\{b=0\}}(u, X_u)\, du = 0 \quad \mathbb{P}\text{-a.s.}$$

(ii) $(X, \mathbb{F})$ is said to be *nonabsorbing* if $\tilde{U}_\infty = \tilde{T}_\infty$ $\mathbb{P}$-a.s.

We notice that $(X, \mathbb{F})$ is nonabsorbing if and only if $\tilde{A}_t < \tilde{A}_\infty$ on $\{\tilde{A}_\infty < +\infty\}$ $\mathbb{P}$-a.s. for every $t \geq 0$.

Lemma 5.2 *Let* $(X, \mathbb{F})$ *be a solution of Eq. (1.1). We then have:*

(i) *For every* $t \geq 0$,

$$\tilde{T}_t \geq \int_0^t |b|^{-\alpha}(\tilde{T}_u, x_0 + S_u)\, du. \tag{5.8}$$

(ii) *If* $(X, \mathbb{F})$ *is a basic solution of Eq. (1.1) then*

$$\tilde{T}_t \wedge \tilde{U}_\infty = \int_0^{t \wedge \tilde{A}_\infty} |b|^{-\alpha}(\tilde{T}_u, x_0 + S_u)\, du \quad \text{for every} \quad t \geq 0 \quad \mathbb{P}\text{-a.s.} \tag{5.9}$$

and if, additionally, $(X, \mathbb{F})$ *is nonabsorbing then*

$$\tilde{T}_t = \int_0^t |b|^{-\alpha}(\tilde{T}_u, x_0 + S_u)\, du \quad \text{for every} \quad t \geq 0 \quad \mathbb{P}\text{-a.s.} \tag{5.10}$$

(iii) *Let*

$$N_b = \{x \in \mathbb{R} : \exists t \geq 0 \quad \text{such that} \quad b(t, x) = 0\} \tag{5.11}$$

and suppose that $\lambda(N_b) = 0$. *If (5.9) holds then* $(X, \mathbb{F})$ *is a basic solution of Eq. (1.1).*

Proof. The continuity of $\tilde{A}$ on $[0, \tilde{T}_\infty)$ and (5.4) yield $\tilde{A}_{\tilde{T}_t} = t \wedge \tilde{A}_\infty$ and, by time change in the integral (see, e.g., [5], Lemma 1.6), we obtain

$$
\begin{aligned}
\int_0^{t \wedge \tilde{A}_\infty} |b|^{-\alpha}(\tilde{T}_u, x_0 + S_u)\, du &= \int_0^{\tilde{A}_{\tilde{T}_t}} |b|^{-\alpha}(\tilde{T}_u, x_0 + S_u)\, du \\
&= \int_0^{\tilde{T}_t} |b|^{-\alpha}(u, X_u)\, d\tilde{A}_u \\
&= \int_0^{\tilde{T}_t} |b|^{-\alpha}(u, X_u)\, |b|^\alpha(u, X_u)\, du \leq \tilde{T}_t.
\end{aligned}
$$

Because $\tilde{T}_t = +\infty$ for $t \geq \tilde{A}_\infty$, this proves (5.8). If $(X, \mathbb{F})$ is a basic solution then the inequality becomes an equality $\mathbb{P}$-a.s. on $\{\tilde{T}_t < \tilde{U}_\infty\} = \{t < \tilde{A}_\infty\}$ which yields (5.9) on this set. Passing to the limit $t \uparrow \tilde{A}_\infty$ we obtain

$$\tilde{U}_\infty = \tilde{T}_{\tilde{A}_\infty -} = \int_0^{\tilde{A}_\infty} |b|^{-\alpha}(\tilde{T}_u, x_0 + S_u)\, du \qquad (5.12)$$

which is (5.9) on $\{\tilde{T}_t \geq \tilde{U}_\infty\} = \{t \geq \tilde{A}_\infty\}$. If, additionally, $(X, \mathbb{F})$ is nonabsorbing then $\tilde{U}_\infty = \tilde{T}_\infty$ and, in particular, $\tilde{U}_\infty = +\infty$ on $\{\tilde{A}_\infty < +\infty\}$. From this and from (5.9) and (5.12) easily follows that (5.10) is satisfied. Finally, we show (iii): Using (5.9), on $\{t < \tilde{A}_\infty\} = \{\tilde{T}_t < \tilde{U}_\infty\}$ we obtain

$$\int_0^{\tilde{T}_t} \mathbf{1}_{\{b=0\}}(u, X_u)\, du \;\leq\; \int_0^{\tilde{T}_t} \mathbf{1}_{N_b}(X_u))\, du = \int_0^t \mathbf{1}_{N_b}(x_0 + S_u))\, d\tilde{T}_u$$

$$= \int_0^t |b|^{-\alpha}(\tilde{T}_u, x_0 + S_u)\, \mathbf{1}_{N_b}(x_0 + S_u)\, du = 0 \quad \mathbb{P}\text{-a.s.}$$

because $x_0 + S$ has $\mathbb{P}$-a.s. occupation time zero in the set N_b of Lebesgue measure zero. Now $\tilde{T}_t \uparrow \tilde{U}_\infty$ as $t \uparrow \tilde{A}_\infty$ and hence $(X, \mathbb{F})$ is a basic solution of Eq. (1.1). $\qquad \square$

For every $N \geq 1$, we introduce the functions $\underline{b}_N$ and $\bar{b}_N$ by

$$\underline{b}_N(x) = \inf_{0 \leq t \leq N} |b(t, x)|, \quad \bar{b}_N(x) = \sup_{0 \leq t \leq N} |b(t, x)|, \quad x \in \mathbb{R}.$$

Let $x_0 \in \mathbb{R}$. We shall need the following condition $(\mathbf{E}(x_0))$:

$$(\mathbf{E}(x_0)) \begin{cases} (E_1) & \text{For } \lambda\text{-a.e. } x \in \mathbb{R}, \ b(t, x) \text{ is continuous in } t. \\ (E_2) & \text{For every } N \geq 1, \\ & \quad \textbf{(a)} \quad \text{if } 1 < \alpha \leq 2, \text{ then } \underline{b}_N^{-\alpha} \text{ is locally integrable,} \\ & \quad \textbf{(b)} \quad \text{if } \alpha = 1, \text{ then } f = \underline{b}_N^{-\alpha}(x_0 + \cdot) \text{ satisfies (2.3),} \\ & \quad \textbf{(c)} \quad \text{if } 0 < \alpha < 1, \text{ then } f = \underline{b}_N^{-\alpha}(x_0 + \cdot) \text{ satisfies (2.5).} \end{cases}$$

Theorem 5.3 *Suppose that the condition* $(\mathbf{E}(x_0))$ *is satisfied. Then there exists a (possibly, exploding) nonabsorbing basic solution* $(X, \mathbb{F})$ *of Eq.* (1.1) *with* $X_0 = x_0$.

Proof. Let $(S, \mathbb{G})$ be a symmetric α-stable process on a probability space $(\Omega, \mathcal{F}, \mathbb{P})$. The condition $(\mathbf{E}(x_0))$ ensures that for $h = |b|^{-\alpha}(\cdot, x_0 + \cdot)$ on $[0, +\infty) \times \mathbb{R}$ the conditions of Theorem 3.1 are satisfied. (Of course, $|b|^{-\alpha}(t, x) = +\infty$ if $b(t, x) = 0$ for $(t, x) \in [0, +\infty) \times \mathbb{R}$.) For this we notice that $(\mathbf{E}(x_0))$ implies $\lambda(N_b) = 0$ where N_b is defined by (5.11). (Note that an x_0-polar set is of Lebesque-measure zero.) By Theorem 3.1, there exists an $\mathbb{F}^S$-adapted solution T of

$$T_t = \int_0^{t+} |b|^{-\alpha}(T_u, x_0 + S_u)\, du \quad \left(= \int_0^{t+} |b|^{-\alpha}(T_u, x_0 + S_{u-})\, du \right), \quad t \geq 0,$$

such that

$$A_t < A_\infty \quad \text{on} \quad \{A_\infty < +\infty\} \quad \mathbb{P}\text{-a.s.} \quad \text{for all} \quad t \geq 0. \qquad (5.13)$$

Here A denotes the right inverse of the increasing process T defined by (3.2). Obviously, T is strictly increasing and continuous on $[0, A_\infty)$ $\mathbb{P}$-a.s. From this follows that A is strictly increasing and continuous on $[0, T_\infty)$ $\mathbb{P}$-a.s., too. In particular, from this we get the identities $T_{A_t} = t \wedge T_\infty$ and $A_{T_t} = t \wedge A_\infty$, $t \geq 0$. Furthermore, A_t is an $\mathbb{F}^S$-stopping time for every $t \geq 0$. We define the time changed process $(X, \mathbb{F})$ by

$$X_t = \begin{cases} x_0 + S_{A_t} & \text{if } t < T_\infty, \\ \Delta & \text{otherwise} \end{cases}$$

and $\mathcal{F}_t = \mathcal{G}_{A_t}$ for every $t \geq 0$. Then $(X, \mathbb{F})$ is a process with paths in D^e. We will show that $(X, \mathbb{F})$ is a solution to Eq. (1.1). Let N_b be defined by (5.11). As mentioned above, the condition $(\mathbf{E}(x_0))$ ensures $\lambda(N_b) = 0$. Consequently, $x_0 + S_t$ has occupation time zero in N_b $\mathbb{P}$-a.s. By time change in the integral (see, e.g., [5], Lemma 1.6), we now obtain

$$\begin{aligned} A_t &= \int_0^{A_t} \mathbf{1}_{N_b^c}(x_0 + S_u)\, du = \int_0^{A_t} |b|^\alpha(T_u, x_0 + S_u)\, |b|^{-\alpha}(T_u, x_0 + S_u)\, du \\ &= \int_0^{A_t} |b|^\alpha(T_u, x_0 + S_u)\, dT_u = \int_0^{T_{A_t}} |b|^\alpha(u, x_0 + S_{A_u})\, du \\ &= \int_0^{t \wedge T_\infty} |b|^\alpha(u, X_u)\, du \end{aligned}$$

and, consequently,

$$A_t = \int_0^{t \wedge T_\infty} |b|^\alpha(u, X_u)\, du \quad \text{for all } t \geq 0 \quad \mathbb{P}\text{-a.s.} \tag{5.14}$$

Furthermore, Proposition 4.3 (i), (ii) implies that the stochastic integral

$$Z_t = \int_0^{A_t} \mathbf{1}_{N_b^c} b^{-1}(T_u, x_0 + S_{u-})\, dS_u, \quad t \geq 0,$$

is well-defined and $(Z, \mathbb{F})$ is a symmetric α-stable process stopped at $U_\infty = T_\infty$. On the other side, changing the time in the stochastic integral with respect to a semimartingale (see, e.g., [8], Theorem 10.21) yields

$$\begin{aligned} Z_t &= \int_0^{A_t} \mathbf{1}_{N_b^c} b^{-1}(T_u, x_0 + S_{u-})\, dS_u = \int_0^t \mathbf{1}_{N_b^c} b^{-1}(T_{A_u}, x_0 + S_{A_u-})\, dS_{A_u} \\ &= \int_0^t \mathbf{1}_{N_b^c} b^{-1}(u, X_u^-)\, dX_u, \quad t < T_\infty, \quad \mathbb{P}\text{-a.s.} \end{aligned} \tag{5.15}$$

From (5.14)

$$\int_0^t |b|^\alpha(u, X_u^-)\, du = \int_0^t |b|^\alpha(u, X_u)\, du = A_t < +\infty, \quad t < T_\infty,$$

and by Proposition 4.3 (iii) (changing the roles of A and T) there exists the stochastic integral $\int_0^t b(u, X_u^-) \, dZ_u$. Because of (5.15) we conclude

$$\int_0^t b(u, X_u^-) \, dZ_u = \int_0^t b(u, X_u^-) b^{-1}(u, X_u^-) \, dX_u$$

$$= \int_0^t 1_{N_b^c}(X_u^-) \, dX_u = X_t - x_0, \quad t < T_\infty, \quad \text{P-a.s.} \quad (5.16)$$

where we have used that $\int_0^t 1_{N_b}(X_u^-) \, dX_u = 0$, $t < T_\infty$, P-a.s. Indeed, again using Theorem 10.21 of [8] we have

$$\int_0^t 1_{N_b}(X_u^-) \, dX_u = \int_0^{A_t} 1_{N_b}(x_0 + S_{u-}) \, dS_u, \quad t < T_\infty, \quad \text{P-a.s.}$$

and it is sufficient to verify that

$$\int_0^t 1_{N_b}(x_0 + S_{u-}) \, dS_u = 0, \quad t \geq 0, \quad \text{P-a.s.}$$

But this follows from Proposition 4.3 (iv) and $\mathbb{E} \int_0^t 1_{N_b}(x_0 + S_{u-}) \, du = 0$: We have $\lambda(N_b) = 0$ and hence $x_0 + S_{u-}$ has no occupation time in N_b. This proves the claim. Enlarging the probability space, we can assume that Z is extended to a full symmetric α-stable process (not only stopped at T_∞) and, by (5.16), $(X, \mathbb{F})$ is a solution of Eq. (1.1) with explosion time T_∞. The construction of $(X, \mathbb{F})$ and Lemma 5.2 (iii) ensure that the solution is basic. Finally, (5.13) yields that $(X, \mathbb{F})$ is nonabsorbing. $\square$

Theorem 5.4 *Suppose that b satisfies the following condition:*

(D) *For every $N \geq 1$,*

    (a) *if $1 \leq \alpha \leq 2$, then $\lambda(\{\bar{b}_N < \infty\}) > 0$,*

    (b) *if $0 < \alpha < 1$, then $\exists c_N > 0 : \mu_0(\{|x| \geq 1, \bar{b}_N(x) > c_N|x|\}) < +\infty$ where μ_0 is defined by (2.6).*

Then every solution $(X, \mathbb{F})$ to Eq. (1.1) does not explode.

Proof. Let $(X, \mathbb{F})$ be a solution of Eq. (1.1). We define $\tilde{A}$ by (5.2) and $\tilde{T}$, the right inverse of the increasing process $\tilde{A}$, by (5.3). By (5.6) we know that $(S, \mathbb{G})$ with

$$S_t = X_{\tilde{T}_t} - x_0 = \int_0^{\tilde{T}_t} b(u, X_u^-) \, dZ_u, \quad \mathbb{G}_t = \mathcal{F}_{\tilde{T}_t}, \quad t \geq 0,$$

is a symmetric α-stable process stopped at $\tilde{A}_{T_\infty-}$. By enlarging the probability space, we can assume that $(S, \mathbb{G})$ is extended to a full symmetric α-stable process. The assumptions on b imply that $h = |b|^{-\alpha}(\cdot, x_0 + \cdot)$ satisfies the condition **(C)** of Theorem 3.2. In view of Lemma 4.3 (i), $\tilde{T}$ is a supersolution of Eq. (3.1) for $h = |b|^{-\alpha}(\cdot, x_0 + \cdot)$ and S and Theorem 3.2 now ensures that $\tilde{T}_\infty = +\infty$ P-a.s., hence $\tilde{A}_t < +\infty$ P-a.s. for all $t \geq 0$ which means that $(X, \mathbb{F})$ is nonexploding. $\square$

We note that the condition $(E(x_0))$ implies the condition **(D)(a)**. Combining Theorem 5.3 and Theorem 5.4 we therefore get the following

Theorem 5.5 *Suppose that b satisfies the condition* $(\mathbf{E}(x_0))$. *If* $0 < \alpha < 1$, *additionally we assume that, for every* $N \geq 1$,

$$\exists c_N > 0 : \quad \mu_0(\{x \in \mathbb{R} : |x| \geq 1, \ \bar{b}_N(x) > c_N|x|\}) < +\infty. \tag{5.17}$$

Then there exists a nonexploding and nonabsorbing basic solution $(X, \mathbb{F})$ *of Eq.* (1.1) *with* $X_0 = x_0$.

We recall that, in case of $0 < \alpha < 1$, for the nonexplosion condition (5.17) each of the following conditions is sufficient:

$$\exists c_N > 0 : \quad \lambda(\{x \in \mathbb{R} : \bar{b}_N(x) > c_N|x|\}) < +\infty. \tag{5.18}$$

$$\bar{b}_N(x) = O(|x|), \quad |x| \to \infty, \quad \lambda\text{-a.e.} \tag{5.19}$$

Hence, if b satisfies the condition $(\mathbf{E}(x_0))$ and (5.18) or (5.19) then the statement of Theorem (5.5) is in force. The following result shows that, in a certain sense, these conditions cannot be improved: Roughly speaking, if $|b|$ is growing faster than $|x|$ explosion does occur $\mathbb{P}$-a.s.

Theorem 5.6 *Let* $0 < \alpha < 1$ *and* $(X, \mathbb{F})$ *be a nonabsorbing basic solution of Eq.* (1.1). *Suppose that* $\underline{b} := \inf_{N \geq 0} \underline{b}_N$ *is such that* $\underline{b}^{-\alpha}(x_0 + \cdot)$ *satisfies condition* (2.5) *and there exists* $\delta > 0$ *with*

$$|x|^{1+\delta} = O(\underline{b}(x)), \quad |x| \to \infty.$$

Then X is exploding $\mathbb{P}$-a.s.

Proof. The solution $(X, \mathbb{F})$ is nonabsorbing, hence $\tilde{U}_\infty = \tilde{T}_\infty$ where $\tilde{U}_\infty$ and $\tilde{T}_\infty$ are defined in (5.5) and (5.3). In view of Lemma 5.2 (ii) we have

$$\tilde{T}_\infty = \int_0^\infty |b|^{-\alpha}(\tilde{T}_u, x_0 + S_u) \, du \leq \int_0^\infty \underline{b}^{-\alpha}(x_0 + S_u) \, du \quad \mathbb{P}\text{-a.s.}$$

Now $\underline{b}^{-\alpha}(x_0 + \cdot)$ satisfies the conditions of Proposition 2.13 and thus the right member is finite $\mathbb{P}$-a.s. Since $\tilde{T}_\infty$ is the explosion time of X the proof is completed. □

Remark 5.7 In the case $\alpha = 2$, A. Rozkosz and L. Słomiński [15] and T. Senf [17], [18] proved existence of solutions for only measurable diffusion coefficients b. Having this in mind, one should think that condition (E_1) stating the continuity of b in t for a.a. x is too stringent. However, the proofs in [15] – [18] are essentially based on Krylov estimates (as also existence results of Krylov himself), a tool which is not available for general $0 < \alpha < 2$. Nevertheless, the existence condition $(\mathbf{E}(x_0))$ is of interest for $\alpha = 2$, too. (Of course, in this case $(\mathbf{E}(x_0))$ does not depend on x_0.) In comparison with the condition of T. Senf [18] who only assumed that b^2 and b^{-2} are locally integrable functions the condition $(\mathbf{E}(x_0))$ has two advantages: Firstly, it guarantess the existence of a *nonexploding* solution. Secondly, $(\mathbf{E}(x_0))$ does not include the local integrability of b^2 at all. We notice that a slightly weaker version of Theorem 5.5 for $\alpha = 2$ is already due to the dissertation of T. Senf [17].

6.   THE HOMOGENEOUS CASE

In this section we assume that the diffusion coefficient does not depend on the time and hence is a Borel (or only Lebesgue) measurable function $b : \mathbb{R} \to \mathbb{R}$. This case was extensively studied by P.A. Zanzotto [20] – [22] not only for symmetric but also for skew strictly stable processes where in the latter case b is assumed to be nonnegative. It should be noted that the approach of the present paper also works for skew strictly stable driving processes and nonnegative b. This is based on the fact that Proposition 4.3 remains valid for nonnegative H in this case. For the sake of simplicity, however, we concentrated on the symmetric case.

In comparison with P.A. Zanzotto [20] – [22] we only consider global solutions (and not local solutions). A local solution is only defined until leaving some (small) interval around the starting point. A global solution (which is defined for all $t \geq 0$) is allowed to explode. An exploding solution is in this sense not local: It is a maximal solution leaving every compact subset of $\mathbb{R}$ in finite time for those paths for which explosion occurs. The phenomenon of explosions for homogeneous coefficients b only occurs for parameters $0 < \alpha < 1$ and it was not studied previously.

First we give a reformulation of the results of the last section.

Theorem 6.1 *Let $0 < \alpha \leq 2$. If $1 < \alpha \leq 2$, suppose that the function $f = |b|^{-\alpha}$ is locally integrable. If $0 < \alpha \leq 1$, we assume that $E(f)$ is x_0-polar and, additionally:*

  (a) *If $\alpha = 1$, then $\int_{0-}^{0+} |\ln|y|| \, |b|^{-1}(x_0 + y) \, dy < +\infty$.*

  (b) *If $0 < \alpha < 1$, then $\int_{0-}^{0+} |y|^{\alpha-1} \, |b|^{-\alpha}(x_0 + y) < +\infty$.*

We then have:

  (i)   *There exists a nonabsorbing basic solution $(X, \mathbb{F})$ with $X_0 = x_0$.*

  (ii) *If $1 \leq \alpha \leq 2$, the solution is nonexploding.*

  (iii) *If $0 < \alpha < 1$ and $\exists c > 0$: $\mu_0(\{x \in \mathbb{R} : |x| \geq 1, |b|(x) > c|x|\}) < +\infty$ where μ_0 is defined by (2.6) then the solution is nonexploding.*

  (iv) *If $0 < \alpha < 1$ and there exists $\delta > 0$ such that $|x|^{1+\delta} = O(|b|(x))$, $|x| \to \infty$, then the solution is exploding $\mathbb{P}$-a.s.*

The notion of a nonabsorbing basic solution (cf. (5.1)) is new. Clearly, every nonabsorbing basic solution is nontrivial but not conversely. (A trivial solution X starting at x_0 satisfies $X_t = x_0$ for all $t \geq 0$ $\mathbb{P}$-a.s.) Under the condition of Theorem 6.1, the existence of a nontrivial (nonexploding) solution was proven in [5] for $\alpha = 2$ and by P.A. Zanzotto [20] for $1 < \alpha < 2$. In the case $\alpha = 1$, P.A. Zanzotto [21] proved the existence of a nontrivial (nonexploding) solution under the stronger condition $H(x_0)$ (cf. [20] – [22]). In the case $0 < \alpha < 1$, in [21] he proved the existence of a nontrivial and nonexploding solution under the more restrictive conditions $H(x_0)$ and (2.10) for $f = |b|^{-\alpha}$. The latter means that, outside of some set of finite Lebesgue measure, b is bounded. The nonexplosion condition (iii) and the explosion condition (iv) in Theorem 6.1 are new. They seem to be sharp as the results of Section 2 suggest (also see Example 6.8 below). For example, a sufficient condition for nonexplosion is that, outside of some set of finite Lebesgue measure, $b(x) = O(|x|)$ as $|x| \to \infty$ and a sufficient condition for explosion is that b is growing faster than $|x|^{1+\delta}$ as $|x| \to \infty$ for some $\delta > 0$.

We recall that the assumption in Theorem 5.1 that $E(f)$ is x_0-polar for $f = |b|^{-\alpha}$ is satisfied if, except for a, possibly, denumerable set of isolated points, $|b|^{-\alpha}$ is integrable over a sufficiently small neighbourhood of $x \in \mathbb{R}$. Thus, if $0 < \alpha \le 1$, $|b|^{-\alpha}$ is allowed to have (at least) a denumerable set of singularities of arbitrary order consisting of isolated points distinct from x_0. However, the condition $H(x_0)$ employed by P.A. Zanzottto [20] – [22] excludes such singularities. Also, for $0 < \alpha \le 1$ the condition that $|b|^{-\delta}$ is locally integrabble for some $\delta > 1$ (cf. [20] for $0 < \alpha < 1$, [21], Proposition (4.13), for $0 < \alpha \le 1$) is much more restrictive than the conditions used in Theorem 6.1 for every $x_0 \in \mathbb{R}$.

The next theorem states the uniqueness in law of the nonabsorbing basic solution and is new. The proof is essentially the same as the uniqueness proof of [6], Theorem 2.

Theorem 6.2 *Suppose that $b : \mathbb{R} \to \mathbb{R}$ is measurable and $x_0 \in \mathbb{R}$. The nonabsorbing basic solution $(X, \mathbb{F})$ of Eq. (1.1) satisfying $X_0 = x_0$ is unique in law.*

Proof. Let $\tilde{A}$, $\tilde{T}$ and $\tilde{U}_\infty$ be defined by (5.2), (5.3) and (5.5). By Lemma 5.2 (ii)

$$\tilde{T}_t = \int_0^t |b|^{-\alpha}(x_0 + S_u)\, du \quad \text{for every} \quad t \ge 0 \quad \mathbb{P}\text{-a.s.} \tag{6.1}$$

where $(S, \mathbb{G})$ is the symmetric α-stable process defined by (5.6). From (6.1) we now see that $\tilde{T}$ and hence $\tilde{A}$ is a well-defined $\mathcal{G}_\infty^S$-measurable functional. Hence the distribution of the pair $(x_0 + S, \tilde{A})$ is uniquely determined. Recalling the representation (5.7) for X, we conclude that the distribution of X on the path space $(D^e, \mathcal{D}^e)$ is unique. □

Next we will describe the situation in the case $1 < \alpha \le 2$ and present necessary and sufficient conditions for existence and uniqueness. Since in this case explosions do not occur, from now on $(X, \mathbb{F})$ always signifies a nonexploding solution of Eq. (1.1). We use the notation $N := N_b = \{x \in \mathbb{R} : b(x) = 0\}$ and $E := E(|b|^{-\alpha})$ (cf. (2.2)).

The following lemma shows that the set E is always absorbing. By $D_E(X)$, we will denote the first entry time of X into E.

Lemma 6.3 *For every solution $(X, \mathbb{F})$, we have*

$$X_t = X_{t \wedge D_E(X)}, \quad t \ge 0, \quad \mathbb{P}\text{-a.s.}$$

Proof. Let $(S, \mathbb{G})$ be the symmetric α-stable process defined by (5.6). In view of Lemma 5.2 (i) and Proposition 2.1 (ii), we get $\tilde{T}_{D_{E-x_0}} = +\infty$ $\mathbb{P}$-a.s. Using the relation $\tilde{A}_{D_E(X)} = D_{F-x_0}$ $\mathbb{P}$-a.s., from this it follows

$$\tilde{A}_\infty \le D_{E-x_0} = \tilde{A}_{D_E(X)} \quad \mathbb{P}\text{-a.s.}$$

and, consequently, $\tilde{A}_\infty = \tilde{A}_{D_E(X)}$ $\mathbb{P}$-a.s. This implies the assertion. □

Definition 6.4 A solution $(X, \mathbb{F})$ is called a *fundamental solution* if

$$\int_0^\infty \mathbf{1}_{N \cap E^c}(X_u)\, du = 0 \quad \mathbb{P}\text{-a.s.}$$

Lemma 6.5 *Let* $(X, \mathbb{F})$ *be a solution. Then* $(X, \mathbb{F})$ *is a fundamental solution if and only if the following conditions are satisfied:*

(i) $(X, \mathbb{F})$ *is a basic solution.*

(ii) $\tilde{U}_\infty = D_E(X)$.

Proof. It is clear that the conditions (i) and (ii) are sufficient. Conversely, if $(X, \mathbb{F})$ is a fundamental solution then, obviously,

$$\int_0^{D_E(X)} \mathbf{1}_N(X_u)\, du = 0 \quad \mathbb{P}\text{-a.s.}$$

and, in particular, $D_E(X) \le \tilde{U}_\infty$ $\mathbb{P}$-a.s. From Lemma 6.3 we observe the inequality $\tilde{U}_\infty \le D_E(X)$ and hence the conditions (i) and (ii) are satisfied. $\qquad\square$

In the following theorem, the equivalence of (i) and (iii) was also shown by P.A. Zanzotto in [21], Theorem (3.5). For our statement, we exploit the notion of a fundamental solution. The proof of the implication (i)$\Rightarrow$(ii) is almost the same as for Theorem 5.3, with only slight modifications. Our proof completely unifies the case $1 < \alpha < 2$ on one side and the case $\alpha = 2$ on the other side and is very close to that given by the first author and W. Schmidt in [6].

Theorem 6.6 *Suppose that* $1 < \alpha \le 2$ *and let* $b : \mathbb{R} \to \mathbb{R}$ *a measurable function. Then the following conditions are equivalent:*

(i) $E \subseteq N$.

(ii) *For every* $x_0 \in \mathbb{R}$, *there exists a fundamental solution* $(X, \mathbb{F})$ *of Eq.* (1.1) *with* $X_0 = x_0$.

(iii) *For every* $x_0 \in \mathbb{R}$, *there exists a solution* $(X, \mathbb{F})$ *of Eq.*(1.1) *with* $X_0 = x_0$.

Proof. For proving (i)$\Rightarrow$(ii), as in the proof of Theorem 5.3, let $(S, \mathbb{G})$ be a symmetric α-stable process and we put

$$T_t = \int_0^{t+} |b|^{-\alpha}(x_0 + S_u)\, du, \quad t \ge 0.$$

Let A be the right inverse of T. Then T is strictly increasing and continuous on $[0, A_\infty)$. The $\mathbb{G}^S$-time change A is finite, strictly increasing on $[0, T_{A_\infty -})$ and continuous $\mathbb{P}$-a.s. By Proposition 2.1, $A_\infty = D_{E-x_0}$ $\mathbb{P}$-a.s. In particular, we have $T_\infty = +\infty$ $\mathbb{P}$-a.s. Precisely as in the proof of Theorem 5.3, we now define $(X, \mathbb{F})$ and the symmetric α-stable process $(Z, \mathbb{F})$ stopped at $U_\infty := T_{A_\infty -}$ by

$$X_t = x_0 + S_{A_t}, \quad t \ge 0,$$

and

$$Z_t := \int_0^{A_t} \mathbf{1}_{N^c} b^{-1}(x_0 + S_{u-}) \, dS_u = \int_0^t \mathbf{1}_{N^c} b^{-1}(X_{u-}) \, dX_u, \quad t \geq 0.$$

Since $A_\infty = D_{E-x_0}$ and A is finite and continuous we can conclude

$$\begin{aligned} U_\infty &= \inf\{t \geq 0 : A_t = A_\infty\} = \inf\{t \geq 0 : A_t = D_{E-x_0}\} \\ &= \inf\{t \geq 0 : x_0 + S_{A_t} \in E\} = \inf\{t \geq 0 : X_t \in E\} = D_E(X). \end{aligned}$$

Using this equality, now we get

$$A_t = \int_0^t |b|^\alpha(X_u) \, du, \quad t \geq 0.$$

The proof is similar to that of (5.14) (and exactly the same as in [6], Theorem 1) and therefore omitted. By Proposition 4.3 (iii) (changing the roles of A and T), the stochastic integrals

$$\int_0^t b(X_{u-}) \, dZ_u, \quad t < T_\infty = +\infty,$$

exist, and we may conclude

$$\int_0^t b(X_{u-}) \, dZ_u = \int_0^t \mathbf{1}_{N^c}(X_{u-}) \, dX_u, \quad t \geq 0,$$

or, equivalently,

$$\int_0^{t \wedge U_\infty} b(X_{u-}) \, dZ_u = \int_0^{t \wedge U_\infty} \mathbf{1}_{N^c}(X_{u-}) \, dX_u, \quad t \geq 0.$$

Enlarging the probability space, without loss of generality we extend $(Z, \mathbb{F})$ to a full symmetric α-stable process again denoted by $(Z, \mathbb{F})$. Since $b(x) = 0$ for $x \in E$, we can write

$$\int_0^t b(X_{u-}) \, dZ_u = \int_0^{t \wedge U_\infty} \mathbf{1}_{N^c \cap E^c}(X_{u-}) \, dX_u, \quad t \geq 0.$$

From the definition of E, $\lambda(N \cap E^c) = 0$ and, as in the proof of Theorem 5.3,

$$\int_0^t \mathbf{1}_{N \cap E^c}(X_{u-}) \, dX_u = 0, \quad t \geq 0, \quad \mathbb{P}\text{-a.s.}$$

This yields

$$\int_0^t b(X_{u-}) \, dZ_u = \int_0^{t \wedge U_\infty} \mathbf{1}_{E^c}(X_{u-}) \, dX_u = \int_0^{t \wedge U_\infty} dX_u = X_{t \wedge U_\infty} - X_0 = X_t - x_0$$

for all $t \geq 0$. It remains to show that $(X, \mathbb{F})$ is fundamental:

$$\int_0^\infty \mathbf{1}_{N \cap E^c}(X_u) \, du = \int_0^\infty \mathbf{1}_{N \cap E^c}(x_0 + S_u) \, dT_u$$

$$= \int_0^\infty |b|^{-\alpha}(x_0 + S_u) \mathbf{1}_{N \cap E^c}(x_0 + S_u)\, du = 0 \quad \mathbb{P}\text{-a.s.}$$

since $\lambda(N \cap E^c) = 0$ and hence $x_0 + S$ has occupation time zero in $N \cap E^c$. The implication (ii)$\Rightarrow$(iii) is trivial. To verify (iii)$\Rightarrow$(i), let $x_0 \in E$. We consider a solution $(X, \mathbb{F})$ of Eq. (1.1) such that $X_0 = x_0$. Using Lemma 5.2 (i) we get

$$\tilde{T}_t \geq \int_0^{t+} |b|^{-\alpha}(x_0 + S_u)\, du, \quad t \geq 0,$$

where $\tilde{T}$ and S are defined by (5.3) and (5.6). From Proposition 2.1, the right hand side is equal to $+\infty$ $\mathbb{P}$-a.s., hence $\tilde{T}_t = +\infty$ $\mathbb{P}$-a.s. for all $t \geq 0$ which implies that for the right inverse $\tilde{A}$,

$$\tilde{A}_t = \int_0^{t+} |b|^\alpha(X_u)\, du, \quad t \geq 0,$$

(see (5.2)) we have $\tilde{A}_t = 0$ $\mathbb{P}$-a.s. for all $t \geq 0$. This is only possible if $X_t = x_0$ $\mathbb{P}$-a.s. for every $t \geq 0$ and hence $b(x_0) = 0$. This proves $E \subseteq N$. $\qquad\square$

Finally, we come to the uniqueness in law for $1 < \alpha \leq 2$. In the next theorem, part (ii) is due to the first author and W. Schmidt [4] (Theorem 2) for $\alpha = 2$ and to P.A. Zanzotto [21] (Theorem (3.21)) for $1 < \alpha < 2$. The proof is exactly the same as in the Brownian case (cf. [6]) and will only be given for the sake of self-containedness. For $\alpha = 2$, part (i) can be found in [5] (Theorem (5.4)) in the case $E = \emptyset$ or in [7] (Theorem (4.22)) in a slightly different context.

Theorem 6.7 *Suppose that $1 < \alpha \leq 2$ and let $b : \mathbb{R} \to \mathbb{R}$ be a measurable function. We then have:*

(i) *The fundamental solution $(X, \mathbb{F})$ of Eq. (1.1) with $X_0 = x_0$, if it exists, is unique in law.*

(ii) *Suppose that $E \subseteq N$. Then, for every $x_0 \in \mathbb{R}$, the solution $(X, \mathbb{F})$ with $X_0 = x_0$ is unique in law if and only if $E = N$.*

Proof. For proving (i), let $(X, \mathbb{F})$ be a fundamental solution and $\tilde{T}$ and $(S, \mathbb{G})$ be defined by (5.3) and (5.6). First we show

$$\tilde{T}_t = \int_0^{t+} |b|^{-\alpha}(x_0 + S_u)\, du \quad \text{for every} \quad t \geq 0 \quad \mathbb{P}\text{-a.s.} \tag{6.2}$$

In view of Lemma 5.2 (i) and Proposition 2.1 this relation holds on $\{t \geq D_{E-x_0}\}$. Furthermore, by Lemma 6.5 we have $D_E(X) = \tilde{U}_\infty$ $\mathbb{P}$-a.s. Using this and Lemma 5.2 (i) and (ii), $\mathbb{P}$-a.s. on $\{D_E(X) = +\infty\}$ we observe

$$\tilde{T}_t = \int_0^{t \wedge \tilde{A}_\infty} |b|^{-\alpha}(x_0 + S_u)\, du \leq \int_0^t |b|^{-\alpha}(x_0 + S_u)\, du \leq \tilde{T}_t$$

which proves (6.2) on this set. Finally, since $\tilde{A}$ is continuous, on $\{D_E(X) < \infty\}$

$$\tilde{A}_{D_E(X)} = \inf\{\tilde{A}_t : X_t \in E\} = \inf\{\tilde{A}_t : x_0 + S_{\tilde{A}_t} \in E\}$$

$$=\inf\{0 \le u \le \tilde{A}_\infty : x_0 + S_u \in E\} = \inf\{u \ge 0 : x_0 + S_u \in E\} = D_{E-x_0}$$

and hence $D_{E-x_0} \le \tilde{A}_\infty$ on this set. This implies

$$\{D_E(X) < +\infty, t < D_{E-x_0}\} \subseteq \{t < \tilde{A}_\infty\}.$$

Once again using Lemma 5.2 (ii) and the identity $\{t < \tilde{A}_\infty\} = \{\tilde{T}_t < \tilde{U}_\infty\}$ we conclude that (6.2) is also true on the set $\{D_E(X) < +\infty, t < D_{E-x_0}\}$. This completes the proof of (6.2). Now (6.2) shows that $\tilde{T}$ is a $\mathcal{G}_\infty^S$-measurable functional. Exactly as in the proof of Theorem 6.2, from this the uniqueness in law of $(X, \mathbb{F})$ with given initial value x_0 is established. Finally, we prove (ii). If $E = N$ then every solution is a fundamental solution. Using part (i) we see that, for every $x_0 \in \mathbb{R}$, the solution with initial value x_0 is unique in law. Conversely, suppose $E \subseteq N$ and $x_0 \in E^c \cap N$. Then there is a trivial solution X with $X_t = x_0$ for all $t \ge 0$ and there is, by Theorem 6.6, a fundamental solution X with $X_0 = x_0$. These solutions have different laws and hence the uniqueness fails. □

In conclusion, we give the following instructive example.

Example 6.8 We consider the power function b with

$$b(x) = |x|^\gamma, \quad x \in \mathbb{R}, \quad x \ne 0, \quad \gamma \in \mathbb{R}.$$

Case 1 $< \alpha \le 2$, $b(0) = 0$: From Theorem 6.6 we get:

1) *For any $x_0 \in \mathbb{R}$ there exists a fundamental solution X starting at x_0. Stopping X at $D_{\{0\}}(X)$ yields also a solution. In particular, there is a trivial solution starting at 0. For every solution, explosion does not occur.*

The uniqueness statement of Theorem 6.7 can be sharpened.

2) *Let $x_0 \in \mathbb{R}$ be fixed. The solution starting at x_0 is unique if and only if $\alpha^{-1} \le \gamma$. In this case, the trivial solution is the (pathwise) unique solution starting at 0.*

For proving this, first we assume $\alpha^{-1} \le \gamma$. Then $E = N = \{0\}$ and the uniqueness of the solution starting at x_0 follows from Theorem 6.7. Conversely, let $\gamma < \alpha^{-1}$. Now let X be the fundamental solution starting at x_0. As the next property **3)** shows, X reaches 0 $\mathbb{P}$-a.s. Stopping X at $D_{\{0\}}(X)$ we obtain a second solution with a different law. Hence uniqueness in law fails.

3) *If $\gamma < \alpha^{-1}$, then every solution X reaches 0 $\mathbb{P}$-a.s.*

Indeed, in view of (5.7) we have the representation

$$X_t = x_0 + S_{\tilde{A}_t} \quad \text{for all} \quad t < \tilde{T}_\infty = +\infty, \tag{6.3}$$

where S is a symmetric α-stable process and $\tilde{A}$ is defined by (5.2). Now, as in the proof of Lemma 5.2, on $\{D_{\{0\}}(X) = +\infty\}$

$$\tilde{T}_t = \int_0^t |b|^{-\alpha}(x_0 + S_u) \, du \quad \text{for every} \quad t \ge 0 \quad \mathbb{P}\text{-a.s.}$$

and, using Corollary 2.2, we observe that $\tilde{T}_t$ is finite for every $t \ge 0$, hence $\tilde{A}_\infty = +\infty$ $\mathbb{P}$-a.s. on this set. Consequently, since $x_0 + S$ reaches 0 in finite time $\mathbb{P}$-a.s. (cf. proof of

Proposition 2.1) and $\tilde{A}$ is finite and continuous, (6.3) yields that this is also true for X on $\{D_{\{0\}}(X) = +\infty\}$ which is, however, only possible if $\mathbb{P}(\{D_{\{0\}}(X) = +\infty\}) = 0$.
This result can be made more precise in the following way:

4) *For any solution starting at $x_0 \neq 0$, the point 0 will be reached with probability 1 if $\gamma < 1$.*
Actually, 0 will be reached with probability 1 if (say, $x_0 > 0$)

$$\int_0^{x_0} |x|^{\alpha-1} |b(x)|^{-\alpha} \, dx < +\infty$$

from which **4)** derives. For $\alpha = 2$, see S. Assing and T. Senf [1]. For the general case $1 < \alpha \leq 2$, a proof will be published elsewhere.

Case 1 $< \alpha \leq 2, \mathbf{b(0)} \neq \mathbf{0}$: There is an essential difference to the case $b(0) = 0$ treated above: The trivial solution starting at 0 is now excluded. Therefore the following property is true:

5) *There exists a solution starting at 0 if and only if $\gamma < \alpha^{-1}$.*
Now let $\tilde{b} = \mathbf{1}_{\{b \neq 0\}} \cdot b$ and consider a fundamental solution X to Eq. (1.1) for $\tilde{b}$ starting at $x_0 \in \mathbb{R}$. If $\gamma < \alpha^{-1}$ and $x_0 \neq 0$ then X reaches 0 $\mathbb{P}$-a.s. (cf. **3)**) but has no occupation time in 0 $\mathbb{P}$-a.s. Hence X is also a solution to Eq. (1.1) for b. This verifies:

6) *There exists a solution starting at $x_0 \neq 0$ if $\gamma < \alpha^{-1}$.*
If $\gamma \in [\alpha^{-1}, 1)$, then every solution X starting at $x_0 \neq 0$ would reach 0 $\mathbb{P}$-a.s. (cf. **4)**) but it cannot be continued after $D_{\{0\}}(X)$, which implies:

7) *There exists no solution starting at $x_0 \neq 0$ if $\gamma \in [\alpha^{-1}, 1)$.*
The case $\gamma \geq 1$ and $x_0 \neq 0$ remains open. The question is whether the fundamental solution X to Eq. (1.1) corresponding to $\tilde{b}$ does not reach 0 $\mathbb{P}$-a.s. or not. In the first case, there is a solution to Eq. (1.1) also for b (indeed, one can take X as solution), otherwise not.

8) *Every solution is fundamental and hence unique in law.*

Case 0 $< \alpha \leq 1, \mathbf{b(0)} = \mathbf{0}$:

9) *For any starting point $x_0 \in \mathbb{R}$, there exists a solution.*
For $x_0 \neq 0$ this follows from Theorem 6.1, for $x_0 = 0$ there is at least the trivial solution.

10) *The trivial solution starting at 0 is the (even pathwise) unique solution if and only if $1 \leq \gamma$.*

11) *There is a second, nonabsorbing and basic solution starting at 0 if and only if $\gamma < 1$. This solution is unique in law.*
Indeed, Lemma 5.2 (i) and Example 2.9 show that the trivial solution is the unique solution starting at 0 if $1 \leq \gamma$. On the other side, from Theorem 6.1 we observe that there is a nonabsorbing and basic solution starting at 0 if $\gamma < 1$. The uniqueness statement in **11)** follows from Theorem 6.2.
Now we consider the behaviour of a solution X starting at $x_0 \neq 0$. Since 0 is x_0-polar for the symmetric α-stable process S $(0 < \alpha \leq 1)$, from (6.3) we see that X does

not reach 0 $\mathbb{P}$-a.s. But $b(x) \neq 0$ for all $x \neq 0$ and, consequently, X is nonabsorbing and basic. Property **9)** and Theorem 6.2 now show:

12) *There is a unique solution starting at $x_0 \neq 0$. This solution is nonabsorbing and basic and it does not reach 0 $\mathbb{P}$-a.s.*

This completely clarifies existence and uniqueness of solutions in the case $0 < \alpha \leq 1$ and $b(0) = 0$. What about explosion or nonexplosion?

13) *If $\alpha = 1$, explosion does not occur $\mathbb{P}$-a.s.*

This follows from Theorem 6.1 together with the above uniqueness conclusions **11)** and **12)**. Now we assume $0 < \alpha < 1$. If $\gamma \leq 1$ then $b(x) = O(|x|)$, $|x| \to \infty$, and by Theorem 6.1 (iii) (together with **11)** and **12)**) every solution is nonexploding. Furthermore, if $1 < \gamma$, then $|x|^{1+\delta} = O(|b(x)|)$, $|x| \to \infty$, for $0 < \delta \leq 1 - \gamma$ and by Theorem 6.1 (iv) the solution starting at $x_0 \neq 0$ is exploding $\mathbb{P}$-a.s. Also, if $1 < \gamma$, the solution starting at 0 is trivial (cf. **10)**). This gives:

14) *If $0 < \alpha < 1$, the solution is exploding $\mathbb{P}$-a.s. if and only if $x_0 \neq 0$ and $1 < \gamma$. Otherwise the solution is nonexploding $\mathbb{P}$-a.s.*

Case $0 < \alpha \leq 1$, $b(0) \neq 0$: The results are similar, with only one essential difference: The trivial solution starting at 0 is now excluded. Consequently, we can state:

15) *There is a solution starting at 0 if and only if $\gamma < 1$. In this case, every solution is nonexploding and fundamental. Hence the solution is unique.*

16) *There is a unique solution starting at $x_0 \neq 0$. This solution does not reach 0 $\mathbb{P}$-a.s. Moreover, the solution is exploding $\mathbb{P}$-a.s. if $0 < \alpha < 1$ and $1 < \gamma$. Otherwise the solution is nonexploding $\mathbb{P}$-a.s.*

REFERENCES

[1] S. Assing, T. Senf, On stochastic differential equations without drift, *Stochastics Stochastics Rep.* **36** (1991), 21–39.

[2] R. M. Blumenthal and R. K. Getoor, *Markov processes and potential theory*, Academic Press, New York, 1968.

[3] E. S. Boylan, Local times for a class of Markov processes, *Illinois J. Math.* **8** (1964), 19–39.

[4] H. J. Engelbert and W. Schmidt, On the behaviour of certain functionals of the Wiener process and applications to stochastic differential equations, *Lecture Notes in Control and Information Sciences* **36**, 47–55, Springer-Verlag, Berlin, 1981.

[5] H. J. Engelbert and W. Schmidt, On solutions of one-dimensional stochastic differential equations without drift, *Z. Wahrscheinlichkeitstheorie verw. Gebiete* **68** (1985), 287–314.

[6] H. J. Engelbert and W. Schmidt, On one-dimensional stochastic differential equations with generalized drift, *Lecture Notes in Control and Information Sciences* **69**, 143–155, Springer-Verlag, Berlin, 1985.

[7] H. J. Engelbert and W. Schmidt, Strong Markov continuous local martingales and solutions of one-dimensional stochastic differential equations, III, *Math. Nachr.* **151** (1991), 149–197.

[8] J. Jacod, *Calcul stochastique et problèmes de martingales*, Lecture Notes in Math. **714**, Springer-Verlag, Berlin, 1979.

[9] H. Kesten, Hitting probabilities of single points for processes with stationary, independent increments, *Mem. Amer. Math. Soc.* **93**, 1–129, Providence RI.

[10] V. P. Kurenok, On the existence of solutions of one-dimensional stochastic differential equations, *Izveztija of Akad. of Sci. of Belarus* N **4** (1989), 38–43 (in Russian).

[11] R. S. Liptser and A. N. Shiryaev, *Theory of martingales*, Nauka, Moscow, 1986 (in Russian).

[12] S.C. Port, Hitting times and potentials for recurrent stable processes, *J. d'Anal. Math.* **20** (1967), 371–395.

[13] D. Revuz, M. Yor, *Continuous martingales and Brownian motion*, Springer-Verlag, 1994.

[14] J. Rosiński, W. Woyczyński, On Itô stochastic integration with respect to p-stable motion: inner clock, integrability of sample paths, double and multiple integrals, *Ann. Probab.* **14** (1986), 271–286.

[15] A. Rozkosz and L. Słomiński, On weak solutions of one-dimensional SDEs with time-dependent coefficients, *Stochastics Stochastics Rep.* **42** (1993), 199–208.

[16] K. Sato, *Lévy processes and infinitely divisible distributions*, Cambridge University Press, Cambridge, 1999.

[17] T. Senf, Stochastische Differentialgleichungen mit inhomogenen Koeffizienten, *Dissertation*, Friedrich-Schiller-Universität Jena, 1992.

[18] T. Senf, On one-dimensional stochastic differential equations without drift and with time-dependent diffusion coefficients, *Stochastics Stochastics Rep.* **43** (1993), 199–220.

[19] Ch. Stone, The set of zeros of a semistable process, *Illinois J. Math.* **7** (1963), 631–637.

[20] P. A. Zanzotto, On solutions of one-dimensional stochastic differential equations driven by stable Lévy motion, *Stochastic Processes Appl.* **68** (1997), 209–228.

[21] P. A. Zanzotto, On stochastic differential equations driven by Cauchy process and the other stable Lévy motions, *Technical Report* 2.312.1127, Dipartimento di Matematica, Università di Pisa, 1998.

[22] P. A. Zanzotto, On stochastic differential equations driven by Cauchy process and the other α-stable motions, *Mini-proceedings*, Conference on Lévy Processes: Theory and Applications, Miscellana No. 11, April 1999, MaPhySto, University of Aarhus, 179–183.

INTEGRAL FUNCTIONALS OF STRONG MARKOV CONTINUOUS LOCAL MARTINGALES*

H.-J. ENGELBERT   and   GUNAR TITTEL

Friedrich-Schiller-Universität Jena
Institut für Stochastik
D-07743 Jena, Germany

Abstract We investigate integral functionals of the type $\int_0^t f(X_s)\, ds$, where f is a nonnegative Borel function and X is a strong Markov continuous local martingale. In dependence of the behaviour of the process X, we give purely analytical conditions on f which are necessary and sufficient for the **P**-a.s. convergence or divergence of the integral functionals. If X is a Brownian motion, the problem was treated in [6].

Key Words Brownian motion, strong Markov continuous local martingales, regular points, absorbing points, local time, integral functionals, speed measure.

1. PRELIMINARIES

We start with a family $(\Omega, \mathcal{F}^0, \mathbf{P}_x, x \in \mathbb{R})$ of probability spaces such that for every $A \in \mathcal{F}^0$ the function $\mathbf{P}_\cdot(A)$ is $\mathcal{B}^u(\mathbb{R})$-measurable, where $\mathcal{B}^u(\mathbb{R})$ denotes the universal completion of $\mathcal{B}(\mathbb{R})$. For every probability measure μ on $(\mathbb{R}, \mathcal{B}(\mathbb{R}))$ we set

$$\mathbf{P}_\mu(A) \;=\; \int_{\mathbb{R}} \mathbf{P}_x(A)\, \mu(dx), \ A \in \mathcal{F}^0,$$

*Research supported by Grant INTAS 97-30204.

111

thus obtaining a new probability measure $\mathbf{P}_\mu$. Let $\mathcal{F}^\mu$ be the completion of $\mathcal{F}^0$ with respect to $\mathbf{P}_\mu$ and $\mathcal{F} := \bigcap_\mu \mathcal{F}^\mu$, where the intersection ranges over all probability measures μ on $(\mathbb{R}, \mathcal{B}(\mathbb{R}))$. Now, the probability measures $\mathbf{P}_\mu$ and, in particular, $\mathbf{P}_x$ can be extended to the σ-algebra $\mathcal{F}$.

Let $\mathbb{F} = (\mathcal{F}_t)_{t\geq 0}$ be a right-continuous filtration of sub-σ-algebras of $\mathcal{F}$ such that $\mathcal{F}_0$ is complete in $\mathcal{F}$ with respect to the family $(\mathbf{P}_\mu)$. Furthermore, let $X = (X_t)_{t\geq 0}$ be a real valued stochastic process. We write $(X, \mathbb{F})$ to indicate that X is $\mathbb{F}$-adapted. By $\mathbb{F}^X$ we denote the filtration generated by X and completed in $\mathcal{F}$ with respect to the family $(\mathbf{P}_\mu)$. We call $(X, \mathbb{F})$ a stochastic process on the family of probability spaces $(\Omega, \mathcal{F}, \mathbf{P}_x, x \in \mathbb{R})$. Finally, we say that an assertion holds almost surely (and write a.s.) if it holds $\mathbf{P}_x$-a.s. for every $x \in \mathbb{R}$.

Definition 1.1 A stochastic process $(X, \mathbb{F})$ is said to be a *strong Markov continuous local martingale* if the following properties are fulfilled:

   (i) $\mathbf{P}_x(X_0 = x) = 1$, $x \in \mathbb{R}$.

   (ii) $(X, \mathbb{F})$ is a continuous local martingale for every $\mathbf{P}_x$.

   (iii) $(X, \mathbb{F}^X)$ possesses the strong Markov property, i.e., for every $\mathbb{F}^X$-stopping time S, for all $t \geq 0$, $x \in \mathbb{R}$ and $A \in \mathcal{B}(\mathbb{R})$

$$\mathbf{P}_x(X_{S+t} \in A \mid \mathcal{F}_S^X) = \mathbf{P}_{X_S}(X_t \in A)$$

holds $\mathbf{P}_x$-a.s. on $\{S < +\infty\}$.

Throughout this paper, unless stated otherwise, we will assume that $(X, \mathbb{F})$ is a strong Markov continuous local martingale.

First we will classify the states of a strong Markov continuous local martingale. An important feature of such processes is the possible occurrence of absorbing points.

Definition 1.2 A point $x \in \mathbb{R}$ is *absorbing* if

$$\mathbf{P}_x(X_t = x, \ t \geq 0) = 1$$

holds.

By E we denote the set of absorbing points. A point $x \in \mathcal{R} := \mathbb{R}\backslash E$ is said to be *regular*. The strong Markov continuous local martingale $(X, \mathbb{F})$ is called regular if $\mathcal{R} = \mathbb{R}$ holds.

The following lemma is a well-known fact, its proof can be found in [5].

Lemma 1.3 *The set E of absorbing points is closed.*

By Lemma 1.3 we can decompose the set of regular points $\mathcal{R} = \mathbb{R}\backslash E$ into open intervals. These intervals are called the *regular components* of the state space. Using this decomposition we can reduce our considerations to some special cases, depending on the starting point x.

On the one hand the process X can start in an absorbing point x. This case is trivial and will therefore be excluded; the process is constant and the corresponding integral functional is deterministic.

So we will always assume that the process X starts in the interior of a regular component (a, b). For the boundary points a and b of the interval we have to distinguish between three possible cases.

In the first case, we have $a = -\infty$ and $b = +\infty$, i.e., the strong Markov continuous local martingale is regular. The Wiener process is an important example for this class of processes.

In the second case, one of the boundary points of the interval (a, b) is finite. We have $a > -\infty$ and $b = +\infty$ (respectively, $a = -\infty$ and $b < +\infty$). In this case, the process cannot reach points $x \in \mathbb{R}$ with $x < a$ (respectively, $x > b$) with probability greater than zero. So it is irrelevant if the process possesses other absorbing points beyond the absorbing boundary point a or b. Without loss of generality we can assume that either a or b is the only absorbing point of the process X.

Finally, both of the boundary points a and b can be finite. In this case the process X cannot leave the interval $[a, b]$ and possible absorbing points outside $[a, b]$ are not important. So we can assume that the absorbing points a and b are the only ones.

Next we will describe the asymptotic behaviour of the process X dependent on the fact whether X possesses absorbing points or not. For this we need some properties of the *quadratic variation* $\langle X \rangle = (\langle X \rangle_t)_{t \geq 0}$ of the process X.

In the following, for any real-valued process Z and $B \in \mathcal{B}(\mathbb{R})$, by $D_B(Z)$ we denote the first entry time of Z into B:

$$D_B(Z) = \inf\{t \geq 0 : Z_t \in B\}.$$

If $B = \{b\}$ consists of a single point then we will also use the notation τ_b^Z for $D_{\{b\}}(Z)$. If $Z = X$ then we simply write D_B and τ_b instead of $D_B(X)$ and τ_b^X, respectively.

Proposition 1.4 *For a strong Markov continuous local martingale the following properties are fulfilled:*
   (i) *For the quadratic variation* $\langle X \rangle$

$$\langle X \rangle_{D_E} = \langle X \rangle_\infty := \sup_{t \geq 0} \langle X \rangle_t$$

holds a.s.
   (ii) *The limit* $X_\infty = \lim_{t \to \infty} X_t$ *a.s. exists on the set* $\{\langle X \rangle_\infty < +\infty\}$ *and is finite.*
   (iii) *On the set* $\{\langle X \rangle_\infty = +\infty\}$ *it holds*

$$\liminf_{t \to \infty} X_t = -\infty \text{ and } \limsup_{t \to \infty} X_t = +\infty \quad a.s.$$

   (iv) *We have* $X_\infty \in E$ *a.s. on the set* $\{\langle X \rangle_\infty < +\infty\}$.

The next proposition illustrates the connection between the existence of absorbing points and the behaviour of the quadratic variation.

Proposition 1.5 *The quadratic variation* $\langle X \rangle$ *of a strong Markov continuous local martingale* X *possesses the following properties:*

(i) *If* $E = \emptyset$ *then*

$$\mathbf{P}_x(\langle X \rangle_\infty = +\infty) = 1$$

holds for all $x \in \mathbb{R}$.
(ii) *If* $E \neq \emptyset$ *then*

$$\mathbf{P}_x(\langle X \rangle_\infty < +\infty) = 1$$

holds for all $x \in \mathbb{R}$.

The proofs of Proposition 1.4 and 1.5 can be found in [5], (1.23) and (1.26). Combining Proposition 1.4 and 1.5 we realize the following asymptotic behaviour of the process X depending on whether X is regular or not.

Corollary 1.6 (i) *If* $E = \emptyset$ *then*

$$\liminf_{t \to \infty} X_t = -\infty \ \text{and} \ \limsup_{t \to \infty} X_t = +\infty \quad a.s.$$

In particular,

$$\mathbf{P}_x(\tau_y < +\infty) = 1, \quad x, y \in \mathbb{R}$$

is fulfilled.
(ii) *If* $E \neq \emptyset$ *then* $\lim_{t \to +\infty} X_t$ *exists and is finite a.s. and*

$$\lim_{t \to +\infty} X_t \in E \quad a.s.$$

As we will see later, in the case that X has absorbing points the process can converge in two different ways. On the one hand the process can reach an absorbing point in finite time and will be absorbed there. On the other hand it is possible that the process does not reach an absorbing point in finite time. So we have to investigate which points of the state space can be attained by X in finite time.

Next we recall some basic facts which are important for our further investigations. First we collect some properties of the local time. The proofs of the assertions can be found, e.g., in [5] or [10].

Proposition 1.7 *Let* $(X, \mathbb{F})$ *be a strong Markov continuous local martingale. Then there exists a version* L^X *of the local time of* X *which possesses the following properties:*
(i) $L^X(t, y)$ *is a.s. increasing for every* $y \in \mathbb{R}$.
(ii) $L^X(t, y)$ *is continuous in* (t, y) *a.s.*
(iii) $L^X(t, y) = L^X(t \wedge D_E, y)$ *for all* $(t, y) \in [0, +\infty) \times \mathbb{R}$ *a.s.*
(iv) $L^X(t, y) = 0$ *for all* $(t, y) \in [0, +\infty) \times E$ *a.s.*
(v) $L^X(t, X_0) > 0$ *for all* $t > 0$ *on* $\{X_0 \in \mathbb{R} \backslash E\}$ *a.s.*
(vi) *For every* $\mathbb{F}$-*stopping time* S

$$L^X(S + t, X_S) > 0 \ \text{for all} \ t > 0 \ \text{on} \ \{X_S \in \mathbb{R} \backslash E, S < \infty\}$$

holds a.s.

Now we recall the notion of the associated Wiener process. In fact, we know that every strong Markov continuous local martingale can be transformed to a Wiener process by a time change. We define $A_t := \langle X \rangle_t$ and

$$T_t := \inf\{s \geq 0 : A_s > t\}$$

to be the right inverse of A. Now we set

$$W = X \circ T := (X_{T_t})_{t \geq 0}, \quad \mathbb{G} = \mathbb{F} \circ T := (\mathcal{F}_{T_t})_{t \geq 0}.$$

Proposition 1.8 *The process* $(W, \mathbb{G})$ *is a strong Markov continuous local martingale with quadratic variation*

$$\langle W \rangle_t = t \wedge D_E(W).$$

So $(W, \mathbb{G})$ is a Wiener process stopped when entering the set E. The process $(W, \mathbb{G})$ is called *the Wiener process associated to the strong Markov continuous local martingale* $(X, \mathbb{F})$. The proof of Proposition 1.8 can be found in [5]. As in [4] an associated Wiener process $(\widetilde{W}, \widetilde{\mathbb{G}})$ can be constructed without absorbing points on a, possibly enlarged, family of probability spaces such that $\widetilde{W}_{t \wedge A_\infty} = W_t$ for all $t \geq 0$. Thus, without loss of generality we will always assume that the associated Wiener process $(W, \mathbb{F})$ is extended to a Wiener process on a, possibly augmented, family of probability spaces such that W does not have absorbing points. The considerations above lead to the following theorem, which is also proved in [5].

Theorem 1.9 *Let* $(X, \mathbb{F})$ *be a strong Markov continuous local martingale and* g *a nonnegative Borel function. Then the following assertions are true:*

(i) *There exists a nonnegative measure* m *on* $(\mathbb{R}, \mathcal{B}(\mathbb{R}))$, *called the speed measure of* X, *which is* σ-*finite on* $\mathbb{R} \backslash E$ *and satisfies* $m(G) > 0$ *for every nonempty open subset* G *of* $\mathbb{R} \backslash E$ *and* $m(\{x\}) = +\infty$ *for every* $x \in E$ *and is such that*

$$T_t \wedge D_E = \int_{\mathbb{R}} L^W(t \wedge A_\infty, y)\, m(dy), \quad t \geq 0 \text{ a.s.}$$

holds.

(ii) *Let* m *be the speed measure of* X. *Then for every nonnegative bounded Borel measurable function* g *it holds*

$$\int_0^{t \wedge D_E} g(X_s)\, ds = \int_{\mathbb{R}} g(y) L^X(t, y)\, m(dy), \quad t \geq 0, \text{ a.s.}$$

The following lemma is often used in the sequel.

Lemma 1.10 *We have*

$$L^W(A_t, y) = L^X(t, y), \quad t \geq 0, \text{ a.s.}$$

Indeed, inserting A_t in the Tanaka formula for W (cf. [10], Chapter VI) we obtain

$$|W_{A_t} - y| = |W_0 - y| + \int_0^{A_t} \text{sgn}(W_s - y)\, dW_s + L^W(A_t, y) \quad \text{a.s.}$$

From this and because of

$$\int_0^{A_t} \text{sgn}(W_s - y)\, dW_s = \int_0^t \text{sgn}(X_s - y)\, dX_s$$

we get the assertion.   □

Finally, we introduce some notations which will be helpful for the formalation of some statements in the following sections.

Definition 1.11 Let μ be an arbitrary measure on $(\mathbb{R}, \mathcal{B}(\mathbb{R}))$.

(i) $\mathbf{L}^1_{\text{loc}}(x+, \mu)$ is the space of all Borel functions f which are locally integrable with respect to μ in the right neighbourhood of x, i.e., there exists some $\varepsilon > 0$ such that

$$\int_x^{x+\varepsilon} |f(y)|\, \mu(dy) < +\infty$$

holds.

(ii) $\mathbf{L}^1_{\text{loc}}(x-, \mu)$ is the space of all Borel functions f which are locally integrable with respect to μ in the left neighbourhood of x, i.e., there exists some $\varepsilon > 0$ such that

$$\int_{x-\varepsilon}^{x} |f(y)|\, \mu(dy) < +\infty$$

holds.

(iii) We define

$$\mathbf{L}^1_{\text{loc}}(x, \mu) := \mathbf{L}^1_{\text{loc}}(x-, \mu) \cap \mathbf{L}^1_{\text{loc}}(x+, \mu).$$

(iv) For any $A \subseteq \mathbb{R}$, we set

$$\mathbf{L}^1_{\text{loc}}(A, \mu) := \bigcap_{x \in A} \mathbf{L}^1_{\text{loc}}(x, \mu).$$

(v) Let m be the speed measure of the process X. Then for any $x \in \mathbb{R}$, we define the measure ν_x by

$$\nu_x(A) := \int_A |x - y|\, m(dy), \quad A \in \mathcal{B}(\mathbb{R}).$$

2.   THE REGULAR CASE

In this section we are going to investigate the integral functionals $\int_0^t f(X_s)\, ds$ for *regular* strong Markov continuous local martingales X. In [6], a 0–1-law was proven in the special case that X is a Wiener process. Our aim is to show that an analogous 0–1-law holds in the general case of regular strong Markov continuous local martingales X. Throughout this section we assume that f is a nonnegative Borel function.

First we prove a lemma which will also be used in the nonregular case and is therefore stated for general strong Markov continuous local martingales.

Lemma 2.1 *Let (a, b) a regular component of the state space. For $x_0 \in (a, b)$ we suppose that there exists a strictly positve stopping time τ such that*

$$\mathbf{P}_{x_0}\left(\int_0^\tau f(X_s) \, ds < +\infty \right) > 0$$

holds. Then $f \in L^1_{\text{loc}}(x_0, m)$, i.e., there exists an open neighbourhood U_{x_0} of x_0 with

$$\int_{U_{x_0}} f(y) \, m(dy) < +\infty .$$

Proof. Without loss of generality we can assume that τ is finite $\mathbf{P}_{x_0}$-a.s. Theorem 1.9 yields

$$\int_0^\tau f(X_s) \, ds = \int_{-\infty}^{+\infty} f(y) L^X(\tau, y) \, m(dy). \tag{2.1}$$

By Proposition 1.7 the local time satisfies

$$L^X(s, X_0) > 0, \quad s > 0, \quad \mathbf{P}_{x_0}\text{-a.s.}$$

It follows that

$$\mathbf{P}_{x_0}(L^X(\tau, x_0) > 0) = 1.$$

But the local time is a.s. continuous at x_0. Hence there is a set $\Omega_0 \subseteq \Omega$ with $\mathbf{P}_{x_0}(\Omega_0) = 1$ such that for all $\omega \in \Omega_0$ there exists an open neighbourhood $U_{x_0}(\omega)$ of x_0 and a constant $c(\omega)$ such that

$$L^X(\tau(\omega), y, \omega) \geq c(\omega) > 0, \quad y \in U_{x_0}(\omega) .$$

Thus, we obtain

$$\int_{-\infty}^\infty f(y) L^X(\tau(\omega), y, \omega) \, m(dy) \geq c(\omega) \int_{U_{x_0}(\omega)} f(y) \, m(dy).$$

In view of the assumption and (2.1) the right hand member is finite with strictly positive $\mathbf{P}_{x_0}$-probability and the proof is complete. □

For regular strong Markov continuous local martingales we have the following 0–1-law. The proof is similar to the proof of the 0–1-law in [6].

Theorem 2.2 *Let $(X, \mathbb{F})$ be a regular strong Markov continuous local martingale. For all $x \in \mathbb{R}$ the following conditions are equivalent:*
(i) $\mathbf{P}_x(\int_0^t f(X_s) \, ds < +\infty, \forall t \geq 0) > 0$,
(ii) $\mathbf{P}_x(\int_0^t f(X_s) \, ds < +\infty, \forall t \geq 0) = 1$,
(iii) $f \in L^1_{\text{loc}}(\mathbb{R}, m)$.

Remark 2.3 We notice that the condition (iii) is equivalent to
(iv) For every compact subset K of $\mathbb{R}$,

$$\int_K f(y) \, m(dy) < +\infty.$$

Proof. We first show the implication (i) $\Rightarrow$ (iii). Let $x_0 \in \mathbb{R}$ and $t > 0$ be fixed. For $\tau_{x_0} := D_{\{x_0\}}$, by Corollary 1.6 we obtain

$$\mathbf{P}_x(\tau_{x_0} < +\infty) = 1.$$

Now the assumption i) yields

$$\mathbf{P}_x\left(\int_0^{\tau_{x_0}+t} f(X_s)\, ds < \infty\right) > 0.$$

If we set

$$\widetilde{X}_s := X_{\tau_{x_0}+s}, \quad s \geq 0,$$

we have

$$\mathbf{P}_x\left(\int_0^t f(\widetilde{X}_s)\, ds < \infty\right) = \mathbf{P}_x\left(\int_{\tau_{x_0}}^{\tau_{x_0}+t} f(X_s)\, ds < \infty\right).$$

Because of the strong Markov property the process $\widetilde{X}$ has the same distribution with respect to $\mathbf{P}_x$ as the process X with respect to $\mathbf{P}_{x_0}$. We thus obtain

$$\mathbf{P}_{x_0}\left(\int_0^t f(X_s)\, ds < \infty\right) > 0.$$

By Lemma 2.1 there exists an open neighbourhood U_{x_0} of x_0 with

$$\int_{U_{x_0}} f(y)\, m(dy) < \infty.$$

Since $x_0 \in \mathbb{R}$ was chosen arbitrarily, this proves (iii).

Next we show the implication (iii) $\Rightarrow$ (ii). By Theorem 1.9 (ii) it holds

$$\int_0^t f(X_s(\omega))\, ds = \int_{\mathbb{R}} f(y) L^X(t, y, \omega)\, m(dy) \quad \text{a.s.}$$

We define

$$M_t := \max_{s \leq t} X_s \quad \text{and} \quad m_t := \min_{s \leq t} X_s. \tag{2.2}$$

The local time of X vanishes for all $y \notin [m_t(\omega), M_t(\omega)]$. Because of the continuity in y, the local time $L^X(t, \cdot)$ is bounded on every compact set K and this yields

$$\int_0^t f(X_s(\omega))\, ds = \int_{m_t(\omega)}^{M_t(\omega)} f(y) L^X(t, y, \omega)\, m(dy)$$

$$\leq \max_{y \in [m_t(\omega), M_t(\omega)]} L^X(y, t, \omega) \int_{m_t(\omega)}^{M_t(\omega)} f(y)\, m(dy) < \infty$$

$\mathbf{P}_x$-a.s. where we have used (iv) which is equivalent to (iii) by Remark 2.3. This proves (ii).

The implication from (ii) to (i) is trivial.   □

Next we investigate the functional $\int_0^\infty f(X_s)\, ds$.

Proposition 2.4 *Let m be the speed measure of X and f be a nonnegative Borel function such that $m(\{y : f(y) > 0\}) > 0$ is fulfilled. Then for all $x \in \mathbb{R}$*

$$\mathbf{P}_x\left(\int_0^\infty f(X_s)\, ds = +\infty\right) = 1$$

holds.

Proof. Let $x \in \mathbb{R}$ be an arbitrary starting point. From Theorem 1.9 it follows

$$\int_0^\infty f(X_s)\, ds = \int_{\mathbb{R}} f(y) L^X(\infty, y)\, m(dy) \quad \mathbf{P}_x\text{-a.s.}$$

Let W be the associated Wiener process. Due to Lemma 1.10 and Proposition 1.5 (i), for all $y \in \mathbb{R}$ we have

$$L^X(\infty, y) = L^W(\langle X \rangle_\infty, y) = L^W(\infty, y) = \infty \quad \mathbf{P}_x\text{-a.s.,}$$

where the last equality is shown in [10], Chapter VI, Corollary 2.4. The theorem of Fubini now yields

$$m(\{y : L^X(\infty, y) < +\infty\}) = 0 \quad \mathbf{P}_x\text{-a.s.}$$

Obviously, from this the assertion follows. $\qquad\square$

3. THE SINGULAR CASE

Now we consider the case when $(X, \mathbb{F})$ possesses absorbing points, which is referred to as the *singular* case. First we look for conditions that an absorbing point will be attained in finite time. In this section we always assume that f is a nonnegative Borel function. For simplicity, we make the additional assumption that the process X starts in $X_0 = 0$. This can always be achieved by a translation of the state space. For our further considerations we will need the following lemma which is due to T. Jeulin [7]. A proof can also be found in W. Schmidt and S. Assing [2] and X.-X. Xue [11].

Lemma 3.1 *Let B be a Borel set and $\{L_y\}_{y \in B}$ a family of nonnegative random variables on a probability space $(\Omega, \mathcal{F}, \mathbf{P})$ such that L_y is $\mathcal{B}(B) \otimes \mathcal{F}$-measurable and $P(L_y = 0) = 0$ holds for all $y \in B$. Additionally, we assume that there exists a measurable function $\varphi : B \to (0, +\infty)$ such that for all $y \in B$ the random variable $\frac{L_y}{\varphi(y)}$ possesses the same distribution as a certain integrable random variable ξ. Furthermore, let μ be a σ-finite measure on $(B, \mathcal{B}(B))$. Then the following properties are equivalent:*

(i) $\mathbf{P}\left(\int_B L_y\, \mu(dy) < \infty\right) > 0,$

(ii) $\mathbf{P}\left(\int_B L_y\, \mu(dy) < \infty\right) = 1,$

(iii) $\int_B \varphi(y)\, \mu(dy) < +\infty.$

Lemma 3.1 turns out to be very useful. For applications to the study of the behaviour of integral functionals associated with Bessel processes the reader is referred to J.W. Pitman and M. Yor [9] and X.-X. Xue [11]. Using this lemma we now prove the following 0–1-law. The result is taken from S. Assing [1]. For similar statements, also see S. Assing and W. Schmidt [2], (A1.8), and S. Assing and T. Senf [3], Lemma 2.

Proposition 3.2 *Let* $(W, \mathbb{G})$, *on a probability space* $(\Omega, \mathcal{F}, \mathbf{P})$, *be a Wiener process and* I *an interval of the type* $[0, r)$ *or* $(r, 0]$. *Furthermore, let* μ *be a* σ-*finite measure on* $(I, \mathcal{B}(I))$. *Then the following conditions are equivalent:*

(i) $\quad \mathbf{P}(\int_I L^W(\tau_r^W, y)\, \mu(dy) < +\infty) > 0,$

(ii) $\quad \mathbf{P}(\int_I L^W(\tau_r^W, y)\, \mu(dy) < +\infty) = 1,$

(iii) $\quad \int_I |r - y|\, \mu(dy) < +\infty.$

Proof. We will only give the proof for the case $I = [0, r)$; the case $I = (r, 0]$ can be treated analogously. We set $L_y = L^W(\tau_r^W, y)$. Now, the Ray–Knight–Theorem (cf. [8], Chapter 6, Theorem 4.7) states that for $r > 0$ the local time $L^W(\tau_r^W, y)$ is distributed like a squared two-dimensional Bessel process in $|r - y|$, i.e., $L^W(\tau_r^W, y) \overset{d}{\sim} B^2_{|r-y|}$. In particular, for $y \neq r$, $P(L^W(\tau_r^W, y) = 0) = 0$ is fulfilled. However, a squared two-dimensional Bessel process is the radius of a two-dimensional Wiener process (W^1, W^2). From this it follows

$$\frac{L_y}{|r - y|} \overset{d}{\sim} \frac{(W^1_{|r-y|})^2 + (W^2_{|r-y|})^2}{|r - y|} \overset{d}{\sim} X^2 + Y^2 =: \xi,$$

where X and Y are independent standard normal variables. So we can apply Lemma 3.1 and the proof is complete. □

Now we are going to investigate the behaviour of the stopping time D_E. If the process X, starting in $X_0 = 0$, possesses only one absorbing point $b > 0$ then Theorem 1.9 yields

$$D_E = \tau_b = T_{\tau_b^W -} = T_{\tau_b^W -} \wedge D_E$$

$$= \int_{\mathbb{R}} L^W(\tau_b^W, y)\, m(dy) \quad \text{a.s.}$$

Now the local time vanishes outside of $[\min_{t \leq \tau_b^W} W_t, b]$ and we can decompose the integral on the right hand side of the equality into

$$\int_{[\min_{t \leq \tau_b^W(\omega)} W_t(\omega), 0)} L^W(\tau_b^W(\omega), y)\, m(dy) + \int_{[0,b)} L^W(\tau_b^W(\omega), y)\, m(dy).$$

The speed measure m is locally finite on $(-\infty, b)$. So the left part of the sum is $\mathbf{P}_0$-a.s. finite and for the right part we can apply Proposition 3.2. The case that X possesses only one absorbing point $b < 0$ can be reduced to the case $b > 0$ by considering the process $\tilde{X} = -X$ and therefore is analogous.

If the process possesses two absorbing points a and b with $-\infty < a < 0 < b < +\infty$ then we can proceed in a similar way, but now we have to distinguish the sets $\{X_{D_E} = a\}$ and $\{X_{D_E} = b\}$, respectively. In the next lemma we calculate the probability of reaching a point $y \in (a, b)$ before another point $z \in (a, b)$ will be attained by the process X.

Lemma 3.3 *Let (a, b) be a regular component of the state space with $-\infty < a$ and $b < +\infty$. For an interval $(y, z) \subseteq (a, b)$ let*

$$T_{(y,z)} = \inf\{t \geq 0 : X_t \notin (y, z)\} \tag{3.1}$$

be the first exit time of the process X from the interval (y, z). Then

$$\mathbf{P}_x(X_{T_{(y,z)}} = y) = \frac{z - x}{z - y}, \quad x \in [y, z],$$

holds.

Proof. We notice that, for any stopping time σ, X_σ is defined as X_∞ on $\{\sigma = \infty\}$ where $X_\infty := \limsup\limits_{t \to \infty} X_t$. If, however, $E \neq \emptyset$, $X_\infty = \lim\limits_{t \to \infty} X_t$ a.s. by Proposition 1.4 and 1.5. The stopped process $X_{t \wedge T_{(y,z)}}$ is a bounded martingale and it holds

$$x = E_x X_0 = E_x X_{T_{(y,z)}} = y \cdot \mathbf{P}_x(X_{T_{(y,z)}} = y) + z \cdot (1 - \mathbf{P}_x(X_{T_{(y,z)}} = y))$$

which leads to the desired result. $\qquad\square$

Summarizing the considerations above we now come to the following proposition.

Proposition 3.4 *Let $(X, \mathbb{F})$ be a strong Markov continuous local martingale starting at $X_0 = 0$ and let m be the speed measure of $(X, \mathbb{F})$.*

(I) *If X possesses only one absorbing point $b > 0$ then the following properties are equivalent:*

(i) $\mathbf{P}_0(\tau_b < +\infty) > 0$,

(ii) $\mathbf{P}_0(\tau_b < +\infty) = 1$,

(iii) $\displaystyle\int_{[0,b)} |b - y|\, m(dy) < +\infty$.

(II) *If the process X possesses two absorbing points $-\infty < a < 0 < b < +\infty$ then the following conditions (i) – (iii) as well as (i') – (iii') are equivalent:*

(i) $\mathbf{P}_0(D_E < +\infty \mid X_{D_E} = b) > 0$,

(ii) $\mathbf{P}_0(D_E < +\infty \mid X_{D_E} = b) = 1$,

(iii) $\displaystyle\int_{[0,b)} |b - y|\, m(dy) < +\infty$,

(i') $\mathbf{P}_0(D_E < +\infty \mid X_{D_E} = a) > 0$,

(ii') $\mathbf{P}_0(D_E < +\infty \mid X_{D_E} = a) = 1$,

(iii') $\displaystyle\int_{(a,0]} |a - y|\, m(dy) < +\infty$.

Remark 3.5 In the notation of Definition 1.11 the conditions (iii) and (iii') mean that $1 \in L^1_{loc}(b-, \nu_b)$ and $1 \in L^1_{loc}(a+, \nu_a)$, respectively.

Now we are going to investigate the behaviour of the integral functionals. First we show the following lemma.

Lemma 3.6 *Let (a, b) be a regular component of the state space with $-\infty \leq a < 0 < b \leq +\infty$ and f a nonnegative Borel function. If*

$$\mathbf{P}_0\left(\int_0^t f(X_s)\, ds < +\infty, \forall t < D_E\right) = 1$$

holds then $f \in L^1_{loc}((a, b), m)$.

Proof. For $x \in (a, b)$, from Lemma 3.3 $\mathbf{P}_0(\tau_x < D_E) > 0$ follows. By

$$\widetilde{\mathbf{P}}_0(A) := \mathbf{P}_0(A \mid \tau_x < +\infty), \quad A \in \mathcal{F}^X_\infty,$$

we define a new probability measure $\widetilde{\mathbf{P}}_0$. From the assumption

$$\mathbf{P}_0\left(\int_0^t f(X_s)\, ds < +\infty, \forall t < D_E\right) = 1$$

it follows

$$\widetilde{\mathbf{P}}_0\left(\int_0^{\tau_x+t} f(X_s)\, ds < +\infty, \forall t < D_E - \tau_x\right) = 1.$$

Because of the strong Markov property the process $\widetilde{X}_t := X_{\tau_x+t}, t \geq 0$, defined on $\{\tau_x < +\infty\}$ has the same distribution with respect to $\widetilde{\mathbf{P}}_0$ as the process X with respect to $\mathbf{P}_x$ and we conclude

$$\mathbf{P}_x\left(\int_0^t f(X_s)\, ds < +\infty, \forall t < D_E\right) = 1.$$

Since $\mathbf{P}_x(D_E > 0) = 1$, we can find $t > 0$ such that $\mathbf{P}_x(t < D_E) > 0$ and hence

$$\mathbf{P}_x\left(\int_0^t f(X_s)\, ds < +\infty\right) > 0.$$

Now from Lemma 2.1 it follows that there exists an open neighbourhood U_x of x such that

$$\int_{U_x} f(y)\, m(dy) < +\infty$$

holds which proves the assertion.   □

Now we can formulate necessary and sufficient conditions for the integral functionals being finite or infinite, respectively. The methods and techniques of the proofs are similar for the different cases we have to distinguish. To illustrate these techniques we give the proofs for the case that X has exactly one absorbing point $b > 0$.

Proposition 3.7 *Let the process X have exactly one absorbing point $b > 0$. Then the following statements hold:*

(i) *We have*

$$\mathbf{P}_0\left(\int_0^t f(X_s)\,ds < +\infty, \forall\, t < D_E\right) = 1$$

if and only if $f \in \mathbf{L}^1_{loc}((-\infty, b), m)$ holds.

(ii) *We have*

$$\mathbf{P}_0\left(\int_0^t f(X_s)\,ds < +\infty, \forall\, t < D_E\right) = 0$$

if and only if $f \notin \mathbf{L}^1_{loc}([0, b), m)$, i.e., there exists a point $x_0 \in [0, b)$ such that

$$\int_{U_{x_0}} f(y)\,m(dy) = +\infty$$

for all open neighbourhoods U_{x_0} of x_0.

Proof. (i) The necessity follows from Lemma 3.6. Conversely, if f is locally integrable in $(-\infty, b)$ with respect to m then, by Theorem 1.9, for $t < D_E$ it follows

$$\begin{aligned}
\int_0^t f(X_s)\,ds &= \int_{\mathbb{R}} f(y)L^X(t, y)\,m(dy) \\
&= \int_{m_t}^{M_t} f(y)L^X(t, y)\,m(dy) < +\infty \quad \mathbf{P}_0\text{-a.s.,}
\end{aligned}$$

(where m_t and M_t are defined by (2.2)), since by Proposition 1.7 the local time is continuous and therefore bounded on the compact set $[m_t, M_t] \subseteq (-\infty, b)$ and vanishes outside $[m_t, M_t]$. So assertion (i) is proved.

(ii) First we show the necessity by contraposition. We assume that for some $y_0 < 0$ the function f is locally integrable with respect to m on the interval $[y_0, b)$. Recalling the definition of $\tau_{(y_0, b)}$ by (3.1) and once again using Theorem 1.9, on the set $\{X_{\tau_{(y_0, b)}} = b\}$ for $t < D_E = \tau_b$ we obtain

$$\begin{aligned}
\int_0^t f(X_s)\,ds &= \int_{\mathbb{R}} f(y)L^X(t, y)\,m(dy) \\
&\leq \int_{y_0}^{M_t} f(y)L^X(t, y)\,m(dy) < +\infty.
\end{aligned}$$

From Lemma 3.3 it follows

$$\mathbf{P}_0(\{X_{\tau_{(y_0, b)}} = b\}) > 0$$

and we thus have

$$\mathbf{P}_0\left(\int_0^t f(X_s)\,ds < +\infty, \forall\, t < D_E\right) > 0,$$

which proves the necessity.

Now we show the sufficiency. Suppose that f is not integrable in any neighbourhood of some point $x_0 \in [0, b)$. For y_0 with $x_0 < y_0 < b$ the stopping time τ_{y_0} satisfies

$$\mathbf{P}_0(\{\tau_{y_0} < +\infty\}) = 1 \quad \text{and} \quad \mathbf{P}_0(\{\tau_{y_0} < \tau_b\}) = 1$$

where the first equality follows from Proposition 1.4 (iv). From this and Theorem 1.9 we derive

$$\int_0^{\tau_{y_0}} f(X_s)\, ds = \int_{\mathbb{R}} f(y) L^X(\tau_{y_0}, y)\, m(dy)$$
$$= \int_{m_{\tau_{y_0}}}^{y_0} f(y) L^X(\tau_{y_0}, y)\, m(dy) \quad \mathbf{P}_0\text{-a.s.}$$

We define the stopping time

$$U(0-) := \inf\{t \geq 0 : X_t < 0\}.$$

The relation

$$\mathbf{P}_0(U(0-) = 0) = 1$$

following from the regularity of 0 (cf., e.g., [5], (1.7)) results in

$$\mathbf{P}_0(m_{\tau_{y_0}} < 0) = 1.$$

Now by 1.7 (vi) and $\mathbf{P}_0(\tau_{x_0} < \tau_{y_0}) = 1$, we get $L^X(\tau_{y_0}, x_0) > 0$ a.s. Due to the continuity of the local time there is a set Ω_0 with $\mathbf{P}_0(\Omega_0) = 1$ such that for $\omega \in \Omega_0$ there exists an open neighbourhood $U_{x_0}(\omega) \subseteq (m_{\tau_{y_0}}(\omega), y_0)$ of x_0 and a constant $c(\omega)$ such that

$$L^X(\tau_{y_0}(\omega), y, \omega) \geq c(\omega) > 0, \quad \forall y \in U_{x_0}(\omega), \ \omega \in \Omega_0$$

is fulfilled. Hence we obtain

$$\int_{m_{\tau_{y_0}}(\omega)}^{y_0} f(y) L^X(\tau_{y_0}(\omega), y, \omega)\, m(dy) \geq c(\omega) \int_{U_{x_0}(\omega)} f(y)\, m(dy)$$
$$= +\infty$$

for all $\omega \in \Omega_0$, and the proof is complete. □

In Proposition 3.7, the condition (ii) is not the negation of condition (i). For this reason, a 0–1-law is not true without additional conditions on the function f. However, the following result is an immediate consequence of Proposition 3.7.

Proposition 3.8 *Let the process X have exactly one absorbing point $b > 0$. Furthermore, let $f \in \mathbf{L}^1_{\text{loc}}((-\infty, 0), m)$. Then the following conditions are equivalent:*
(i) $\mathbf{P}_0(\int_0^t f(X_s\, ds < +\infty, \forall t < D_E) > 0$,
(ii) $\mathbf{P}_0(\int_0^t f(X_s\, ds < +\infty, \forall t < D_E) = 1$,
(iii) $f \in \mathbf{L}^1_{\text{loc}}([0, b), m)$.

Next we investigate the functionals at the time $D_E = \tau_b$. For the definition of the measure ν_x, cf. Definition 1.11.

Proposition 3.9 *Let the process X have exactly one absorbing point $b > 0$. Then the following assertions are true:*

(I) *It holds*

$$\mathbf{P}_0\left(\int_0^{\tau_b} f(X_s)\, ds < +\infty\right) = 1$$

if and only if the following two conditions are fulfilled:

(i) $f \in \mathbf{L}^1_{\mathrm{loc}}((-\infty, b), m)$,
(ii) $f \in \mathbf{L}^1_{\mathrm{loc}}(b-, \nu_b)$.

(II) *It holds*

$$\mathbf{P}_0\left(\int_0^{\tau_b} f(X_s)\, ds = +\infty\right) = 1$$

if and only if one of the following two conditions is fulfilled:

(i) $f \notin \mathbf{L}^1_{\mathrm{loc}}([0, b), m)$,
(ii) $f \notin \mathbf{L}^1_{\mathrm{loc}}(b-, \nu_b)$.

Proof. It holds

$$\int_0^{\tau_b} f(X_s)\, ds \tag{3.2}$$

$$= \int_{[m_{\tau_b}, 0)} f(y) L^X(\tau_b, y)\, m(dy) + \int_{[0,b)} f(y) L^X(\tau_b, y)\, m(dy) \quad \mathbf{P}_0\text{-a.s.}$$

(I) We first show the necessity. Let

$$\mathbf{P}_0\left(\int_0^{\tau_b} f(X_s)\, ds < +\infty\right) = 1$$

be fulfilled. From Proposition 3.7 it follows that f is locally integrable on $(-\infty, b)$ with respect to m. So we only have to show $f \in \mathbf{L}^1_{\mathrm{loc}}(b-, \nu_b)$, i.e.,

$$\int_{[0,b)} |b - y| f(y)\, m(dy) < +\infty.$$

From (3.2) and

$$\int_0^{\tau_b} f(X_s)\, ds < +\infty$$

we obtain

$$\int_{[0,b)} f(y) L^X(\tau_b, y)\, m(dy) < +\infty.$$

In view of

$$L^X(\tau_b, y) = L^W(\tau_b^W, y),$$

where W is the Wiener process associated to X, this yields

$$\int_{[0,b)} f(y) L^W(\tau_b^W, y)\, m(dy) = \int_{[0,b)} f(y) L^X(\tau_b, y)\, m(dy) < +\infty \quad \mathbf{P}_0\text{-a.s.}$$

Since f is nonnegative, by $d\mu = f\,dm$ a σ-finite measure is defined. Proposition 3.2 yields

$$\int_{[0,b)} |b - y| f(y)\, m(dy) \;<\; +\infty,$$

which proves the assertion.

To show the sufficiency we again consider the decomposition (3.2). By Proposition 1.7 the local time is continuous and, therefore, bounded on every compact interval of $(-\infty, b)$. From this and the local integrability of f on $(-\infty, b)$ with respect to m it follows $\mathbf{P}_0$-a.s.

$$\int_{[m_{\tau_b},0)} f(y) L^X(\tau_b, y)\, m(dy) \;\le\; \max_{y \in [m_{\tau_b},0]} L^X(\tau_b, y) \int_{[m_{\tau_b},0)} f(y)\, m(dy) < +\infty.$$

On the other side, the integral

$$\int_{[0,b)} f(y) L^X(\tau_b, y)\, m(dy) = \int_{[0,b)} f(y) L^W(\tau_b^W, y)\, m(dy)$$

is finite by Proposition 3.2. This finishes the proof of (I).

(II) First we prove the necessity. If f on $(-\varepsilon, b)$ is not locally integrable with respect to m for all $\varepsilon > 0$ then, obviously, condition (i) is fulfilled. So let f be locally integrable with respect to m on $(-\varepsilon, b)$ for some $\varepsilon > 0$. By Lemma 3.3 we get

$$\mathbf{P}_0(X_{\tau_{(-\varepsilon,b)}} = b) > 0.$$

In particular,

$$\mathbf{P}_0(m_{\tau_b} > -\varepsilon) > 0.$$

From this and from

$$\int_{[m_{\tau_b},0)} f(y) L^X(\tau_b, y)\, m(dy) \;\le\; \max_{y \in [m_{\tau_b},0]} L^X(\tau_b, y) \int_{[m_{\tau_b},0)} f(y)\, m(dy) \quad \mathbf{P}_0\text{-a.s.}$$

we conclude

$$\mathbf{P}_0\Big(\int_{[m_{\tau_b},0)} f(y) L^X(\tau_b, y)\, m(dy) < +\infty\Big) > 0.$$

Therefore, from the assumption and decomposition (3.2) we have

$$\mathbf{P}_0\Big(\int_{[0,b)} f(y) L^X(\tau_b, y)\, m(dy) = +\infty\Big) > 0.$$

Using $L^X(\tau_b, y) = L^W(\tau_b^W, y)$ and applying Proposition 3.2 for the σ-finite measure μ with $d\mu = f\,dm$ we obtain

$$\int_{[0,b)} |b - y| f(y)\, m(dy) \;=\; +\infty.$$

However, this is only true if at least one of the conditions (i) or (ii) is fulfilled.

Now we show the sufficiency. If $f \notin \mathbf{L}^1_{\text{loc}}([0,b), m)$ then Proposition 3.7 (ii) immediately yields (note that $D_E = \tau_b$)

$$\mathbf{P}_0\left(\int_0^{\tau_b} f(X_s)\, ds = +\infty\right) = 1.$$

Now we assume that $f \notin \mathbf{L}^1_{\text{loc}}(b-, \nu_b)$. Using Proposition 3.2 for the σ-finite measure μ with $d\mu := f\, dm$, we obtain

$$\int_{[0,b)} f(y) L^W(\tau_b^W, y) m(dy) = +\infty \quad \mathbf{P}_0\text{-a.s.}$$

But this is equal to the second summand of the right hand side of the decomposition (3.2) and, consequently, we observe that

$$\mathbf{P}_0\left(\int_0^{\tau_b} f(X_s)\, ds = +\infty\right) = 1.$$

This proves the proposition. $\qquad\qquad\qquad\qquad\qquad\qquad\qquad\qquad\qquad\square$

Here again the conditions in (II) are not the negation of the conditions in (I). So we cannot get a 0–1-law without additional conditions on the function f. The following Proposition, however, is an immediate consequence of Proposition 3.9.

Proposition 3.10 *Let the process X have exactly one absorbing point $b > 0$. Furthermore, let $f \in \mathbf{L}^1_{\text{loc}}((-\infty, 0), m)$. Then the following conditions are equivalent:*
  (i) $\mathbf{P}_0\left(\int_0^{\tau_b} f(X_s\, ds < +\infty\right) > 0$,
  (ii) $\mathbf{P}_0\left(\int_0^{\tau_b} f(X_s\, ds < +\infty\right) = 1$,
  (iii) $f \in \mathbf{L}^1_{\text{loc}}([0,b), m)$ and $f \in \mathbf{L}^1_{\text{loc}}(b-, \nu_b)$.

Now we consider the functional at time $t = \infty$. Here we have to distinguish if the process reaches the absorbing point b in finite time or not. In the first case, the functional will be infinite if $f(b) > 0$ holds. In the second case, we already have the result from the last proposition above.

Corollary 3.11 *Let the process X have exactly one absorbing point $b > 0$. Then the following assertions are true:*
  (I) *Let $1 \in \mathbf{L}^1_{\text{loc}}(b-, \nu_b)$ be fulfilled. Then*

$$\mathbf{P}_0\left(\int_0^\infty f(X_s)\, ds < +\infty\right) = 1$$

holds if and only if the following conditions are satisfied:
  (i) $f \in \mathbf{L}^1_{\text{loc}}((-\infty, b), m)$,
  (ii) $f \in \mathbf{L}^1_{\text{loc}}(b-, \nu_b)$,
  (iii) $f(b) = 0$.
  (II) *Let $1 \in \mathbf{L}^1_{\text{loc}}(b-, \nu_b)$ be fulfilled. Then*

$$\mathbf{P}_0\left(\int_0^\infty f(X_s)\, ds = +\infty\right) = 1$$

holds if and only if at least one of the following conditions is satisfied:
  (i) $f \notin \mathbf{L}_{\mathrm{loc}}^1([0, b), m)$,
  (ii) $f \notin \mathbf{L}_{\mathrm{loc}}^1(b-, \nu_b)$,
  (iii) $f(b) > 0$.

(III) *Let* $1 \notin \mathbf{L}_{\mathrm{loc}}^1(b-, \nu_b)$ *be fulfilled. Then*

$$\mathbf{P}_0\Big(\int_0^\infty f(X_s)\, ds < +\infty\Big) = 1$$

holds if and only if the following conditions are satisfied:
  (i) $f \in \mathbf{L}_{\mathrm{loc}}^1((-\infty, b), m)$,
  (ii) $f \in \mathbf{L}_{\mathrm{loc}}^1(b-, \nu_b)$.

(IV) *Let* $1 \notin \mathbf{L}_{\mathrm{loc}}^1(b-, \nu_b)$ *be fulfilled. Then*

$$\mathbf{P}_0\Big(\int_0^\infty f(X_s)\, ds = +\infty\Big) = 1$$

holds if and only if at least one of the following conditions is satisfied:
  (i) $f \notin \mathbf{L}_{\mathrm{loc}}^1([0, b), m)$,
  (ii) $f \notin \mathbf{L}_{\mathrm{loc}}^1(b-, \nu_b)$.

Finally, we consider the case that $(X, \mathbb{F})$ possesses two absorbing points a, b sucht that $-\infty < a < 0 < b < +\infty$ holds. In this case, by Proposition 1.4 and Lemma 3.3 the process X converges with strictly positive probability to the point a as well as to the point b. Additionally, due to Lemma 3.3 each point $y \in (a, b)$ will be reached with strictly positive probability before the process reaches the point a or b, respectively. This leads to the following three propositions for the times $t < D_E$, $t = D_E$ and $t = \infty$. The proofs use the same techniques as that of the assertions above, we only have to handle the sets $\{X_{D_E} = a\}$ and $\{X_{D_E} = b\}$ separately.

Proposition 3.12 *Let the process X have exactly two absorbing points a and b satisfying $-\infty < a < 0 < b < +\infty$. Then the following assertions are true:*
  (I) *It holds*

$$\mathbf{P}_0\Big(\int_0^t f(X_s)\, ds < +\infty, \ \forall t < D_E\Big) = 1$$

if and only if $f \in \mathbf{L}_{\mathrm{loc}}^1((a, b), m)$.
  (II) *It holds*

$$\mathbf{P}_0\Big(\int_0^t f(X_s)\, ds < +\infty, \ \forall t < D_E\Big) = 0$$

if and only if the following two conditions are satisfied:
  (i) $f \notin \mathbf{L}_{\mathrm{loc}}^1((a, 0], m)$,
  (ii) $f \notin \mathbf{L}_{\mathrm{loc}}^1([0, b), m)$.

Proposition 3.13 *Let the process X have exactly two absorbing points a and b satisfying $-\infty < a < 0 < b < +\infty$. Then the following assertions are true:*

(I) *It holds*

$$\mathbf{P}_0\Big(\int_0^{D_E} f(X_s)\, ds < +\infty\Big) = 1$$

if and only if the following conditions are satisfied:
   (i) $f \in \mathbf{L}^1_{\mathrm{loc}}((a,b), m),$
   (ii) $f \in \mathbf{L}^1_{\mathrm{loc}}(a+, \nu_a),$
   (iii) $f \in \mathbf{L}^1_{\mathrm{loc}}(b-, \nu_b).$

(II) *It holds*

$$\mathbf{P}_0\Big(\int_0^{D_E} f(X_s)\, ds = +\infty\Big) = 1$$

if and only if the following conditions are satisfied:
   (i) $f \notin \mathbf{L}^1_{\mathrm{loc}}((a,0], m) \cap \mathbf{L}^1_{\mathrm{loc}}(a+, \nu_a),$
   (ii) $f \notin \mathbf{L}^1_{\mathrm{loc}}([0,b), m) \cap \mathbf{L}^1_{\mathrm{loc}}(b-, \nu_b).$

Remarks 3.14 (1) The conditions (i) – (iii) of Proposition 3.13 (I) can be written equivalently in form of the following two conditions:
   (i′) $f \in \mathbf{L}^1_{\mathrm{loc}}((a,0], m) \cap \mathbf{L}^1_{\mathrm{loc}}(a+, \nu_a),$
   (ii′) $f \in \mathbf{L}^1_{\mathrm{loc}}([0,b), m) \cap \mathbf{L}^1_{\mathrm{loc}}(b-, \nu_b).$

   (2) For any nonnegative Borel function f, we have

$$f \in \mathbf{L}^1_{\mathrm{loc}}((a,0], m) \cap \mathbf{L}^1_{\mathrm{loc}}(a+, \nu_a)$$

if and only if there exists $\varepsilon > 0$ such that

$$\int_a^\varepsilon |y - a|\, f(y)\, m(dy) < +\infty.$$

   (3) Analogously, for any nonnegative Borel function f, we have

$$f \in \mathbf{L}^1_{\mathrm{loc}}([0,b), m) \cap \mathbf{L}^1_{\mathrm{loc}}(b-, \nu_b)$$

if and only if there exists $\varepsilon > 0$ such that

$$\int_{-\varepsilon}^b |y - b|\, f(y)\, m(dy) < +\infty.$$

Proposition 3.15 *Let the process X have exactly two absorbing points a and b satisfying $-\infty < a < 0 < b < +\infty$. Then the following assertions are true:*
   (I) *Let $1 \in \mathbf{L}^1_{\mathrm{loc}}(a+, \nu_a) \cap \mathbf{L}^1_{\mathrm{loc}}(b-, \nu_b)$ be fulfilled. Then it holds*

$$\mathbf{P}_0\Big(\int_0^\infty f(X_s)\, ds < +\infty\Big) = 1$$

if and only if the following conditions are satisfied:
   (i) $f \in \mathbf{L}^1_{\mathrm{loc}}((a,0], m) \cap \mathbf{L}^1_{\mathrm{loc}}(a+, \nu_a),$
   (ii) $f \in \mathbf{L}^1_{\mathrm{loc}}([0,b), m) \cap \mathbf{L}^1_{\mathrm{loc}}(b-, \nu_b),$

(iii) $f(a) = f(b) = 0$.

It holds

$$\mathbf{P}_0\left(\int_0^\infty f(X_s)\, ds = +\infty\right) = 1$$

if and only if the following conditions are satisfied:

    (i) $f \in \mathbf{L}^1_{\text{loc}}((a, 0], m) \cap \mathbf{L}^1_{\text{loc}}(a+, \nu_a) \Longrightarrow f(a) > 0$,

    (ii) $f \in \mathbf{L}^1_{\text{loc}}([0, b), m) \cap \mathbf{L}^1_{\text{loc}}(b-, \nu_b) \Longrightarrow f(b) > 0$.

  (II) *Let* $1 \notin \mathbf{L}^1_{\text{loc}}(a+, \nu_a) \cup \mathbf{L}^1_{\text{loc}}(b-, \nu_b)$ *be fulfilled. Then it holds*

$$\mathbf{P}_0\left(\int_0^\infty f(X_s)\, ds < +\infty\right) = 1$$

if and only if the following conditions are satisfied:

    (i) $f \in \mathbf{L}^1_{\text{loc}}((a, 0], m) \cap \mathbf{L}^1_{\text{loc}}(a+, \nu_a)$,

    (ii) $f \in \mathbf{L}^1_{\text{loc}}([0, b), m) \cap \mathbf{L}^1_{\text{loc}}(b-, \nu_b)$.

It holds

$$\mathbf{P}_0\left(\int_0^\infty f(X_s)\, ds = +\infty\right) = 1$$

if and only if the following conditions are satisfied:

    (i) $f \notin \mathbf{L}^1_{\text{loc}}((a, 0], m) \cap \mathbf{L}^1_{\text{loc}}(a+, \nu_a)$,

    (ii) $f \notin \mathbf{L}^1_{\text{loc}}([0, b), m) \cap \mathbf{L}^1_{\text{loc}}(b-, \nu_b)$.

  (III) *Let* $1 \in \mathbf{L}^1_{\text{loc}}(a+, \nu_a)$ *and* $1 \notin \mathbf{L}^1_{\text{loc}}(b-, \nu_b)$ *be fulfilled. Then it holds*

$$\mathbf{P}_0\left(\int_0^\infty f(X_s)\, ds < +\infty\right) = 1$$

if and only if the following conditions are satisfied:

    (i) $f \in \mathbf{L}^1_{\text{loc}}((a, 0], m) \cap \mathbf{L}^1_{\text{loc}}(a+, \nu_a)$,

    (ii) $f \in \mathbf{L}^1_{\text{loc}}([0, b), m) \cap \mathbf{L}^1_{\text{loc}}(b-, \nu_b)$,

    (iii) $f(a) = 0$.

It holds

$$\mathbf{P}_0\left(\int_0^\infty f(X_s)\, ds = +\infty\right) = 1$$

if and only if the following conditions are satisfied:

    (i) $f \in \mathbf{L}^1_{\text{loc}}((a, 0], m) \cap \mathbf{L}^1_{\text{loc}}(a+, \nu_a) \Longrightarrow f(a) > 0$,

    (ii) $f \notin \mathbf{L}^1_{\text{loc}}([0, b), m) \cap \mathbf{L}^1_{\text{loc}}(b-, \nu_b)$.

  (IV) *Let* $1 \notin \mathbf{L}^1_{\text{loc}}(a+, \nu_a)$ *and* $1 \in \mathbf{L}^1_{\text{loc}}(b-, \nu_b)$ *be fulfilled. Then it holds*

$$\mathbf{P}_0\left(\int_0^\infty f(X_s)\, ds < +\infty\right) = 1$$

if and only if the following conditions are satisfied:

    (i) $f \in \mathbf{L}^1_{\text{loc}}((a, 0], m) \cap \mathbf{L}^1_{\text{loc}}(a+, \nu_a)$,

    (ii) $f \in \mathbf{L}^1_{\text{loc}}([0, b), m) \cap \mathbf{L}^1_{\text{loc}}(b-, \nu_b)$,

    (iii) $f(b) = 0$.

It holds

$$\mathbf{P}_0\left(\int_0^\infty f(X_s)\,ds = +\infty\right) = 1$$

if and only if the following conditions are satisfied:
   (i) $f \notin \mathbf{L}_{\mathrm{loc}}^1((a, 0], m) \cap \mathbf{L}_{\mathrm{loc}}^1(a+, \nu_a)$,
   (ii) $f \in \mathbf{L}_{\mathrm{loc}}^1([0, b), m) \cap \mathbf{L}_{\mathrm{loc}}^1(b-, \nu_b) \Longrightarrow f(b) > 0$.

Summarizing the results of the present paper, we have given purely analytical conditions on a nonnegative Borel function f for the convergence or divergence of the integral functionals $\int_0^t f(X_s)\,ds$ for $t \geq 0$, $\int_0^{D_E} f(X_s)\,ds$, and $\int_0^\infty f(X_s)\,ds$ for arbitrary strong Markov continuous local martingales X.

We remark that the above results can also be extended to (one-dimensional) strong Markov continuous processes transforming them into strong Markov continuous local martingales by the method of space transformation, the results of the present paper being a first step for this purpose. However, we will not deal with these problems in the present paper. It will be the subject of a forthcoming paper which will be published elsewhere.

REFERENCES

[1] S. Assing, *Homogene stochastische Differentialgleichungen mit gewöhnlicher Drift*, Promotionsschrift, Friedrich-Schiller-Universität Jena, 1994.

[2] S. Assing, W. Schmidt, *Continuous strong Markov processes in dimension one: A stochastic calculus approach*, Lecture Notes in Math. **1688**, Springer-Verlag, Berlin, 1998.

[3] S. Assing, T. Senf, On stochastic differential equations without drift, *Stochastics Stochastics Rep.* **36** (1991), 21–39.

[4] H.-J. Engelbert, J. Hess, Stochastic integrals of continuous local martingales, I, *Math. Nachr.* **97** (1990), 325–343.

[5] H.-J. Engelbert, W. Schmidt, Strong Markov continuous local martingales and solutions of one-dimensional stochastic differential equations (Part I–III), *Math. Nachr.* **143** (1989), 167–184; **144** (1989), 241–281; **151** (1991), 149–197.

[6] H.-J. Engelbert, W. Schmidt, On the behaviour of certain functionals of the Wiener process and applications to stochastic differential equations, *Stochastic Differential Systems*, Proceedings of the IFIP-WG 7/1 Working Conference Visegrád, Hungary, September 15–20, 1980, Lecture Notes in Control and Information Sciences **36**, 47–55, Springer-Verlag, Berlin, 1981.

[7] T. Jeulin, *Semi-martingales et grossissement d'une filtration*, Lecture Notes in Math. **833**, Springer-Verlag, Berlin, 1980.

[8] I. Karatzas, S.E. Shreve, *Brownian motion and stochastic calculus*, 2nd ed., Graduate Texts in Mathematics **113**, Springer-Verlag, 1991.

[9] J.W. Pitman, M. Yor, Some divergent integrals of Brownian motion, *Adv. Appl. Probab.*, Spec. Suppl. (1986), 109-116.

[10] D. Revuz, M. Yor, *Continuous martingales and Brownian motion*, Grundlehren der mathematischen Wissenschaften **293**, Springer-Verlag, 1991.

[11] X.-X. Xue, A zero-one law for integral functionals of the Bessel process, Séminaire de Probab. XXIV 1988/89, 137–153, Lecture Notes in Math. **1426**, Springer-Verlag, Berlin, 1990.

ON THE APPROXIMATION OF STOCHASTIC INTEGRALS AND WEIGHTED BMO

STEFAN GEISS

University of Jyväskylä
Department of Mathematics
FIN-40351 Jyväskylä, Finland

Abstract We approximate certain stochastic integrals with respect to the geometric Brownian motion by stochastic integrals over simple integrands, where deterministic time nets are used and the approximation error is measured with respect to a weighted BMO-space.

Key Words Geometric Brownian motion, stochastic integral, approximation, weighted BMO.

1. INTRODUCTION

The question of the best possible L_2-approximation of stochastic integrals by integrals over simple integrands is strongly motivated by Stochastic Finance if one wishes to replace continuously adjusted portfolios by discretely adjusted ones. So corresponding approximation rates for deterministic time nets are studied in [11], [6], [4], and [10], whereas random time nets are treated in [7]. For the European Call Option in the Black–Scholes model it is shown in [5] that replacing the L_2-estimates by much more restrictive BMO_2^S-estimates, where $S = (S_t)_{t \in [0,T]}$ is the price process, one obtains the same optimal approximation rate if the discretely adjusted portfolios are based on deterministic time nets.

In this note we consider a certain wider class of pay-off functions than the pay-off function of the European Call Option and characterize in Theorem 3.4 those pay-off functions which allow the same upper bounds for the approximation with respect to BMO_2^S as the European Call Option. Moreover, we give additional information for the risk process in Proposition 2.1.

Let us fix a probability space $(\Omega, \mathcal{F}, \mathbb{P})$ and a right-continuous filtration $(\mathcal{F}_t)_{t \in [0,T]}$ such that $\mathcal{F}_T = \mathcal{F}$ and $\mathcal{F}_0$ contains all $\mathcal{F}$ null sets, which is obtained in the usual way by a standard Brownian motion $W = (W_t)_{t \in [0,T]}$ with $W_0 \equiv 0$ (for convenience we assume that all paths are continuous). As price process we take the geometric Brownian motion $S = (S_t)_{t \in [0,T]}$ defined by $S_t = \exp\left(W_t - \frac{t}{2}\right)$. We consider Borel-measurable $f : (0, \infty) \to [0, \infty)$ such that $f(S_T) \in L_2(\Omega, \mathcal{F}, \mathbb{P})$. It is known that

$$\mathbb{E}\left(f(S_T)|\mathcal{F}_t\right) = \mathbb{E}f(S_T) + \int_{(0,t]} \varphi(u, S_u) dS_u \quad \text{for} \quad t \in [0, T] \text{ a.s.,} \qquad (1.1)$$

where one can take as continuous modification of $(\mathbb{E}(f(S_T)|\mathcal{F}_t))_{t \in [0,T]}$ the process $(F(t, S_t))_{t \in [0,T]}$ with

$$F(u, y) := \mathbb{E}f\left(yS_{T-u}\right) \quad \text{and} \quad \varphi(u, y) := \frac{\partial F}{\partial y}(u, y)$$

for $(u, y) \in [0, T) \times (0, \infty)$, and $\varphi(T, y) := 0$. Note that $f(S_T) \in L_2(\Omega, \mathcal{F}, \mathbb{P})$ implies that there is some $\varepsilon > 0$ such that

$$F \in C^{\infty, \infty}\left((-\varepsilon, T) \times (0, \infty)\right) \quad \text{with} \quad \frac{\partial F}{\partial t} + \frac{y^2}{2}\frac{\partial^2 F}{\partial y^2} = 0; \qquad (1.2)$$

cf. [4].

2. THE RISK PROCESS $R(\tau)$

Given a deterministic time net $\tau = (t_i)_{i=0}^n$ with $0 = t_0 < t_1 < \cdots < t_n = T$ we shall approximate $f(S_T)$ by

$$\mathbb{E}f(S_T) + \sum_{i=1}^n \varphi(t_{i-1}, S_{t_{i-1}})\left(S_{t_i} - S_{t_{i-1}}\right).$$

The corresponding error process is given by

$$C_t(\tau) := \int_{(0,t]}\left(\varphi(u, S_u) - \sum_{i=1}^n \chi_{(t_{i-1}, t_i]}(u)\varphi\left(t_{i-1}, S_{t_{i-1}}\right)\right) dS_u$$

for $t \in [0, T]$, where we can assume that all paths of $C(\tau) := (C_t(\tau))_{t \in [0,T]}$ are continuous and $C_0(\tau) \equiv 0$. Furthermore, we introduce the risk process $R(\tau) := (R_t(\tau))_{t \in [0,T]}$ to be

$$R_t(\tau) := \mathbb{E}\left(|C_T(\tau) - C_t(\tau)|^2 \mid \mathcal{F}_t\right) \quad \text{for} \quad t \in [0, T].$$

The following general decomposition of the risk process $R(\tau)$ extends the consideration for the European Call Option from [5] to the maybe right and natural form: The decomposition below shows that, in general, the risk process consists of two *non-negative* parts: Firstly the part $X(\tau)$, which depends on φ and vanishes at each time-knot t_i of the net τ. Secondly the part $Y(\tau)$, which depends on $\frac{\partial \varphi}{\partial y}$ and which is related to the function $H(u)$ from [4] used to compute the global L_2-approximation error there. To formulate the decomposition we let $M = (M_t)_{t \in [0,T)}$, throughout this note, be given by

$$M_t := S_t^2 \frac{\partial \varphi}{\partial y}(t, S_t).$$

Furthermore, for a net $\tau = (t_i)_{i=0}^n \subset \mathbb{R}, 0 = t_0 \leq t_1 \leq \cdots \leq t_n = T$, we use

$$\|\tau\|_\infty := \sup_{1 \leq i \leq n} |t_i - t_{i-1}|.$$

Finally, given $A, B \geq 0$ and $c > 0$, $A \sim_c B$ is standing for $\frac{1}{c} A \leq B \leq cA$.

Proposition 2.1 (i) $(M_t)_{t \in [0,T)}$ *is a martingale.*
(ii) $(M_t)_{t \in [0,T)} \subseteq L_2(\Omega, \mathcal{F}, \mathbb{P})$ *and* $\int_{[0,T)} (T - t) \, (\mathbb{E} M_t^2) \, dt < \infty$.
(iii) *For a deterministic time net* $\tau = (t_i)_{i=0}^n$ *with* $0 = t_0 < t_1 < \cdots < t_n = T$ *and for* $0 \leq a < T$ *with* $t_{i_0-1} \leq a < t_{i_0}$ *one has*

$$R_a(\tau) \sim_{c(\|\tau\|_\infty)} X_a(\tau) + Y_a(\tau) \quad a.s.$$

with

$$X_a(\tau) := (t_{i_0} - a) S_a^2 \left(\varphi(a, S_a) - \varphi \left(t_{i_0-1}, S_{t_{i_0-1}} \right) \right)^2,$$

$$Y_a(\tau) := \mathbb{E} \left(\int_{(a,T]} \left(\left(\sum_{i=1}^n t_i \chi_{(t_{i-1}, t_i]}(u) \right) - u \right) M_u^2 du \mid \mathcal{F}_a \right)$$

where $c : [0, T] \to [1, \infty)$ *depends at most on* T, *is increasing, and satisfies the condition* $\lim_{\varepsilon \downarrow 0} c(\varepsilon) = 1$ *and where* $M_T :\equiv 0$.

Proof. (ii) was shown in [4]. To prove (i) we first observe that

$$\frac{\partial}{\partial t} \left(y^2 \frac{\partial \varphi}{\partial y}(t, y) \right) + \frac{y^2}{2} \frac{\partial^2}{\partial y^2} \left(y^2 \frac{\partial \varphi}{\partial y}(t, y) \right) = 0$$

on $(-\varepsilon, T) \times (0, \infty)$ as a consequence of equation (1.2). Itô's formula gives

$$M_t = M_0 + \int_0^t H_u dS_u \quad \text{for} \quad t \in [0, T) \text{ a.s.}$$

with

$$H_u := \frac{\partial}{\partial y} \left(y^2 \frac{\partial \varphi}{\partial y}(u, y) \right) |_{y=S_u}.$$

Given a sequence $(u_n)_{n=1}^\infty$, $u_n \in [0, T)$, with $\sup_n u_n < T$, it is shown in [4] that $\mathbb{E} \sup_n |M_{u_n}|^2 < \infty$ (in [4] it was assumed that $(u_n)_{n=1}^\infty$ is converging, but the argument remains true for $\sup_n u_n < T$). Hence, by the continuity of the paths,

$$\mathbb{E} \sup_{u \in [0,t]} |M_u|^2 < \infty \quad \text{whenever} \quad t \in [0, T)$$

so that $(M_t)_{t\in[0,T)}$ is a martingale.

(iii) Because of

$$C_T(\tau) - C_a(\tau) = \int_{(a,t_{i_0}]} \left(\varphi(u, S_u) - \varphi\left(t_{i_0-1}, S_{t_{i_0-1}}\right)\right) dS_u$$

$$+ \sum_{i=i_0+1}^{n} \int_{(t_{i-1},t_i]} \left(\varphi(u, S_u) - \varphi\left(t_{i-1}, S_{t_{i-1}}\right)\right) dS_u \quad \text{a.s.}$$

we obtain

$$\mathbb{E}\left(|C_T(\tau) - C_a(\tau)|^2 \mid \mathcal{F}_a\right)$$

$$= \mathbb{E}\left(\int_{(a,t_{i_0}]} \left|\varphi(u, S_u) - \varphi\left(t_{i_0-1}, S_{t_{i_0-1}}\right)\right|^2 S_u^2 du \mid \mathcal{F}_a\right)$$

$$+ \sum_{i=i_0+1}^{n} \mathbb{E}\left(\int_{(t_{i-1},t_i]} \left|\varphi(u, S_u) - \varphi\left(t_{i-1}, S_{t_{i-1}}\right)\right|^2 S_u^2 du \mid \mathcal{F}_a\right) \quad \text{a.s.}$$

Given $0 \leq a_0 \leq a \leq b \leq T$, from [5] it follows that

$$\mathbb{E}\left(\int_{(a,b]} |\varphi(t, S_t) - \varphi(a_0, S_{a_0})|^2 S_t^2 dt \mid \mathcal{F}_a\right)$$

$$\sim_{c(b-a)} (b-a)\left(\varphi(a, S_a) - \varphi(a_0, S_{a_0})\right)^2 S_a^2$$

$$+ \mathbb{E}\left(\int_{(a,b]} (b-u)\left(S_u^2 \frac{\partial\varphi}{\partial y}(u, S_u)\right)^2 du \mid \mathcal{F}_a\right) \quad \text{a.s.}$$

where $c : [0, T] \to [1, \infty)$ is a function satisfying the desired properties of our assertion. Hence

$$\mathbb{E}\left(|C_T(\tau) - C_a(\tau)|^2 \mid \mathcal{F}_a\right)$$

$$\sim_{c(\|\tau\|_\infty)} (t_{i_0} - a)\left(\varphi(a, S_a) - \varphi\left(t_{i_0-1}, S_{t_{i_0-1}}\right)\right)^2 S_a^2$$

$$+ \mathbb{E}\left(\int_{(a,t_{i_0}]} (t_{i_0} - u)\left(S_u^2 \frac{\partial\varphi}{\partial y}(u, S_u)\right)^2 du \mid \mathcal{F}_a\right)$$

$$+ \sum_{i=i_0+1}^{n} \mathbb{E}\left(\int_{(t_{i-1},t_i]} (t_i - u)\left(S_u^2 \frac{\partial\varphi}{\partial y}(u, S_u)\right)^2 du \mid \mathcal{F}_a\right) \quad \text{a.s.}$$

$\square$

Remark 2.2 One consequence of the martingale property of $(M_t)_{t\in[0,T)}$ is that the function $u \to (\mathbb{E}M_u^2)^{1/2}$, which is the main tool in [4] in order to compute the L_2-approximation rates for the stochastic integrals, is non-decreasing in u.

3. THE RESULT

As pay-off functions we are going to use a class of continuous non-decreasing functions whose (right-hand side) derivative is monotone.

Definition 3.1 A Borel measurable function $f : (0, \infty) \to [0, \infty)$ belongs to the class C provided that $f(S_T) \in L_2(\Omega, \mathcal{F}, \mathbb{P})$ and $f(y) = \int_0^y K(x)dx$ for a non-constant Borel-function $K : (0, \infty) \to [0, \infty)$ which is either increasing or decreasing on $(0, \infty)$ (not necessarily strictly).

Remark 3.2 Obviously the pay-off function $f(y) = (y - K)^+$, $K > 0$, of the European Call Option belongs to C by taking $K(x) = \chi_{[K, \infty)}(x)$.

Now we define the weighted BMO-spaces we shall use to measure the approximation error.

Definition 3.3 Assume that $\Phi = (\Phi_t)_{t \in [0,T]}$ is $(\mathcal{F}_t)_{t \in [0,T]}$-adapted and has right-continuous paths with left-hand side limits for all $\omega \in \Omega$ and that $\Phi_t(\omega) > 0$ for all $t \in [0, T]$ and $\omega \in \Omega$. Then, given $C(\tau) = (C_t(\tau))_{t \in [0,T]}$ as before, we let $\|C(\tau)\|_{\mathrm{BMO}_2^\Phi} := \inf c$, where the infimum is taken over all $c > 0$ such that

$$\mathbb{E}\left(\left| \frac{C_T(\tau) - C_\sigma(\tau)}{\Phi_\sigma} \right|^2 \Big| \mathcal{F}_\sigma \right) \leq c^2 \text{ a.s.}$$

for all stopping times $\sigma : \Omega \to [0, T]$. In the case there is no such $c > 0$, then we let $\|C(\tau)\|_{\mathrm{BMO}_2^\Phi} := \infty$.

There is an extensive list of literature concerning weighted BMO with [8] as one of the starting points. Behind the probabilistic approach we are going to exploit here stands the Garsia-Neveu lemma (see for example [3], [9], [2], [1]). Replacing the exponent 2 in our definition by 1 one gets a more standard form of weighted BMO. However, under certain regularity assumptions on Φ, which are not discussed here, one ends up with the same weighted BMO. We have taken the exponent 2 since the basic upper BMO-estimate, which is coming from [5], is a local L_2-estimate rather than an L_1-estimate. Let us turn now to the main result of this note.

Theorem 3.4 *Assume that* $f \in C$ *with kernel* K *and* $\Phi_t = \psi(S_t)$ *for* $t \in [0, T]$ *with* $\psi : (0, \infty) \to (0, \infty)$ *being continuous and increasing. Then the following assertions are equivalent:*

(i) There exist a deterministic net $\tau = (t_i)_{i=0}^n$, $0 = t_0 < t_1 < \cdots < t_n = T$, $n \geq 2$, *an* $a \in [0, T] \backslash \{t_0, ..., t_n\}$, *and a constant* $c_1 > 0$ *such that*

$$X_a(\tau) \leq c_1^2 \, \psi(S_a)^2 \text{ a.s.}$$

where $X_a(\tau)$ *is defined in Proposition 2.1.*

(ii) *There exists a constant $c_2 > 0$ such that*

$$\|C(\tau)\|_{\mathrm{BMO}_2^{\Phi}} \leq c_2 \sqrt{\|\tau\|_{\infty}}$$

for all deterministic $\tau = (t_i)_{i=0}^n$ with $0 = t_0 < t_1 < \cdots < t_n = T$.
(iii) $\sup_{x>0} K(x) < \infty$ *and* $\inf_{y>0} \frac{\psi(y)}{y} > 0.$

Let us comment the above result. The implication (iii) $\Rightarrow$ (ii) is a natural extension of the corresponding result proved in [5] for the European Call Option. Here we would like to point out (i) $\Rightarrow$ (iii): Even if one weakens the BMO–estimate drastically by replacing condition (ii) by condition (i), only functions f generated by bounded K give rise to condition (i). Some consequences of the above BMO-estimate are discussed in [5].

Lemma 3.5 *For $f \in C$ with kernel K and $t \in [0, T)$ one has*
  (i) $\varphi(t, y) = \mathbb{E}\left(K(yS_{T-t})S_{T-t}\right)$ *for $y > 0$,*
  (ii) $\sup_{y>0} \varphi(t, y) = \sup_{x>0} K(x),$
  (iii) $\inf_{y>0} \varphi(t, y) = \inf_{x>0} K(x),$
where both sides of (ii) *may be infinite.*

Proof. (i) follows by a standard computation. The monotonicity of K implies that

$$\sup_{y>0} \varphi(t, y) = \mathbb{E}\left(\sup_{y>0} K(yS_{T-t})S_{T-t}\right) = \sup_{x>0} K(x)$$

and (ii). The same argument applies to (iii).   □

Proof of Theorem 3.4. (ii) $\Rightarrow$ (i) is clear.
  (i) $\Rightarrow$ (iii). Assume $0 < t_{i_0-1} < a < t_{i_0}$ and

$$(t_{i_0} - a) S_a^2 \left(\varphi(a, S_a) - \varphi\left(t_{i_0-1}, S_{t_{i_0-1}}\right)\right)^2 = X_a(\tau) \leq c_1^2 \psi(S_a)^2$$

a.s. Then

$$(t_{i_0} - a) y_a^2 \left(\varphi(a, y_a) - \varphi\left(t_{i_0-1}, y_{t_{i_0-1}}\right)\right)^2 \leq c_1^2 \psi(y_a)^2$$

for all $y_a, y_{t_{i_0-1}} > 0$ and

$$\sup_{y>0} |\varphi\left(t_{i_0-1}, y\right)| < \infty$$

so that the first part of (iii) follows as a consequence of Lemma 3.5. For the second part we consider $y_{t_{i_0-1}} \downarrow 0$ and $y_{t_{i_0-1}} \uparrow \infty$ and conclude, again by Lemma 3.5, that

$$c_1^2 \geq (t_{i_0} - a) \frac{y_a^2}{\psi(y_a)^2} \left(\varphi(a, y_a) - \sup_{x>0} K(x)\right)^2$$

and

$$c_1^2 \geq (t_{i_0} - a) \frac{y_a^2}{\psi(y_a)^2} \left(\varphi(a, y_a) - \inf_{x>0} K(x)\right)^2$$

for $y_a > 0$. Hence

$$4 c_1^2 \geq (t_{i_0} - a) \frac{y_a^2}{\psi(y_a)^2} \left(\sup_{x>0} K(x) - \inf_{x>0} K(x) \right)^2$$

for $y_a > 0$, which implies $\inf_{y>0} \frac{\psi(y)}{y} > 0$ (note that $\inf_x K(x) < \sup_x K(x)$ by assumption).

(iii) $\Rightarrow$ (ii). First we observe that it is sufficient to consider the case that K is increasing. Assume K to be decreasing we let

$$K_0(x) := \sup_{z>0} K(z) - K(x)$$

and obtain an increasing K_0. Since

$$f_0(y) = \int_0^y K_0(x) dx = cy - f(y) \quad \text{with} \quad c := \sup_{z>0} K(z)$$

we obtain for the corresponding error processes

$$C_t^{(f_0)} = -C_t^{(f)} \quad \text{for} \ t \in [0, T] \ \text{a.s.}$$

So let us assume that K is increasing and that $\sup_{x>0} K(x) < c < \infty$. For $m = 1, 2, ...$ we find increasing Borel functions $K^{(m)} : (0, \infty) \to [0, \infty)$ taking only finitely many values such that

$$0 \leq K^{(m)}(x) \leq K(x) \leq K^{(m)}(x) + \frac{c}{m} \quad \text{for} \ \ x > 0.$$

Lemma 3.5 implies that

$$0 \leq \varphi^{(m)}(t, y) \leq \varphi(t, y) \leq \varphi^{(m)}(t, y) + \frac{c}{m}, \ y > 0, \ t \in [0, T) \qquad (3.1)$$

where φ and $\varphi^{(m)}$ are the integrands from equation (1.1) for

$$f(y) = \int_0^y K(x) dx \quad \text{and} \quad f^{(m)}(y) := \int_0^y K^{(m)}(x) dx,$$

respectively. Since

$$f^{(m)}(y) = \sum_{i=1}^{N_m} \theta_i^{(m)} (y - K_i^{(m)})^+$$

for some $\theta_i^{(m)} \geq 0$ and $K_i^{(m)} \geq 0$ with $\sum_{i=1}^{N_m} \theta_i^{(m)} \leq c$, from [5] one gets that

$$\left\| C^{(m)}(\tau) \right\|_{\text{BMO}_2^S} \leq \left(\sum_{i=1}^{N_m} \theta_i^{(m)} \right) c_T \sqrt{\|\tau\|_\infty} \leq c \, c_T \sqrt{\|\tau\|_\infty}, \qquad (3.2)$$

where $C^{(m)}(\tau)$ is the error process of $f^{(m)}$, since the multiplicative constant $c_T > 0$ in [5] for the upper BMO-estimate for the European Call Option does not depend on the strike price. On the other hand, equation (3.1) gives

$$\left| \varphi(u, S_u) - \varphi^{(m)}(u, S_u) - \sum_{i=1}^{n} \chi_{(t_{i-1}, t_i]}(u) \left(\varphi(t_{i-1}, S_{t_{i-1}}) - \varphi^{(m)}(t_{i-1}, S_{t_{i-1}}) \right) \right| \leq \frac{c}{m}$$

for $u \in [0, T)$, so that

$$\left\| C(\tau) - C^{(m)}(\tau) \right\|_{\text{BMO}_2^S} \leq d_T \frac{c}{m} \tag{3.3}$$

by a standard computation, where $d_T > 0$ depends on T only. Combining equations (3.2) and (3.3) yields

$$c_\psi \left\| C(\tau) \right\|_{\text{BMO}_2^\Phi} \leq \left\| C(\tau) \right\|_{\text{BMO}_2^S} \leq c\, c_T \sqrt{\|\tau\|_\infty} + d_T \frac{c}{m}$$

with $c_\psi := \inf_{y>0} \frac{\psi(y)}{y} > 0$ so that by $m \to \infty$ the desired result follows. $\square$

REFERENCES

[1] N.L. Bassily and J. Mogyoródi, On the $\mathcal{K}_\Phi$-spaces with general Young function φ, *Ann. Univ. Sci. Budap. Rolando Eötvös, Sect. Math.*, **27** (1984), 205–214.

[2] C. Dellacherie and P.-A. Meyer, *Probabilities and Potential B*, Mathematics Studies 72, North–Holland, 1982.

[3] A.M. Garsia, *Martingale Inequalities*, Seminar Notes on Recent Progress, Benjamin, Reading, 1973.

[4] S. Geiss, Quantitative approximation of certain stochastic integrals, 2000.

[5] S. Geiss, Weighted BMO and discrete time hedging within the Black–Scholes model, 2000.

[6] E. Gobet and E. Temam, Discrete time hedging errors for options with irregular payoffs, *Preprint 177 of Ecole Nationale des Ponts et Chaussées*, Paris, July 1999.

[7] C. Martini and C. Patry, Variance optimal hedging in the Black–Scholes model for a given number of transactions, *Preprint 3767 of INRIA*, Paris, September 1999.

[8] B. Muckenhoupt and R.L. Wheeden, Weighted bounded mean oscillation and the Hilbert transform, *Studia Math.* **54** (3) (1976), 221–237.

[9] J. Neveu, *Martingales a temps discret*, Masson CIE, 1972.

[10] E. Temam, Rate of convergence of the discrete time hedging strategy in a complete multi-dimensional model, *Preprint 190 of Ecole Nationale des Ponts et Chaussées*, Paris, March 2000.

[11] R. Zhang, *Couverture approchée des options Européennes*, PhD thesis, Ecole Nationale des Ponts et Chaussées, Paris, 1998.

MINIMAL DISTANCE MARTINGALE MEASURES AND OPTIMAL PORTFOLIOS CONSISTENT WITH OBSERVED MARKET PRICES

THOMAS GOLL and LUDGER RÜSCHENDORF

Universität Freiburg
Institut für Mathematische Stochastik
D-79104 Freiburg i. Br., Germany

Abstract In this paper we study derivative pricing with information on observed market prices of some derivatives. To this purpose we restrict the class of martingale measures to those which yield derivative prices which are consistent with the observed market prices. As pricing measure we propose the minimal distance martingale measure consistent with the observed market prices. We characterize this pricing measure with respect to f-divergence distances and obtain as a result some necessary and some sufficient conditions for utility-optimal portfolios of the underlyings and the derivatives with observed market prices.

Key Words Derivative pricing, f-divergences, observed market prices, portfolio optimization.

1. INTRODUCTION

A common approach to derivative pricing in incomplete markets is to base the prices on minimal distance martingale measures with respect to certain distances like the L^2-distance (see [26] and [6]), the Hellinger distance (see [19]), the entropy distance (see [9]) and others.

This approach to derivative pricing takes into consideration the probabilistic model of the future behaviour of the underlyings, but not the information on derivative prices observed in the market. In order to include this information in the model we consider only those martingale measures which yield derivative prices consistent with the observed market prices. For derivative pricing we propose the minimal distance martingale measure consistent with observed market prices. This notion is equivalent to the least favourable consistent pricing measure which is proposed independently of this paper in [18]. Related ideas of consistent derivative pricing are studied by Kallsen (see [16] and [17]) and Avellaneda (see [2]). In [16] and [17] derivative prices are considered which are derived from utility maximization and which are consistent with given demand vectors in some derivatives of the market. In [2] probability measures are characterized which minimize the relative entropy distance of the pricing measure to the class of all probability measures consistent with the observed market prices. However, the calibrated pricing measure obtained in this way is not necessarily a martingale measure.

We consider the class of all f-divergence distances defined by strictly convex, differentiable functions f which includes the distances above and further examples (see [21]). We obtain some necessary and some sufficient conditions for projections of the underlying measure on the set of martingale measures consistent with the observed prices of a finite number of derivatives in a general semimartingale market model.

The paper is organized as follows. In Section 2 we recall a characterization of f-projections on classes of distributions determined by inequality constraints. Based on this result we obtain, in Section 3, some necessary and some sufficient conditions for minimal distance martingale measures under constraints. In Section 4 we recall the notion of a minimax measure with respect to concave utility functions and convex sets of probability measures. Minimax martingale measures are equivalent to minimal distance martingale measures with respect to f-divergence distances induced by the convex conjugate of the utility function. As a consequence the characterizations of minimal distance martingale measures consistent with observed market prices are closely related to the determination of optimal portfolios, if one additionally allows constant positions in the derivatives with observed market prices.

Acknowledgement The authors thank J. Kallsen for fruitful discussions. The authors also thank H.-J. Engelbert and the referee for their helpful remarks.

2. f-DIVERGENCES AND MINIMAL DISTANCE MEASURES

In the following we recall a characterization of projections with respect to f-divergence distances on classes of distributions determined by inequality constraints. For a detailed discussion of f-divergence distances we refer to [21] or [28].

Let $(\Omega, \mathcal{F}, P)$ be a probability space.

Definition 2.1 Let $Q \ll P$ and let $f : (0, \infty) \to \mathbb{R}$ be a convex function. Then the f-divergence distance between Q and P is defined as:

$$f(Q\|P) := \begin{cases} \int f(\frac{dQ}{dP})dP, & \text{if the integral exists} \\ \infty, & \text{else,} \end{cases}$$

where $f(0) = \lim_{x \downarrow 0} f(x)$.

Examples of f-divergence distances are the Kullback-Leibler or entropy distance for $f(x) = x \log x$, the total variation distance for $f(x) = |x - 1|$, the Hellinger distance for $f(x) = -\sqrt{x}$, the reverse relative entropy distance for $f(x) = -\log(x)$ and many others.

In the following we assume that f is a continuous, strictly convex and differentiable function. Let $\mathcal{K}$ be a convex set of probability measures on $(\Omega, \mathcal{F})$ dominated by P. A measure $Q^* \in \mathcal{K}$ is called an f-projection of P on $\mathcal{K}$ if

$$f(Q^*||P) = \inf_{Q \in \mathcal{K}} f(Q||P) =: f(\mathcal{K}||P).$$

Let F be a convex cone of real valued random variables on $(\Omega, \mathcal{F})$, i.e.,

$$\left\{ \sum_{i=1}^{k} \alpha_i f_i : \alpha_i \geq 0, f_i \in F \right\} = F,$$

and define the moment family determined by inequality constraints with respect to F as

$$\mathcal{K}_F := \{ Q \ll P : F \subset L^1(Q) \text{ and } E_Q f \geq 0 \text{ for all } f \in F \}.$$

The following result was given in [23], Theorem 2:

Theorem 2.2 *Let $Q^* \ll P$ satisfy $f(Q^*||P) < \infty$ and assume that $c := E_{Q^*} f'(\frac{dQ^*}{dP})$ is finite.*

(i) $Q^* \in \mathcal{K}$ *is the f-projection of P on $\mathcal{K}$ if and only if*

$$E_{Q^*} f'\left(\frac{dQ^*}{dP}\right) \leq E_Q f'\left(\frac{dQ^*}{dP}\right) \text{ for all } Q \in \mathcal{K} \text{ with } f(Q||P) < \infty.$$

(ii) *If $Q^* \in \mathcal{K}_F$ is the f-projection on $\mathcal{K}_F$, then*

$$f'\left(\frac{dQ^*}{dP}\right) - c \in \bar{F}, \text{ the } L^1(\Omega, \mathcal{F}, Q^*)\text{-closure of } F.$$

(iii) *If $Q^* \in \mathcal{K}_F$ such that $f'(\frac{dQ^*}{dP}) - c \in F$, then Q^* is the f-projection on $\mathcal{K}_F$.*

Proposition 8.5 in [21] yields the existence of an f-projection of P on a convex class $\mathcal{K}$ under the assumptions that $\mathcal{K}$ is closed with respect to the variational distance and $\lim_{x \to \infty} \frac{f(x)}{x} = \infty$. By Corollary 2.3 in [11] an f-projection is equivalent to P if $f'(0) = -\infty$ and if there exists a measure $Q \in \mathcal{K}$ with $Q \sim P$ and with a finite distance $f(Q||P) < \infty$.

3. CHARACTERIZATION OF MINIMAL DISTANCE MARTINGALE MEASURES UNDER CONSTRAINTS

In the following we apply Theorem 2.2 to characterize f-projections on the set of martingale measures which fulfill some additional constraints. Our mathematical framework is as follows: $(\Omega, \mathcal{F}, (\mathcal{F}_t)_{0 \leq t \leq T}, P)$ is a filtered probability space, where the filtration is assumed to be right-continuous and $\mathcal{F} = \mathcal{F}_T$. Let S be a $\mathbb{R}^d$-valued semimartingale with deterministic S_0. Vector stochastic integrals are written as $\varphi \cdot S_t = \int_0^t \varphi_s dS_s$. (For the definition of a vector stochastic integral see [14].)

Let $\mathcal{M}$ ($\mathcal{M}_{\mathrm{loc}}$) be the set of P-absolutely continuous (local) martingale measures and $\mathcal{M}^e$ ($\mathcal{M}^e_{\mathrm{loc}}$) the subset of $\mathcal{M}$ ($\mathcal{M}_{\mathrm{loc}}$) consisting of probability measures which are equivalent to P. Let $H_1, \ldots, H_n$ be a finite set of $\mathcal{F}_T$-measurable random variables and $r \in \{0, \ldots, n\}$.

We define the set of martingale measures with constraints as

$$\widehat{\mathcal{M}} := \{Q \in \mathcal{M} : H_1, \ldots, H_n \in L^1(Q), \ E_Q H_i \geq 0 \text{ for } 1 \leq i \leq r$$
$$\text{and } E_Q H_i = 0 \text{ for } r + 1 \leq i \leq n\}.$$

The class $\widehat{\mathcal{M}}$ stands for the set of martingale measures consistent with some information on the prices of derivatives $H_1, \ldots, H_n$. Notice that price information of the form $E_Q H_i \in [q_i, p_i]$ can also be described by inequality constraints as in the definition of $\widehat{\mathcal{M}}$. The sets $\widehat{\mathcal{M}}_{\mathrm{loc}}, \widehat{\mathcal{M}}^e, \widehat{\mathcal{M}}^e_{\mathrm{loc}}$ are defined analogously to $\widehat{\mathcal{M}}$. We assume throughout that

$$\widehat{\mathcal{M}}^e \neq \varnothing.$$

For a $\mathbb{R}^d$-valued local martingale N the class $L^1_{\mathrm{loc}}(N)$ of predictable integrands is defined in [14]. For $Q \in \mathcal{M}_{\mathrm{loc}}$, by $L^1_{\mathrm{loc}}(S, Q)$ we denote the class of integrands $L^1_{\mathrm{loc}}(S)$ which is defined with respect to the measure Q. The following theorem gives a necessary condition for the f-projection of P on $\widehat{\mathcal{M}}$.

Theorem 3.1 *Let $Q^* \in \widehat{\mathcal{M}}$ satisfy $f(Q^* \| P) < \infty$. If Q^* is the f-projection of P on $\widehat{\mathcal{M}}$ and $c := E_{Q^*} f'(\frac{dQ^*}{dP})$ is finite, then*

$$f'\left(\frac{dQ^*}{dP}\right) = c + \varphi \cdot S_T + \sum_{i=1}^n \mu_i H_i \quad Q^*\text{-a.s.} \tag{3.1}$$

with $\mu_1, \ldots, \mu_n \in \mathbb{R}$ such that $\mu_i \geq 0$ and $\mu_i (E_{Q^*} H_i) = 0$ for $1 \leq i \leq r$, and with some process $\varphi \in L^1_{\mathrm{loc}}(S, Q^*)$ such that $\varphi \cdot S$ is a martingale under Q^*.

Proof. First we introduce a set G of random variables which determines $\widehat{\mathcal{M}}$ as a moment family. We define the set G' as

$$G' := \{\varphi \cdot S_T : \varphi^i = Y^i 1_{]s_i, t_i]}, s_i < t_i, Y^i \text{ bounded } \mathcal{F}_{s_i}\text{-measurable}\}.$$

Let G be the convex cone generated by the set

$$G' \cup \{H_i : 1 \leq i \leq n\} \cup \{-H_i : r + 1 \leq i \leq n\}.$$

Then we have the following characterization of $\widehat{\mathcal{M}}$:

$$\widehat{\mathcal{M}} = \{Q \ll P : G \subset L^1(Q) \text{ and } E_Q g \geq 0 \text{ for all } g \in G\}.$$

The necessary condition in Theorem 2.2 (ii) yields: $f'(\frac{dQ^*}{dP}) - c \in \bar{G}$. Corollary 2.5.2 in [29] (for a multidimensional version see Theorem 1.6 in [7]) implies that the $L^1(Q^*)$-closure of the vector space generated by G' is contained in

$$\{\varphi \cdot S_T : \varphi \in L^1_{\mathrm{loc}}(S, Q^*) \text{ such that } \varphi \cdot S \text{ is a } Q^*\text{-martingale}\}.$$

According to Proposition 1.1 in [13] this result is valid without the assumption of a complete filtration. Extending Proposition I.3.3 in [25] from vector spaces to the class of closed convex cones one gets

$$f'\left(\frac{dQ^*}{dP}\right) = c + \varphi \cdot S_T + \sum_{i=1}^{n} \mu_i H_i \quad Q^*\text{-a.s.,}$$

where $\mu_i \geq 0$ for $1 \leq i \leq r$. This implies by the definition of c that $\mu_i(E_{Q^*} H_i) = 0$ for $1 \leq i \leq r$. □

The following theorem is a variant of Theorem 3.1. It shows that the necessary condition in Theorem 3.1 is also valid for the set $\widehat{\mathcal{M}}_{\text{loc}}$ under the additional assumption that S is locally bounded.

Theorem 3.2 *Let S be locally bounded. Let $Q^* \in \widehat{\mathcal{M}}_{\text{loc}}$ satisfy $f(Q^*\|P) < \infty$. If Q^* is the f-projection of P on $\widehat{\mathcal{M}}_{\text{loc}}$ and $c := E_{Q^*} f'(\frac{dQ^*}{dP})$ is finite, then*

$$f'\left(\frac{dQ^*}{dP}\right) = c + \varphi \cdot S_T + \sum_{i=1}^{n} \mu_i H_i \quad Q^*\text{-a.s.} \tag{3.2}$$

with $\mu_1, \ldots, \mu_n \in \mathbb{R}$ such that $\mu_i \geq 0$ and $\mu_i(E_{Q^*} H_i) = 0$ for $1 \leq i \leq r$, and with some process $\varphi \in L^1_{\text{loc}}(S, Q^*)$ such that $\varphi \cdot S$ is a martingale under Q^*.

Proof. Let G_{loc} be the convex cone generated by

$$\{\varphi \cdot S_T : \varphi^i = Y^i 1_{]s_i, t_i]} 1_{[0, \widehat{T}^i]}, s_i < t_i, Y^i \text{ bounded } \mathcal{F}_{s_i}\text{-measurable}, \widehat{T}^i \in \gamma^i\}$$
$$\cup \{H_i : 1 \leq i \leq n\} \cup \{-H_i : r + 1 \leq i \leq n\},$$

where $\gamma^i := \{\widehat{T}^i \text{ stopping time} : (S^i)^{\widehat{T}^i} \text{ is bounded}\}$. Then the convex cone G_{loc} determines $\widehat{\mathcal{M}}_{\text{loc}}$ as a moment family

$$\widehat{\mathcal{M}}_{\text{loc}} = \{Q \ll P : \widehat{G}_{\text{loc}} \subset L^1(Q) \text{ and } E_Q g \geq 0 \quad \forall g \in G_{\text{loc}}\}.$$

The presentation (3.2) is then obtained as in the proof of Theorem 3.1. □

Remark Theorems 3.1 and 3.2 are generalizations of the corresponding results without constraints, see [11], [12], [20] and [24].

From Theorem 2.2 (i) we obtain the following sufficient condition for f-projections of P on $\widehat{\mathcal{M}}_{\text{loc}}$ ($\widehat{\mathcal{M}}$). By $L(S)$ we denote the set of predictable, S-integrable processes with respect to P (see [13]).

Theorem 3.3 *Let $Q^* \in \widehat{\mathcal{M}}_{\text{loc}}$ with $f(Q^*\|P) < \infty$ such that for some process $\varphi \in L(S)$ and $\mu_1, \ldots, \mu_n \in \mathbb{R}$ the following conditions hold:*

$$\text{(i)} \quad f'\left(\frac{dQ^*}{dP}\right) = c + \varphi \cdot S_T + \sum_{i=1}^{n} \mu_i H_i \quad P\text{-a.s.,}$$

(ii) $-\varphi \cdot S$ is bounded from below P-a.s.,

(iii) $E_{Q^*}(\varphi \cdot S_T) = 0$,

(iv) $\mu_i \geq 0$, $\mu_i(E_{Q^*} H_i) = 0$ for $1 \leq i \leq r$.

Then Q^* is the f-projection of P on $\widehat{\mathcal{M}}_{\mathrm{loc}}$.

Proof. It follows from Corollary 3.5 in [1] and condition (ii) that $-\varphi \cdot S$ is a Q-local martingale and hence a Q-supermartingale for any $Q \in \widehat{\mathcal{M}}_{\mathrm{loc}}$. Therefore,

$$
E_Q f'\left(\frac{dQ^*}{dP}\right) = c + E_Q(\varphi \cdot S_T) + \sum_{i=1}^{n} \mu_i E_Q H_i
$$

$$
\geq c = E_{Q^*} f'\left(\frac{dQ^*}{dP}\right).
$$

Now the result follows from Theorem 2.2 (i).   □

The following proposition shows that one can transform the minimization problem $\inf_{Q \in \widehat{\mathcal{M}}} f(Q \| P)$ with respect to $\widehat{\mathcal{M}}$ into a minimization problem with respect to $\mathcal{M}$ including some penalty terms for violating the constraints. The coefficients γ_i in the penalty terms can be interpreted as Lagrange multipliers.

Proposition 3.4 Let S and $H_1, \ldots, H_n$ be bounded and $\inf_{Q \in \widehat{\mathcal{M}}} E f(\frac{dQ}{dP}) < \infty$. Assume the existence of a measure $Q_0 \in \widehat{\mathcal{M}}$ such that $E_{Q_0} H_i > 0$ for $1 \leq i \leq r$ and $f(Q_0 \| P) < \infty$. Furthermore assume that there exists a neighbourhood V of $(0, \ldots, 0) \in \mathbb{R}^{n-r}$ such that for all $v \in V$ there exists an element $Q \in \mathcal{M}$ with $f(Q \| P) < \infty$ and $(E_Q H_{r+1}, \ldots, E_Q H_n) = v$.

Then $Q^* \in \widehat{\mathcal{M}}$ is a f-projection of P on $\widehat{\mathcal{M}}$, i.e.,

$$
f(Q^* \| P) = \inf_{Q \in \widehat{\mathcal{M}}} f(Q \| P),
$$

if and only if there are $\gamma_1, \ldots, \gamma_n \in \mathbb{R}$ such that

$$
f(Q^* \| P) + \sum_{i=1}^{n} \gamma_i E_{Q^*} H_i = \inf_{Q \in \mathcal{M}} \left\{ f(Q \| P) + \sum_{i=1}^{n} \gamma_i E_Q H_i \right\}
$$

and $\gamma_i \leq 0$, $\gamma_i(E_{Q^*} H_i) = 0$ for $1 \leq i \leq r$.

Proof. 1. Since S is bounded, it follows that the set $\mathcal{M}$ of absolutely continuous martingale measures is closed in variation. As $H_1, \ldots, H_n$ are bounded this is also true for $\mathcal{M}' := \{Q \in \mathcal{M} : E_Q H_i = 0 \text{ for } r+1 \leq i \leq n\}$. Since $\mathcal{M}, \mathcal{M}'$ are convex, they are also closed in $\sigma(L^1, L^\infty)$, if one identifies $Q \in \mathcal{M}$ with its Radon-Nikodym density with respect to P (see, for example, Proposition IV.3.1 in [25]). We define a function $B : \mathcal{M}' \to \mathbb{R}^r$ by

$$
B(Q) = (-E_Q H_1, \ldots, -E_Q H_r).
$$

Obviously, the component mappings of B are convex and continuous with respect to $\sigma(L^1, L^\infty)$. The optimization problem is to minimize the lower semicontinuous functional $f(\cdot\|P)$ (see Theorem 1.47 in [21]) over $\widehat{\mathcal{M}}$,

$$\inf_{Q\in\widehat{\mathcal{M}}} f(Q\|P).$$

This problem can be written in the form

$$\inf_{\substack{Q\in\mathcal{M}' \\ B(Q)\leq 0}} f(Q\|P),$$

where $B(Q) \leq 0$ is understood componentwise.

The assumption on the existence of an inner point Q_0 in $\mathcal{M}'$ allows to apply a Lagrange multiplier theorem (see [8], Theorem III.5.1), which results in the following equivalence:

$Q^* \in \widehat{\mathcal{M}}$ is a f-projection of P on $\widehat{\mathcal{M}}$ if and only if there are $\gamma_i \leq 0$ such that

$$f(Q\|P) + \sum_{i=1}^{r} \gamma_i E_{Q^*} H_i = \inf_{Q\in\mathcal{M}'}\left\{f(Q\|P) + \sum_{i=1}^{r}\gamma_i E_Q H_i\right\}, \qquad (3.3)$$

and $\gamma_i(E_{Q^*} H_i) = 0$ for $1 \leq i \leq r$.

2. Next we follow a similar line of argument to handle the equality constraints on the right-hand side of (3.3). Since the component mappings of B are continuous and linear, the mapping $J : \mathcal{M} \to \mathbb{R}$, defined by $J(Q) := f(Q\|P) + \sum_{i=1}^{r}\gamma_i E_Q H_i$, is lower semicontinuous and convex. We define a function $B' : \mathcal{M} \to \mathbb{R}^{n-r}$ by

$$B'(Q) := (E_Q H_{r+1}, \ldots, E_Q H_n).$$

The optimization problem

$$\inf_{Q\in\mathcal{M}'} J(Q)$$

can be written as

$$\inf_{\substack{Q\in\mathcal{M} \\ B'(Q)=0}} J(Q).$$

The perturbation function $\Phi : \mathcal{M} \times \mathbb{R}^{n-r} \to \bar{\mathbb{R}}$ is chosen as

$$\Phi(Q, v) := \begin{cases} J(Q), & \text{if } Q \in \mathcal{M} \text{ and } B'(Q) = v \\ \infty, & \text{otherwise.} \end{cases}$$

For $v \in \mathbb{R}^{n-r}$ we define $h(v) := \inf_{Q\in\mathcal{M}} \Phi(Q, v)$. By assumption, $h(0)$ is finite. We observe that $\mathcal{M}$ is closed, J is lower semicontinuous and convex, and B' is linear and continuous. Therefore, the function h is convex (see [8], Lemma III.5.2 and Lemma III.2.1). Theorem 23.4 in [22] shows that h is subdifferentiable in 0. Hence by Proposition III.3.2 in [8] we obtain the following equivalence:

$Q^* \in \mathcal{M}'$ solves $\inf_{Q \in \mathcal{M}'} J(Q)$ if and only if there are $\gamma_{r+1}, \ldots, \gamma_n \in \mathbb{R}$ such that

$$f(Q^*\|P) + \sum_{i=1}^{n} \gamma_i E_{Q^*} H_i \;=\; \inf_{Q \in \mathcal{M}} \{ f(Q\|P) + \sum_{i=1}^{n} \gamma_i E_Q H_i \}.$$

$\square$

Remark Following the line of arguments of the proof of Theorem 7.1 in [11] one verifies that under the additional assumption $f'(\frac{dQ^*}{dP}) \in L^1(Q^*)$ the coefficients $\gamma_1, \ldots, \gamma_n$ in Proposition 3.4 correspond to $-\mu_1, \ldots, -\mu_n$, where μ_i are the coefficients of the characterization of Q^* in Theorem 3.1.

4. RELATIONSHIP TO PORTFOLIO OPTIMIZATION

Minimal distance martingale measures are closely related to minimax martingale measures and hence to utility maximization problems. We briefly restate the notion of a minimax measure and some results about minimax measures for general convex models as introduced in [11], Section 4, in order to point out the relationship to portfolio optimization. In the following, as in Section 2, we denote by $\mathcal{K}$ a general convex set of probability measures on $(\Omega, \mathcal{F})$ dominated by P.

A utility function $u: \mathbb{R} \to \mathbb{R} \cup \{-\infty\}$ is assumed to be strictly increasing, strictly concave, continuously differentiable in $\mathrm{dom}(u) := \{x \in \mathbb{R} \mid u(x) > -\infty\}$ and to satisfy

$$u'(\infty) \;=\; \lim_{x \to \infty} u'(x) = 0, \tag{4.1}$$

$$u'(\bar{x}) \;=\; \lim_{x \downarrow \bar{x}} u'(x) = \infty \tag{4.2}$$

for $\bar{x} := \inf\{x \in \mathbb{R} \mid u(x) > -\infty\}$. This implies that either $\mathrm{dom}(u) = (\bar{x}, \infty)$ or $\mathrm{dom}(u) = [\bar{x}, \infty)$. By I we denote the inverse of the derivative of u. Assumption (4.1) implies that $I(0) = \infty$. The convex conjugate function $u^*: \mathbb{R}_+ \to \mathbb{R}$ of u is defined by

$$u^*(y) \;:=\; \sup_{x \in \mathbb{R}} \{ u(x) - xy \} = u(I(y)) - y I(y).$$

For $Q \in \mathcal{K}$ and $x > \bar{x}$ we define

$$U_Q(x) := \sup\{ E u(Y) : Y \in L^1(Q), E_Q Y \leq x, E u(Y)^- < \infty \}. \tag{4.3}$$

The value $U_Q(x)$ can be interpreted as the maximal expected utility which can be achieved with endowment x, if the market prices are computed by Q. If $E_Q(I(\lambda \frac{dQ}{dP}))$ is finite for all $\lambda > 0$, then $U_Q(x)$ has the following well-known representation (see, for example, [11], Lemma 4.1):

$$U_Q(x) \;=\; E \Big[u \Big(I \Big(\lambda_Q(x) \frac{dQ}{dP} \Big) \Big) \Big], \tag{4.4}$$

where $\lambda_Q(x)$ is chosen so that $E I(\lambda_Q(x) \frac{dQ}{dP}) = x$. The random variable $I(\lambda_Q(x) \frac{dQ}{dP})$ can be interpreted as optimal contingent claim which is financeable under the pricing measure Q.

Definition 4.1 A measure $Q^* = Q^*(x) \in \mathcal{K}$ is called minimax measure with respect to endowment x and model $\mathcal{K}$ if it minimizes $Q \mapsto U_Q(x)$ over all $Q \in \mathcal{K}$, i.e.,

$$U_{Q^*}(x) = U(x) := \inf_{Q \in \mathcal{K}} U_Q(x).$$

We refer to [3], [10] and [11] for further information about minimax measures. By $u^*_{\lambda_0}(\cdot||\cdot)$ we denote the f-divergence distance corresponding to $f(x) = u^*(\lambda_0 x)$.

Under the assumptions:

$$\exists x > \bar{x} \text{ with } U(x) < \infty, \tag{4.5}$$

$$E_Q I\left(\lambda \frac{dQ}{dP}\right) < \infty \quad \forall \lambda > 0 \quad \forall Q \in \mathcal{K} \tag{4.6}$$

one gets the following result (see Proposition 4.3 in [11]):

Proposition 4.2 *Let* $x > \bar{x}$.
(i) *Let* $\lambda_0 > 0$ *such that* $\lambda_0 \in \partial U(x)$. *If* $Q^* \in \mathcal{K}$ *is a* $u^*_{\lambda_0}$*-projection of* P *on* $\mathcal{K}$, *then* Q^* *is a minimax measure for* $\mathcal{K}$ *and* $\lambda_0 = \lambda_{Q^*}(x)$.
(ii) *If* $Q^* \in \mathcal{K}$ *is a minimax measure, then* Q^* *is a* $u^*_{\lambda_{Q^*}(x)}$*-projection of* P *on* $\mathcal{K}$ *and* $\lambda_{Q^*}(x) \in \partial U(x)$.

This result shows that minimax measures can be determined by distance minimization and conversely a $u^*_{\lambda_0}$-projection has an alternative interpretation in the sense of utility maximization. Notice that for the standard utility functions like $u(x) = \frac{x^p}{p}$ ($p \in (-\infty, 1) \setminus \{0\}$), $u(x) = \log x$ and $u(x) = 1 - e^{-px}$ ($p \in (0, \infty)$) the minimax measure does not depend on x and the u^*_λ-projection does not depend on λ, respectively. Since $U(x)$ is typically not known explicitly it is of interest to be able to determine some $\lambda_0 \in \partial U(x)$. In the next proposition we give a sufficient condition to imply $\lambda_0 \in \partial U(x)$.

According to the relationship between minimax measures and u^*_λ-projections in Proposition 4.2 the preceding results also induce necessary and sufficient conditions for a minimax measure in $\widehat{\mathcal{M}}$ and in $\widehat{\mathcal{M}}_{\text{loc}}$, respectively.

Assume that the conditions 4.5, 4.6 hold true for $\widehat{\mathcal{M}}$ ($\widehat{\mathcal{M}}_{\text{loc}}$).

Proposition 4.3 *Let* $Q^* \in \widehat{\mathcal{M}}$ ($\widehat{\mathcal{M}}_{\text{loc}}$), $\lambda > 0$ *satisfy* $u^*_\lambda(Q^*||P) < \infty$. *Assume that for some* $\varphi \in L(S)$ *and constants* $\mu_1, \ldots, \mu_n \in \mathbb{R}$ *the following conditions hold:*

$$\text{(i)} \quad I\left(\lambda \frac{dQ^*}{dP}\right) = x + \varphi \cdot S_T + \sum_{i=1}^{n} \mu_i (H_i - E_{Q^*} H_i) \quad P\text{-a.s.,}$$

(ii) $\quad \varphi \cdot S$ *is bounded from below* P*-a.s.,*

(iii) $\quad E_{Q^*}(\varphi \cdot S_T) = 0$,

(iv) $\quad \mu_i \leq 0$ *and* $\mu_i(E_{Q^*} H_i) = 0$ *for* $1 \leq i \leq r$.

Then Q^* *is a minimax measure for* $\widehat{\mathcal{M}}$ ($\widehat{\mathcal{M}}_{\text{loc}}$) *and* x *and* $\lambda \in \partial U(x)$.

Proof. Since $E_Q \cdot I(\lambda \frac{dQ^*}{dP}) = x$ one gets that $\lambda = \lambda_{Q^*}(x)$. Moreover, since $(u^*_\lambda)'(x) = -\lambda I(\lambda x)$, we conclude from Theorem 3.3 that Q^* is the u^*_λ-projection of P on $\widehat{\mathcal{M}}$ ($\widehat{\mathcal{M}}_{\text{loc}}$) and that the condition of Theorem 2.2 (i) is fulfilled. Hence for all measures $Q \in \widehat{\mathcal{M}}$ ($\widehat{\mathcal{M}}_{\text{loc}}$) satisfying $u_\lambda(Q||P) < \infty$ one gets $E_Q I(\lambda \frac{dQ^*}{dP}) \leq x$. This implies that

$$U_{Q^*}(x) = E\left[u\left(I\left(\lambda\left(\frac{dQ^*}{dP}\right)\right)\right)\right] \leq U_Q(x).$$

Assumption (4.6) implies that (see Lemma 4.1 in [11])

$$\{Q \in \widehat{\mathcal{K}} : u^*_\lambda(Q||P) < \infty\} = \{Q \in \mathcal{K} : U_Q(x) < \infty\}.$$

Therefore, Q^* is a minimax measure for x and $\widehat{\mathcal{M}}$ ($\widehat{\mathcal{M}}_{\text{loc}}$). From Proposition 4.2 we conclude that $\lambda = \lambda_{Q^*}(x) \in \partial U(x)$.   □

In the following we point out the relationship between minimal distance martingale measures under constraints and portfolio optimization. We consider a market model which consists of $d + 1$ assets. We assume that the assets $0, \ldots, d$ are modeled by the $\mathbb{R}^{d+1}$-valued semimartingale $S = (S^0, \ldots, S^d)$ and suppose without loss of generality that the price of asset 0 is constant, i.e., $S^0_t \equiv 1$. Assume that additionally to the price process S the prices $\{p_1, \ldots, p_n\}$ of a finite set of contingent claims $\{H_1, \ldots, H_n\}$ are known at time $t = 0$. We suppose that one can buy or sell these contingent claims in $t = 0$.

We call a pair (φ, μ) of a predictable, S-integrable, $\mathbb{R}^{d+1}$-valued process φ and a vector $\mu \in \mathbb{R}^n$ an admissible strategy if

$$\sum_{i=0}^{d} \varphi^i_t S^i = x - \sum_{i=1}^{n} \mu_i p_i + \varphi \cdot S_t$$

for any $t \in \mathbb{R}_+$ and the process $\varphi \cdot S$ is bounded from below. The set of admissible strategies is denoted by $\mathcal{A}$.

We define an *optimal portfolio strategy* as a strategy $(\widehat{\varphi}, \widehat{\mu}) \in \mathcal{A}$ which maximizes

$$(\varphi, \mu) \mapsto E\left(u\left(x + \varphi \cdot S_T + \sum_{i=1}^{n} \mu_i(H_i - p_i)\right)\right) \tag{4.7}$$

over all $(\varphi, \mu) \in \mathcal{A}$. By $\widehat{\mathcal{M}}$ we denote the class of martingale measures consistent with the observed market prices of the derivatives $H_1, \ldots, H_n$,

$$\widehat{\mathcal{M}} := \{Q \in \mathcal{M} : H_i \in L^1(Q) \text{ and } E_Q H_i = p_i, 1 \leq i \leq n\}.$$

Optimal portfolio strategies can be obtained from a representation of the $u^*_{\lambda_0}$-projection Q^* of P on $\widehat{\mathcal{M}}$ as in Theorem 3.1. Assume that the conditions 4.5, 4.6 hold true for $\widehat{\mathcal{M}}$.

Theorem 4.4 *Let $Q^* \in \widehat{\mathcal{M}}^e$ be the $u^*_{\lambda_0}$-projection of P on $\widehat{\mathcal{M}}$ for some $\lambda_0 \in \partial U(x)$, $\lambda_0 > 0$.*

(i) *There exist constants* $\widehat{\mu}_1, \ldots, \widehat{\mu}_n \in \mathbb{R}$ *and a process* $\widehat{\varphi} \in L(S)$ *such that:*

$$I\left(\lambda_0 \frac{dQ^*}{dP}\right) = x + \widehat{\varphi} \cdot S_T + \sum_{i=1}^{n} \widehat{\mu}_i(H_i - p_i). \tag{4.8}$$

(ii) *If the representation (4.8) holds and if the stochastic integral* $\widehat{\varphi} \cdot S$ *is bounded from below, then* $(\widehat{\varphi}, \widehat{\mu})$ *is an optimal portfolio strategy (where* $\widehat{\varphi}_t^0 := x + \widehat{\varphi} \cdot S_t - \sum_{i=1}^{n} \widehat{\mu}_i p_i - \sum_{i=1}^{d} \widehat{\varphi}_t^i S_t).$

Proof. 1. Theorem 3.1 and the identity $\lambda_0 = \lambda_{Q^*}(x)$ (see Proposition 4.2) imply the existence of the representation in (4.8).

2. By the definition of $\widehat{\varphi}_t^0$ and observing that S^0 is assumed to be identical 1 it holds that $\sum_{i=0}^{d} \widehat{\varphi}_t^i S_t^i = x - \sum_{i=1}^{n} \widehat{\mu}_i p_i + \widehat{\varphi} \cdot S_t$ for any $t \in \mathbb{R}_+$ and hence $(\widehat{\varphi}, \widehat{\mu}) \in \mathcal{A}$. It follows from Corollary 3.5 in [1] that $\varphi \cdot S$ is a Q^*-local martingale and hence a Q^*-supermartingale for any $(\varphi, \mu) \in \mathcal{A}$. Therefore,

$$E_{Q^*}\left(x + \varphi \cdot S_T + \sum_{i=1}^{n} \mu_i(H_i - p_i)\right) \leq x.$$

Since $I(\lambda_0 \frac{dQ^*}{dP})$ is the optimal contingent claim which is financeable under the pricing measure Q^* and the endowment x (see (4.4)) we conclude that $(\widehat{\varphi}, \widehat{\mu})$ is an optimal portfolio strategy. $\qquad\square$

Remarks 1. If the utility-function is finite on $(-\infty, \infty)$, i.e., $\bar{x} = -\infty$, then the condition that $\widehat{\varphi} \cdot S$ is bounded from below is not fulfilled in general. In this case one has to choose a suitable extended concept of admissible strategies in order to solve utility maximization problems. This issue is discussed in [24], [5] and [15].

2. Theorem 4.4 shows that if derivative prices are computed by the minimax (respectively, minimal distance) measure Q^* for $\widehat{\mathcal{M}}$ and x, then the optimal contingent claim can be duplicated by a strategy $(\widehat{\varphi}, \widehat{\mu})$. Hence, although it may be profitable to invest in the derivatives $H_1, \ldots, H_n$, it turns out that one cannot increase the maximal expected utility in comparison to the strategy $(\widehat{\varphi}, \widehat{\mu})$ by trading in further derivatives. This shows that Q^* yields consistent derivative prices in the sense of [18].

3. There is a close relationship of the portfolio optimization problem (4.7) to utility-based hedging of one contingent claim H_1:

$$\sup_{(\varphi, -1) \in \mathcal{A}} E(u(x_0 + \varphi \cdot S_T - H_1)). \tag{4.9}$$

If for $k = 1$ and $\widehat{\mu} = -1$ the strategy $\widehat{\varphi}$ in representation (4.8) is such that $\widehat{\varphi} \cdot S$ is bounded from below, then $\widehat{\varphi}$ turns out to be the optimal portfolio strategy for the utility-based hedging problem (4.9) with initial endowment $x_0 = x - p_1$. Hence problem (4.7) is closely related to problem (4.9). In the portfolio optimization problem (4.7) we have a fixed initial price of the derivative, in the utility-based hedging problem (4.9) we have a fixed number of derivatives in the set of allowed portfolios. Due to this relationship, results of this paper are closely related to results on utility-based hedging. The problem of utility-based hedging has been studied recently in [4], [5], [11] and [15].

Example We consider a discrete-time market model. Let $S = (S_0, S_1, \ldots, S_T)$ be the price process of a risky asset with $S_0 \in \mathbb{R}$, $S_0 > 0$ and $S_t = S_0 \prod_{s=1}^{t} X_s$ where $X_1, \ldots, X_T$ are $(0, \infty)$-valued random variables. Moreover the price p of one derivative H is given.

Theorem 3.3 gives as sufficient conditions for a measure $Q^* \in \widehat{\mathcal{M}}$, where

$$\widehat{\mathcal{M}} := \{Q \in \mathcal{M} : E_Q H = p\},$$

to minimize the relative entropy (corresponding to $f(x) = x \log x$) over all measures in $\widehat{\mathcal{M}}$:

$$\text{(i)} \qquad \frac{dQ^*}{dP} = \frac{e^{\varphi \cdot S + \mu H}}{E\left(e^{\varphi \cdot S + \mu H}\right)},$$

$$\text{(ii)} \qquad \varphi \cdot S \text{ is bounded from below.}$$

In the discrete-time setting condition (ii) implies condition (iii) in Theorem 3.3. Condition (i) can be rewritten in the following way:

$$\frac{dQ^*}{dP} = \frac{e^{\sum_{i=1}^{T} \gamma_i (X_i - 1) + \mu H}}{E\left(e^{\sum_{i=1}^{T} \gamma_i (X_i - 1) + \mu H}\right)},$$

where γ_i is $\mathcal{F}_{i-1}$-measurable and $\mu \in \mathbb{R}$. The random variable γ_i describes the amount of money invested at time i in the risky asset.

The condition that $Q^* \in \widehat{\mathcal{M}}$ leads to $T + 1$ recursive nonlinear equations for the parameters γ_i, μ. First γ_T is determined dependent on the parameter μ by the equation:

$$E\left((X_T - 1)\, e^{\gamma_T (X_T - 1) + \mu H} \middle| \mathcal{F}_{T-1}\right) = 0.$$

Then γ_{T-1} is determined by

$$E\left((X_{T-1} - 1)\, e^{\gamma_{T-1}(X_{T-1} - 1) + \gamma_T (X_T - 1) + \mu H} \middle| \mathcal{F}_{T-2}\right) = 0.$$

Finally γ_1 is determined by

$$E\left((X_1 - 1)\, e^{\sum_{i=2}^{T} \gamma_i (X_i - 1) + \gamma_1 (X_1 - 1) + \mu H} \middle| \mathcal{F}_0\right) = 0.$$

According to a generalized Bayes formula (see, for example, [27], page 438–439) this procedure ensures that Q^* is a martingale measure.

The parameter μ is determined by the moment constraint $E_{Q^*} H = p$. One finally has to check condition (ii) for $\varphi_i := \frac{\gamma_i}{S_0 \prod_{s \leq i-1} X_s}$. Then as consequence Q^*, as constructed above, is the minimal distance measure for $\widehat{\mathcal{M}}$ and $f(x) = x \log x$. Moreover, (φ, μ) is an optimal portfolio strategy for the exponential utility function $u(x) = -e^{-x}$. Notice that for the exponential utility function the optimal portfolio strategy is independent of the initial endowment, as $e^{y+z} = e^y e^z$.

REFERENCES

[1] J. Ansel and C. Stricker, Couverture des actifs contingents et prix maximum, *Ann. Inst. H. Poincaré Probab. Statist.* **30** (1994), 303–315.

[2] M. Avellaneda, Minimum-entropy calibration of asset-pricing models, *Int. J. Theor. Appl. Finance* **1** (1998), 447–472.

[3] F. Bellini and M. Frittelli, On the existence of minimax martingale measures, Technical Report 2, Universita di Milano, 1998.

[4] J. Cvitanić, W. Schachermayer, and H. Wang, Utility maximization in incomplete markets with random endowment, Preprint, to appear in *Finance Stoch.*, 2000.

[5] F. Delbaen, P. Grandits, Th. Rheinländer, D. Samperi, M. Schweizer, and Ch. Stricker, Exponential hedging and entropic penalties, Preprint, Technical University of Berlin, 2000.

[6] F. Delbaen and W. Schachermayer, The variance optimal martingale measure for continuous processes, *Bernoulli* **9** (1996), 81–105.

[7] F. Delbaen and W. Schachermayer, A compactness principle for bounded sequences of martingales, in R. C. Dalang, M. Dozzi, and F. Russo, editors, *Seminar on Stochastic Analysis, Random Fields and Applications*, Progress in Probability **45**, 137–173, Birkhäuser, Basel, 1999.

[8] I. Ekeland and R. Temam, *Convex analysis and variational problems*, North-Holland publishing company, Amsterdam, 1976.

[9] M. Frittelli, The minimal entropy martingale measure and the valuation problem in incomplete markets, *Math. Finance* **10** (2000), 39–52.

[10] M. Frittelli, Optimal solutions to utility maximization and to the dual problem, Preprint, University of Milano, 2000.

[11] T. Goll and L. Rüschendorf, Minimax and minimal distance martingale measures and their relationship to portfolio optimization, Preprint, to appear in *Finance Stoch.*, 2000.

[12] P. Grandits and T. Rheinländer, On the minimal entropy martingale measure, Preprint, Technical University of Berlin, 1999.

[13] J. Jacod, *Calcul stochastique et problèmes de martingales*, Lecture Notes in Math. **714**, Springer-Verlag, 1979.

[14] J. Jacod, Intégrales stochastiques par rapport à une semi-martingale vectorielle et changements de filtration, in *Séminaire de Probabilités XIV, 1978/79*, Lecture Notes in Math. **784**, 161–172, Springer-Verlag, 1980.

[15] Y. Kabanov and C. Stricker, On the optimal portfolio for the exponential utility maximization: remarks to the six-author paper, Preprint, University of Franche-Comte, 2001.

[16] J. Kallsen, Duality links between portfolio optimization and derivative pricing, Technical Report 40/1998, Mathematische Fakultät Universität Freiburg i. Br., 1998.

[17] J. Kallsen, *Semimartingale modelling in finance*, Dissertation Universität Freiburg i. Br., 1998.

[18] J. Kallsen, Utility-based derivative pricing in incomplete markets, Preprint, to appear in *Selected Proceedings of the First Bachelier Congress*, held in Paris, 2000, edited by H. Geman, D. Madan, S. Pliska, and T. Vorst, Springer-Verlag, Berlin, 2001.

[19] U. Keller, *Realistic modelling of financial derivatives*, Dissertation, Universität Freiburg i. Br., 1997.

[20] D. Kramkov and W. Schachermayer, A condition on the asymptotic elasticity of utility functions and optimal investment in incomplete markets, *Ann. Appl. Probab.* **9** (1999), 904–950.

[21] F. Liese and I. Vajda, *Convex statistical distances*, Teubner, Leipzig, 1987.

[22] R.T. Rockafellar, *Convex analysis*, Princeton University Press, Princeton, 1970.

[23] L. Rüschendorf, Projections of probability measures, *Statistics* **18** (1987), 123–129.

[24] W. Schachermayer, Optimal investment in incomplete markets when wealth may become negative, Preprint, to appear in *Ann. Appl. Probab.*, 1999.

[25] H.H. Schaefer, *Topological vector spaces*, Springer, Berlin, 1971.

[26] M. Schweizer, Approximation pricing and the variance-optimal martingale measure, *Ann. Probab.* **24** (1996), 206–236.

[27] A. Shiryaev, *Essentials of stochastic finance*, World Scientific, Singapore, 1999.

[28] I. Vajda, *Theory of statistical inference and information*, Kluwer Academic Publishers, Dordrecht, 1989.

[29] M. Yor, Sous-espaces denses dans L^1 ou H^1 et représentation des martingales, in *Séminaire de Probabilités XII, 1978*, Lecture Notes in Math. **649**, 265–309, Springer-Verlag, 1978.

ON GENERALIZED z-DIFFUSIONS[*]

BRONIUS GRIGELIONIS

Institute of Mathematics and Informatics
Akademijos 4, 2600 Vilnius, Lithuania
Vilnius University
Naugarduko 24, 2600 Vilnius, Lithuania

Abstract The ergodic one-dimensional diffusion processes on an open interval with the predetermined one-dimensional marginal distributions are considered. The analogue of the classical extreme value theory is presented. Several examples, including well-known models from population genetics, population growth, neurobiology, mathematical finance etc., are discussed. Special attention is paid to the class of the generalized z-distributions and related generalized z-diffusion processes.

Key words Ergodic diffusion process, H-diffusions, extreme value theory, generalized z-distributions, generalized z-diffusions.

1. INTRODUCTION

Considering stationary diffusion models in many applications it is desirable to incorporate some of the statistical aspects of the empirical data. In Barndorff-Nielsen [1] a number of key features are formulated for observational series from finance and turbulence. In Bibby and Sørensen [5] and Rydberg [30] one-dimensional ergodic diffusions with stationary generalized hyperbolic distributions are introduced and applications to the mathematical finance are analysed. A class of ergodic diffusions with positive values, having the generalized inverse Gaussian distributions as stationary distributions was defined in Borkovec and Klüppelberg [7], where the extremal behaviour of diffusion models in finance is also investigated.

The purpose of this paper is to turn one's attention to the class of the generalized z-distributions, defined and investigated in Grigelionis [5], as an alternative to the generalized hyperbolic distributions. The generalized z-diffusions are defined as strictly

---

[*]This work was partially supported by the Lithuanian State Science and Studies Foundation Grant # K-014 and NSF Grant # DMS-98-02423

stationary diffusions on R^1 with one-dimensional generalized z-distributions.

Considering extremal behaviour of such diffusion processes in Section 4, we apply general extreme value theory for the ergodic diffusions on an open interval, which is developed in Section 2, basing on the known extreme value properties of one-dimensional diffusions (see Newell [27], Berman [3], Davis [10]) and the classical extreme value theory (see, e.g., Gnedenko [15], Galambos [14], Leadbetter *et al.* [24], Berman [4], Resnick [29], Embrechts *et al.* [11]). Finally, several examples, including well-known models from population genetics, population growth, neurobiology, mathematical finance etc., are presented in Section 3.

2. H-DIFFUSIONS

We consider strictly positive measurable functions $h(x)$ and $\sigma(x)$, $x \in (l,r) \subseteq R^1$, such that

$$\int_l^r h(x)\,dx = 1 \tag{1}$$

and the function $g(x) := (h(x)\sigma^2(x))^{-1}$, $x \in (l,r)$, is differentiable and denote $H(\,dx) = h(x)\,dx$.

Definition 1 *A strictly stationary diffusion process* $X = \{X(t), t \geq 0\}$ *on* (l,r) *is called an H-diffusion, if the probability law* $\mathcal{L}(X(t)) \equiv H$.

Let

$$\mu(x) = -\frac{\sigma^2(x)g'(x)}{2g(x)}, \quad G(x) = \int_{x_0}^x g(v)\,dv, \quad x \in (l,r), \tag{2}$$

where x_0, $l < x_0 < r$, is a fixed point, and we are given a standard Brownian motion $W(t)$, $t \geq 0$, defined on a probability space $(\Omega, \mathcal{F}, \mathbf{P})$.

We assume that

$$G(x) \to \infty \text{ as } x \uparrow r, \quad G(x) \to -\infty \text{ as } x \downarrow l, \tag{3}$$

and for each $x \in (l,r)$ there exists $\varepsilon > 0$ such that $(x - \varepsilon, x + \varepsilon) \subseteq (l,r)$ and

$$\int_{x-\varepsilon}^{x+\varepsilon} \left(\frac{1}{\sigma^2(v)} + \frac{|g'(v)|}{g(v)} \right) dv < \infty. \tag{4}$$

Theorem 1 *Under the assumptions* (1) – (4) *there exists the unique strictly stationary weak solution for the stochastic differential equation*

$$\begin{cases} dX(t) = \mu(X(t))\,dt + \sigma(X(t))\,dW(t), & t > 0, \\ \mathcal{L}(X(0)) = H, \end{cases} \tag{5}$$

which is an ergodic H-diffusion.

Proof. From (1) and (2) it follows that

$$s'(x) := \exp\left\{ -2\int_{x_0}^{x} \frac{\mu(v)}{\sigma^2(v)}\, dv \right\} = \frac{g(x)}{g(x_0)}, \quad x \in (l,r), \tag{6}$$

$$m'(x) := \frac{2}{\sigma^2(x)s'(x)} = 2h(x)g(x_0), \quad x \in (l,r), \tag{7}$$

$$|m| := \int_{l}^{r} m'(x)\, dx = 2g(x_0) < \infty$$

and $|m|^{-1} m'(x) = h(x)$, $x \in (l,r)$.

So the functions $g(x)$ and $h(x)$ are proportional to the densities of the scale and speed measures, respectively, of the diffusion on (l,r) with the drift coefficient $\mu(x)$ and the diffusion coefficient $\sigma^2(x)$.

Now it remains to apply known results on existence, uniqueness and ergodicity of weak solutions to (5) (see, e.g., Mandl [25], Engelbert and Schmidt [12], Karatzas and Shreve [21]). □

We denote

$$M_T^X = \max_{0 \le s \le T} X(s), \quad \underline{M}_T^X = \min_{0 \le s \le T} X(s), \quad T \ge 0,$$

the Gumbel distributions

$$\Lambda(x) = \exp\{-e^{-x}\}, \quad \underline{\Lambda}(x) = 1 - \Lambda(-x), \quad x \in R^1,$$

the Fréchet distributions

$$\Phi_\alpha(x) = \exp\{-x^{-\alpha}\} \mathrm{I\!I}(0,\infty), \quad \underline{\Phi}_\alpha(x) = 1 - \Phi_\alpha(-x), \quad x \in R^1, \, \alpha > 0,$$

and the Weibull distributions

$$\Psi_\alpha(x) = \begin{cases} \exp\{-(-x)^\alpha\} & \text{for } x \le 0, \\ 1 & \text{for } x > 0, \, \alpha > 0, \end{cases}$$
$$\underline{\Psi}_\alpha(x) = 1 - \Psi_\alpha(-x), \quad x \in R^1.$$

Definition 2 *A stochastic process X belongs to the maximum domain of attraction of Q, denoted by $X \in MDA(Q)$, if there exist constants $a_T > 0$ and $b_T \in R^1$ such that $\mathcal{L}(a_T(M_T^X - b_T))$ weakly converges to Q as $T \to \infty$, and X belongs to the minimum domain of attraction of $\underline{Q}$, denoted by $X \in \underline{MDA}(\underline{Q})$, if there exist constants $\underline{a}_T > 0$ and $\underline{b}_T \in R^1$ such that $\mathcal{L}(\underline{a}_T(\underline{M}_T^X - \underline{b}_T))$ weakly converges to $\underline{Q}$ as $T \to \infty$, where Q and $\underline{Q}$ are probability distributions on R^1.*

We shall denote $c_T \sim \tilde{c}_T$, if $c_T \tilde{c}_T^{-1} \to 1$ as $T \to \infty$.

Let us consider an H-diffusion X and define the values γ_T and $\underline{\gamma}_T$ from the equalities $2G(\gamma_T) = T$ and $2G(-\underline{\gamma}_T) = -T$, respectively.

Theorem 2 *Under the assumptions* (1) – (4) *the following criteria hold true:*

i) $X \in MDA(\Lambda)$ *iff there exists a function* $a(x) > 0$, $x \in (x_0, r)$, *such that, for each* $x \in R^1$,

$$\lim_{y \uparrow r} \frac{G(y)}{G(y + a(y)x)} = e^{-x};$$

ii) $X \in MDA(\Phi_\alpha)$ *iff* $r = \infty$ *and, for each* $x > 0$,

$$\lim_{y \uparrow \infty} \frac{G(y)}{G(xy)} = x^{-\alpha}, \quad \alpha > 0;$$

iii) $X \in MDA(\Psi_\alpha)$ *iff* $r < \infty$ *and, for each* $x > 0$,

$$\lim_{y \downarrow 0} \frac{G(r - y)}{G(r - xy)} = x^\alpha, \quad \alpha > 0.$$

Moreover, in the case i)

$$\int_{x_0}^{r} (G(v))^{-1}\, dv < \infty$$

and we can take

$$a(x) = G(x) \int_{x}^{r} (G(v))^{-1}\, dv,$$

$$a_T \sim \frac{2}{T} \left(\int_{\gamma_T}^{r} (G(v))^{-1} \right)^{-1}, \quad b_T = \gamma_T + \chi_T,$$

where χ_T *are any constants such that* $a_T \chi_T \to 0$ *as* $T \to \infty$. *In the case* ii)

$$a_T \sim \gamma_T^{-1}, \quad b_T = 0$$

and in the case iii)

$$a_T \sim (r - \gamma_T)^{-1}, \quad b_T = r.$$

Proof. Under the assumptions (1) – (4) from Davis [10] we have that for any constants $u_T \uparrow r$ as $T \to \infty$,

$$\lim_{T \to \infty} \left| \mathbf{P}\left\{ M_T^X \leq u_T \right\} - F^T(u_T) \right| = 0,$$

where $F(x) = e^{-(2G(x))^{-1}}$, $x \in (l, r)$. Let

$$\widehat{F}(x) = \begin{cases} 0 & \text{for } x < \widehat{x}_0, \\ 1 - (2G(x))^{-1} \mathrm{I\!I}(\widehat{x}_0, r) & \text{for } x \geq \widehat{x}_0, \end{cases}$$

where $G(\widehat{x}_0) = \frac{1}{2}$. Because $1 - F(x) \sim 1 - \widehat{F}(x)$ as $x \uparrow r$, the statement of the Theorem 2 now follows from the classical extreme value theory (see, e.g., Embrechts *et al.* [11]). □

Theorem 2' *Under the assumptions* (1) – (4) *the following criteria hold true:*

i') $X \in MDA(\underline{\Lambda})$ *iff there exists a function* $\underline{a}(x) > 0$, $x \in (l, x_0)$, *such that, for each* $x \in R^1$,

$$\lim_{y \Downarrow l} \frac{G(y)}{G(y - \underline{a}(y)x)} = e^{-x};$$

ii') $X \in MDA(\underline{\Phi}_\alpha)$ *iff* $l = -\infty$ *and, for each* $x > 0$,

$$\lim_{y \downarrow -\infty} \frac{G(y)}{G(xy)} = x^{-\alpha}, \quad \alpha > 0;$$

iii') $X \in MDA(\underline{\Psi}_\alpha)$ *iff* $l > -\infty$ *and, for each* $x > 0$,

$$\lim_{y \downarrow 0} \frac{G(l + y)}{G(l + xy)} = x^\alpha, \quad \alpha > 0.$$

Moreover, in the case i')

$$\int\limits_l^{x_0} (G(v))^{-1} \, dv > -\infty$$

and we can take

$$\underline{a}(x) = G(x) \int\limits_l^x (G(v))^{-1} \, dv,$$

$$\underline{a}_T \sim \frac{-2}{T} \left(\int\limits_l^{-\underline{\gamma}_T} (G(v))^{-1} \, dv \right)^{-1}, \quad \underline{b}_T = -\underline{\gamma}_T + \underline{\chi}_T,$$

where $\underline{\chi}_T$ *are any constants such that* $\underline{a}_T \underline{\chi}_T \to 0$ *as* $T \to \infty$. *In the case* ii')

$$\underline{a}_T \sim \underline{\gamma}_T^{-1}, \quad \underline{b}_T = 0$$

and in the case iii')

$$\underline{a}_T \sim (-l - \underline{\gamma}_T)^{-1}, \quad \underline{b}_T = l.$$

Proof. This statement follows directly from Theorem 2, noting that

$$\underline{M}_T^X = -M_T^{-X}, \quad T \geq 0,$$

and $-X$ is the diffusion process with drift coefficient $-\mu(-x)$, diffusion coefficient $\sigma^2(-x)$ and stationary density $h(-x)$, $x \in (-r, -l)$. $\qquad \square$

Because

$$\hat{f}(x) := \hat{F}'(x) = \frac{g(x)}{2G^2(x)}, \quad x \in (\hat{x}_0, r),$$

and

$$\widehat{f}'(x) = \frac{1}{2}\frac{g'(x)}{G^2(x)} - \frac{g^2(x)}{G^3(x)}, \quad x \in (\widehat{x}_0, r),$$

we shall have the following analogues of the classical von Mises theorems (cf. von Mises [34], Galambos [14], Leadbetter *et al.* [24], Embrechts *et al.* [11]).

Theorem 3 *Under the assumptions* (1) – (4) *the following statements hold true:*
   i) *If*

$$\lim_{x\uparrow r} \frac{g'(x)G(x)}{g^2(x)} = 1,$$

then $X \in MDA(\Lambda)$;
   ii) *if* $r = \infty$ *and*

$$\lim_{x\uparrow\infty} \frac{xg(x)}{G(x)} = \alpha > 0,$$

then $X \in MDA(\Phi_\alpha)$;
   iii) *if* $r < \infty$ *and*

$$\lim_{x\uparrow r} \frac{(r-x)g(x)}{G(x)} = \alpha > 0,$$

then $X \in MDA(\Psi_\alpha)$.

Theorem 3′ *Under the assumptions* (1) – (4) *the following statements hold true:*
   i′) *If*

$$\lim_{x\downarrow l} \frac{g'(x)G(x)}{g^2(x)} = 1,$$

then $X \in MDA(\underline{\Lambda})$;
   ii′) *if* $l = -\infty$ *and*

$$\lim_{x\downarrow -\infty} \frac{xg(x)}{G(x)} = \alpha > 0,$$

then $X \in MDA(\underline{\Phi}_\alpha)$;
   iii′) *if* $l > -\infty$ *and*

$$\lim_{x\downarrow l} \frac{(l-x)g(x)}{G(x)} = \alpha > 0,$$

then $X \in MDA(\underline{\Psi}_\alpha)$.

Analogously to the classical case, under the above assumptions asymptotic independence of M_T^X and $\underline{M}_T^X$ is known.

Theorem 4 (Davis [10]) *If* $X \in MDA(Q)$ *and* $X \in MDA(\underline{Q})$, *then, for each* $x, y \in R^1$,

$$\lim_{T\to\infty} \mathbf{P}\left\{a_T(M_T^X - b_T) \leq x, \ \underline{a}_T(\underline{M}_T^X - \underline{b}_T) \leq y\right\} = Q(x)\underline{Q}(y),$$

where Q *equals to* Λ, Φ_α *or* Ψ_α *and* $\underline{Q}$ *equals to* $\underline{\Lambda}$, $\underline{\Phi}_\alpha$ *or* $\underline{\Psi}_\alpha$.

Remark that if $r = \infty$,

$$G(x) \sim C_1(\ln x)^{\gamma_1}, \quad C_1 > 0, \quad \gamma_1 > 0, \quad \text{as } x \uparrow \infty,$$

or if $r < \infty$,

$$G(x) \sim C_1(\ln \frac{1}{r-x})^{\gamma_1}, \quad C_1 > 0, \quad \gamma_1 > 0, \quad \text{as } x \uparrow r,$$

then there don't exist constants $a_T > 0$ and $b_T \in R^1$ with a nondegenerated weak limit for $\mathcal{L}(a_T(M_T^X - b_T))$ as $T \to \infty$.

Analogously, if $l = -\infty$,

$$G(x) \sim C_2(\ln |x|)^{\gamma_2}, \quad C_2 < 0, \quad \gamma_2 > 0, \quad \text{as } x \downarrow -\infty,$$

or if $l > -\infty$,

$$G(x) \sim C_2(\ln(\frac{1}{|l-x|}))^{\gamma_2}, \quad C_2 < 0, \quad \gamma_2 > 0, \quad \text{as } x \downarrow l,$$

then there don't exist constants $\underline{a}_T > 0$ and $\underline{b}_T \in R^1$ with a nondegenerated weak limit for $\mathcal{L}(\underline{a}_T(\underline{M}_T^X - \underline{b}_T))$ as $T \to \infty$.

Indeed, under the above assumptions

$$\int_{x_0}^{r} (G(v))^{-1} dv = \infty, \quad \int_{l}^{x_0} (G(v))^{-1} dv = -\infty,$$

and, for each $x > 0$,

$$\lim_{y \uparrow \infty} \frac{G(y)}{G(xy)} = 1 \quad \text{as } r = \infty,$$

$$\lim_{y \downarrow 0} \frac{G(r-y)}{G(r-xy)} = 1 \quad \text{as } r < \infty,$$

$$\lim_{y \downarrow -\infty} \frac{G(y)}{G(xy)} = 1 \quad \text{as } l = -\infty,$$

$$\lim_{y \downarrow 0} \frac{G(l+y)}{G(l+xy)} = 1 \quad \text{as } l > -\infty.$$

Thus, using Theorems 2 and 2′ and the classical extreme value theory (see, e.g., Embrechts et al. [11]), we find that $X \notin MDA(Q)$ or $X \notin MDA(\underline{Q})$ for any possible in this situation nondegenerated limiting laws of types Λ, Φ_α, Ψ_α, $\underline{\Lambda}$, $\underline{\Phi}_\alpha$ or $\underline{\Psi}_\alpha$.

3. EXAMPLES

For the sake of shortness in the below considered examples we shall omit elementary technical details, mainly reducing to an application of the l'Hospital rule, and we shall not discuss the explicit expressions of the norming constants for the related extreme values. We shall denote norming constants for the stationary distributions by C.

Example 1 Let $(l, r) = (0, 1)$,

$$\sigma^2(x) = \sigma^2 x^{\alpha_1}(1 - x)^{\alpha_2} e^{\mu x},$$
$$h(x) = C x^{\beta_1 - 1}(1 - x)^{\beta_2 - 1} e^{\lambda x},$$
$$\alpha_1, \alpha_2, \lambda, \mu \in R^1, \ \sigma^2 > 0, \ \beta_1, \ \beta_2 > 0.$$

Then

$$\mu(x) = \frac{\sigma^2}{2}\left[(\alpha_1 + \beta_1 - 1)x^{\alpha_1 - 1}(1 - x)^{\alpha_2} - (\alpha_2 + \beta_2 - 1)x^{\alpha_1}(1 - x)^{\alpha_2 - 1} + (\lambda + \mu)x^{\alpha_1}(1 - x)^{\alpha_2}\right] e^{\mu x}.$$

The assumptions (1) – (4) are satisfied iff $\alpha_1 + \beta_1 \geq 2$ and $\alpha_2 + \beta_2 \geq 2$. If $\alpha_1 + \beta_1 \geq 2$ and $\alpha_2 + \beta_2 > 2$, then $X \in MDA(\underline{\Psi}_{\alpha_2 + \beta_2 - 2})$. If $\alpha_1 + \beta_1 > 2$ and $\alpha_2 + \beta_2 \geq 2$, then $X \in MDA(\underline{\Psi}_{\alpha_1 + \beta_1 - 2})$.

Taking $\alpha_1 = \alpha_2 = 1$, $\mu = 0$, we have the Wright–Fisher gene frequency model with mutation and selection in the population genetics (see, e.g., Crow and Kimura [9], Karlin and Taylor [22]).

Example 2 Let $(l, r) = \left(-\frac{\pi}{2}, \frac{\pi}{2}\right)$,

$$\sigma^2(x) = \sigma^2(\cos x)^\gamma, \quad h(x) = C(\cos x)^\lambda, \quad \lambda > -1, \ \gamma \in R^1, \ \sigma^2 > 0.$$

Then

$$\mu(x) = -\frac{\lambda \sigma^2}{2}(\cos x)^\gamma \mathrm{tg}\, x$$

and the assumptions (1) – (4) are satisfied iff $\lambda + \gamma \geq 1$. If $\lambda + \gamma > 1$, then $X \in MDA(\underline{\Psi}_{\lambda + \gamma - 1})$ and $X \in MDA(\underline{\Psi}_{\lambda + \gamma - 1})$. Taking $\gamma = 0$ we have the Ornstein–Uhlenbeck process on a finite interval (see Kessler and Sørensen [21]).

Example 3 Let $(l, r) = (0, \infty)$,

$$\sigma^2(x) = \sigma^2 x^\gamma, \quad h(x) = C x^{\lambda - 1} \exp\left\{-(\psi_1 x^{\beta_1} + \psi_2 x^{-\beta_2})\right\},$$
$$\sigma^2 > 0, \ \beta_1 > 0, \ \beta_2 > 0, \ \psi_1 \geq 0, \ \psi_2 \geq 0, \ \lambda, \gamma \in R^1,$$

assuming that $\lambda < 0$, if $\psi_1 = 0$, and $\lambda > 0$, if $\psi_2 = 0$. Then

$$\mu(x) = \frac{\sigma^2}{2}\left[(\gamma + \lambda - 1)x^{\gamma - 1} - \psi_1 \beta_1 x^{\gamma + \beta_1 - 1} + \psi_2 \beta_2 x^{\gamma - \beta_2 - 1}\right].$$

The following implications hold true:

i) $\psi_1 > 0, \psi_2 > 0, \lambda, \gamma \in R^1 \Rightarrow X \in MDA(\Lambda)$ and $X \in MDA(\underline{\Lambda})$;

ii) $\psi_1 = 0, \lambda < 0, \psi_2 > 0, \lambda + \gamma < 2 \Rightarrow X \in MDA(\Phi_{2 - \lambda - \gamma})$;

iii) $\psi_1 = 0, \lambda < 0, \psi_2 > 0, \gamma \in R^1 \Rightarrow X \in MDA(\underline{\Lambda})$;

iv) $\psi_2 = 0, \psi_1 > 0, \lambda > 0, \gamma \in R^1 \Rightarrow X \in MDA(\Lambda)$;

v) $\psi_2 = 0, \psi_1 > 0, \lambda > 0, \lambda + \gamma > 2 \Rightarrow X \in MDA(\underline{\Psi}_{\lambda + \gamma - 2})$.

If $\gamma = 2$, $\beta_1 = 1$, $\psi_2 = 0$, $\lambda > 0$, we find that

$$\mu(x) = \frac{\sigma^2}{2}(\lambda + 1)x - \psi_1 x^2,$$
$$\sigma^2(x) = \sigma^2 x^2$$

and

$$h(x) = C x^{\lambda - 1}\, e^{-\psi_1 x},$$

giving us a diffusion version of the Pearl–Verhulst logistic population growth model (see, e.g., Karlin and Taylor [22]). This class of diffusions also contains the Cox–Ingersoll–Ross model for short interest rates in bond markets and its generalizations (see, e.g., Cox et al. [8], Borkovec and Klüppelberg [7]).

Example 4 Let $(l, r) = (0, \infty)$,

$$\sigma^2(x) = \sigma^2 x^\gamma, \quad h(x) = C x^{\delta\beta_1 - 1}(1 + \psi x^\delta)^{-(\beta_1 + \beta_2)},$$
$$\beta_1 > 0, \ \beta_2 > 0, \ \delta > 0, \ \psi > 0, \ 2 - \beta_1\delta \le \gamma \le \delta\beta_2 + 2.$$

Then

$$\mu(x) = \frac{\sigma^2}{2}\big[(\beta_1\delta + \gamma - 1)x^{\gamma - 1} - (\beta_1 + \beta_2)\psi\delta x^{\gamma + \delta - 1}\big](1 + \psi x^\delta)^{-1}$$

and the assumptions (1) – (4) are satisfied. If $2 - \beta_1\delta \le \gamma < \beta_2\delta + 2$ then $X \in MDA(\Phi_{\beta_2\delta + 2 - \gamma})$. If $2 - \beta_1\delta < \gamma \le \beta_2\delta + 2$, then $X \in MDA(\underline{\Psi}_{\beta_1\delta - 2 + \gamma})$.

Example 5 Let $(l, r) = (0, \infty)$,

$$\sigma^2(x) = \sigma^2 x^\gamma, \quad h(x) = C x^\lambda \exp\{-\psi(\ln x)^2\}, \quad \sigma^2 > 0, \ \psi > 0, \ \gamma, \lambda \in R^1.$$

Then

$$\mu(x) = \sigma^2 x^{\gamma - 1}\left(\frac{\lambda + \gamma}{2} - \psi \ln x\right),$$

$X \in MDA(\Lambda)$ and $X \in MDA(\underline{\Lambda})$. If $\gamma = 2$,

$$\mu(x) = \sigma^2 x\left(\frac{\lambda + 2}{2} - \psi \ln x\right)$$

and we have the Black–Derman–Toy diffusion model for short interest rates (see Black et al. [6]).

Example 6 Let $(l, r) = (b, \infty)$,

$$\sigma^2(x) = \sigma^2(x - b)^\gamma, \quad h(x) = C x^{-m}, \quad b > 0, \ \sigma^2 > 0, \ \gamma \ge 1, \ m \ge \gamma - 1.$$

The assumptions (1) – (4) are satisfied. We find that

$$\mu(x) = \frac{\sigma^2}{2}\big(\gamma(x - b)^{\gamma - 1} - m(x - b)^\gamma x^{-1}\big).$$

If $\gamma \ge 1, m > 1, m > \gamma - 1$, then $X \in MDA(\Phi_{m - \gamma + 1})$. If $\gamma > 1, m > 1, m \ge \gamma - 1$, then $X \in MDA(\underline{\Psi}_{\gamma - 1})$.

Example 7 (see Borkovec and Klüppelberg [7]) Let $(l, r) = (-\infty, \infty)$, $x_0 = 0$, $\sigma^2(x) = \sigma^2(h(x))^{-1}$, $\sigma^2 > 0$, and $h(x)$ be an arbitrary strictly positive probability density function on R^1. Then

$$g(x) \equiv \sigma^{-2}, \quad \mu(x) \equiv 0, \quad \gamma_T = \underline{\gamma}_T = \frac{\sigma^2}{2} T$$

and, for each $x, y \in R^1$,

$$\lim_{T \to \infty} P\left\{ \frac{2}{\sigma^2 T} M_T^X \leq x, \frac{2}{\sigma^2 T} \underline{M}_T^X \leq y \right\} = \Phi_1(x) \underline{\Phi}_1(y).$$

Example 8 Let $(l, r) = (-\infty, \infty)$,

$$\sigma^2(x) = \sigma^2 \left(1 + \left(\frac{x - \mu}{\alpha} \right)^2 \right)^\gamma,$$

$$h(x) = C \left(1 + \left(\frac{x - \mu}{\alpha} \right)^2 \right)^\lambda \times e^{-\varkappa \operatorname{arctg}\left(\frac{x - \mu}{\alpha} \right)},$$

$$\sigma^2 > 0, \ \alpha > 0, \ \mu, \gamma, \varkappa \in R^1, \ \lambda < -\frac{1}{2}.$$

Then

$$\mu(x) = \frac{\sigma^2}{\alpha} \left(1 + \left(\frac{x - \mu}{\alpha} \right)^2 \right)^{\gamma-1} \left[(\lambda + \gamma) \left(\frac{x - \mu}{\alpha} \right) - \frac{\varkappa}{2} \right]$$

and the assumptions (1) – (4) are satisfied iff $\lambda + \gamma \leq \frac{1}{2}$. If $\lambda + \gamma < \frac{1}{2}$ then $X \in MDA(\Phi_{1-2(\lambda+\gamma)})$ and $X \in MDA(\underline{\Phi}_{1-2(\lambda+\gamma)})$. Taking $\gamma = 1$, we have the Johannesma diffusion model for the stochastic activity of neurons (see Johannesma [19], Hanson and Tuckwell [18], Kallianpur and Wolpert [20]) and one of the Föllmer–Schweizer models for stock returns (see Föllmer and Schweizer [13]).

Example 9 Let $(l, r) = (-\infty, \infty)$,

$$\sigma^2(x) = \sigma^2(1 + \psi x^{2k})^\gamma, \quad h(x) = (2\pi \alpha^2)^{-\frac{1}{2}} \times \exp\left\{ -\frac{1}{2} \left(\frac{x - \mu}{\alpha} \right)^2 \right\},$$

$$\sigma^2 > 0, \ \alpha > 0, \ \psi \geq 0, \ \gamma, \mu \in R^1, \ k = 0, 1, 2, \ldots.$$

Then

$$\mu(x) = \sigma^2(1 + \psi x^{2k})^{\gamma-1} \left[\gamma k \psi x^{2k-1} - \frac{1}{2\alpha^2}(1 + \psi x^{2k})(x - \mu) \right],$$

$X \in MDA(\Lambda)$ and $X \in MDA(\underline{\Lambda})$.

Example 10 Let $(l, r) = (-\infty, \infty)$,

$$\sigma^2(x) = \sigma^2(1 + \psi_1 e^{\chi_1 x} + \psi_2 e^{-\chi_2 x})^\gamma,$$

$$h(x) = (2\pi\alpha^2)^{-\frac{1}{2}} \exp\left\{ -\frac{1}{2}\left(\frac{x-\mu}{\alpha}\right)^2 \right\},$$

$$\sigma^2 > 0, \ \alpha > 0, \ \psi_1 \geq 0, \ \psi_2 \geq 0, \ \chi_1 \geq 0, \ \chi_2 \geq 0, \ \gamma, \mu \in R^1.$$

We find that

$$\begin{aligned}
\mu(x) &= \frac{\sigma^2}{2}\left(1 + \psi_1 e^{\chi_1 x} + \psi_2 e^{-\chi_2 x}\right)^{\gamma-1}\left[\gamma\psi_1\chi_1 e^{\chi_1 x} - \gamma\psi_2\chi_2 e^{-\chi_2 x}\right. \\
&\quad \left. - \frac{1}{\alpha^2}\left(1 + \psi_1 e^{\chi_1 x} + \psi_2 e^{-\chi_2 x}\right)(x - \mu)\right],
\end{aligned}$$

$X \in MDA(\Lambda)$ and $X \in MDA(\underline{\Lambda})$.

As special cases, the last two examples include the Ornstein–Uhlenbeck process and the Vasicek diffusion model from the mathematical finance (see Vasicek [33]).

4. GENERALIZED z-DISTRIBUTIONS AND GENERALIZED z-DIFFUSIONS

Definition 3 (Grigelionis [16]) *A probability distribution H on R^1 is called a generalized z-distribution, denoted by $GZD(\alpha, \beta_1, \beta_2, \delta, \mu)$, $\alpha > 0$, $\beta_1 > 0$, $\beta_2 > 0$, $\delta > 0$, $\mu \in R^1$, if*

$$\int_{-\infty}^{\infty} e^{izx} H(dx) = \left(\frac{B\left(\beta_1 + \frac{i\alpha z}{2\pi}, \beta_2 - \frac{i\alpha z}{2\pi}\right)}{B(\beta_1, \beta_2)}\right)^{2\delta} e^{i\mu z}, \quad z \in R^1, \tag{8}$$

where $B(z_1, z_2)$ is the Euler beta function.

In Grigelionis [16] it is proved that $GZD(\alpha, \beta_1, \beta_2, \delta, \mu)$ is self-decomposable with the Lévy characteristics $(a, 0, \pi(dx))$, where

$$a = \frac{\alpha\delta}{\pi} \int_0^{\frac{2\pi}{\alpha}} \frac{\exp\{-\beta_2 x\} - \exp\{-\beta_1 x\}}{1 - e^{-x}}\, dx + \mu,$$

$$\pi(dx) = v(x)\, dx,$$

$$v(x) = \begin{cases} \dfrac{2\delta \exp\{-\frac{2\pi\beta_2}{\alpha}x\}}{x\left(1 - \exp\{-\frac{2\pi}{\alpha}x\}\right)}, & \text{if } x > 0, \\[3mm] \dfrac{2\delta \exp\{\frac{2\pi\beta_1}{\alpha}x\}}{|x|\left(1 - \exp\{\frac{2\pi}{\alpha}x\}\right)}, & \text{if } x < 0, \end{cases}$$

and have semiheavy tails, i.e., the density function $h(x)$, which is smooth and unimodal (see Yamazoto [35]), have the following behaviour:

$$h(x) \sim C_\pm |x|^{\varrho_\pm} e^{-\sigma_\pm |x|} \quad \text{as} \quad x \to \pm\infty, \tag{9}$$

where

$$\varrho_\pm = 2\delta - 1, \quad \sigma_+ = \frac{2\pi\beta_2}{\alpha}, \quad \sigma_- = \frac{2\pi\beta_1}{\alpha},$$

$$C_\pm = \left(\frac{2\pi}{\alpha B(\beta_1, \beta_2)}\right)^{2\delta} \frac{\exp\{\pm\mu\sigma_\pm\}}{\Gamma(2\delta)}.$$

If $\delta = \frac{1}{2}$, we have the class of z-distributions (see Prentice [28], Barndorff-Nielsen et al. [2]), with densities

$$h(x) = \frac{2\pi \exp\left\{\frac{2\pi\beta_1}{\alpha}(x-\mu)\right\}}{\alpha B(\beta_1, \beta_2)\left(1 + \exp\left\{\frac{2\pi}{\alpha}(x-\mu)\right\}\right)^{\beta_1+\beta_2}}, \quad x \in R^1. \tag{10}$$

If $\beta_1 = 1 - \beta_2 = \frac{1}{2} + \frac{\beta}{2\pi}, -\pi < \beta < \pi$, then

$$\int_{-\infty}^{\infty} e^{izx} H(dx) = \left(\frac{\cos\frac{\beta}{2}}{\cosh\left(\frac{\alpha z - i\beta}{2}\right)}\right)^{2\delta} e^{i\mu z}, \quad z \in R^1,$$

giving us the class of Meixner distributions (see Meixner [26], Schoutens and Teugels [32], Grigelionis [17], Schoutens [31]) with the density functions

$$h(x) = \frac{\left(2\cos\frac{\beta}{2}\right)^{2\delta}}{2\pi\alpha\Gamma(2\delta)}\left|\Gamma\left(\delta + i\frac{x-\mu}{\alpha}\right)\right|^2 \exp\left\{\beta\left(\frac{x-\mu}{\alpha}\right)\right\}, \quad x \in R^1,$$

where $\Gamma(z)$ is the Euler gamma function.

Definition 4 (Grigelionis [16]) *An H-diffusion X on R^1 is called a generalized z-diffusion, denoted by $GZDP(\alpha, \beta_1, \beta_2, \delta, \mu)$, if $H = GZD(\alpha_1, \beta_1, \beta_2, \delta, \mu)$.*

We shall formulate two propositions concerning the extremal behaviour of the generalized z-diffusion X, the proofs of which are based on the Theorems 3 and 3' and formula (9). We again shall omit the technical details of the elementary analysis.

Proposition 1 *Let X be a $GZDP(\alpha, \beta_1, \beta_2, \delta, \mu)$,*

$$\sigma^2(x) = \sigma^2(1 + \psi x^{2k})^\gamma, \quad \sigma^2 > 0, \ \psi \geq 0, \ \gamma \in R^1, \ k = 0, 1, \ldots.$$

Then the assumptions (1) – (4) are satisfied, $X \in MDA(\Lambda)$ and $X \in MDA(\underline{\Lambda})$.

Proposition 2 *Let X be a $GZDP(\alpha, \beta_1, \beta_2, \delta, \mu)$,*

$$\sigma^2(x) = \sigma^2\left(1 + \psi_1 e^{\chi_1 x} + \psi_2 e^{-\chi_2 x}\right)^\gamma,$$
$$\sigma^2 > 0, \ \psi_1 > 0, \ \psi_2 > 0, \ \chi_1 \geq 0, \ \chi_2 \geq 0, \ \gamma \in R^1, \ \sigma_+ \geq \gamma\chi_1, \ \sigma_- \geq \gamma\chi_2,$$

assuming that $\delta \leq 1$ if $\sigma_+ = \gamma\chi_1$ or $\sigma_- = \gamma\chi_2$. Then the assumptions (1) – (4) are satisfied and the following implications hold true:
    i) $\sigma_+ > \gamma\chi_1 \Rightarrow X \in MDA(\Lambda)$;
    ii) $\sigma_+ = \gamma\chi_1, \delta < 1 \Rightarrow X \in MDA(\Phi_{2-2\delta})$;
    iii) $\sigma_- > \gamma\chi_2 \Rightarrow X \in MDA(\underline{\Lambda})$;
    iv) $\sigma_- = \gamma\chi_2, \delta < 1 \Rightarrow X \in MDA(\underline{\Phi}_{2-2\delta})$.

Example 11 Let $(l, r) = (-\infty, \infty)$,

$$\sigma^2(x) = \sigma^2 \cosh^2\left(\pi\gamma\left(\frac{x-\mu}{\alpha}\right)\right), \quad \sigma^2 > 0, \ \gamma > 0, \quad H = GZD(\alpha, \beta_1, \beta_2, \delta, \mu).$$

Then the assumptions (1) – (4) are satisfied iff $\beta_1 \geq \gamma$ and $\beta_2 \geq \gamma$, assuming that $\delta \leq 1$ if $\beta_1 = \gamma$ or $\beta_2 = \gamma$. The following implications hold true:

 i) $\beta_2 > \gamma \Rightarrow X \in MDA(\Lambda)$;
 ii) $\beta_2 = \gamma, \delta < 1 \Rightarrow X \in MDA(\Phi_{2-2\delta})$;
 iii) $\beta_1 > \gamma \Rightarrow X \in MDA(\underline{\Lambda})$;
 iv) $\beta_1 = \gamma, \delta < 1 \Rightarrow X \in MDA(\underline{\Phi}_{2-2\delta})$.

If $\delta = \frac{1}{2}, \sigma^2 = \frac{\alpha^2}{\pi^2}, \gamma = 1$, from (2) and (10) we find that

$$\mu(x) = \frac{\alpha}{2\pi}\cosh\left(\pi\left(\frac{x-\mu}{\alpha}\right)\right)\left[(\beta_1 - \beta_2)\cosh\left(\pi\left(\frac{x-\mu}{\alpha}\right)\right)\right.$$
$$\left. - (\beta_1 + \beta_2 - 2)\sinh\left(\pi\left(\frac{x-\mu}{\alpha}\right)\right)\right], \quad x \in R^1,$$

giving us the generalized logistic diffusion process (cf. Kessler and Sørensen [23], Grigelionis [16]).

REFERENCES

[1] O.E. Barndorff-Nielsen, Processes of normal inverse Gaussian type, *Finance Stoch.* **2** (1998), 41–68.

[2] O.E. Barndorff-Nielsen, J. Kent and M. Sørensen, Normal variance-mean mixtures and z distributions, *Int. Statist. Rev.* **50** (1982), 145–159.

[3] S.M. Berman, Limiting distribution of the maximum of a diffusion process, *Ann. Math. Statist.* **35** (1964), 319–329.

[4] S.M. Berman, *Sojourns and extremes of stochastic processes*, Wadsworth, Belmond, California, 1992.

[5] B.M. Bibby, M. Sørensen, A hyperbolic diffusion model for stock prices, *Finance Stoch.* **1** (1997), 25–41.

[6] F. Black, E. Derman and W. Toy, A one-factor model of interest rates and its application to treasury bond options, *Financial Analysts J.* **11** (1990), 33–39.

[7] M. Borkovec and C. Klüppelberg, Extremal behaviour of diffusion models in finance, *Extremes* 1:1 (1998), 47–80.

[8] J. Cox, J. Ingersoll and S. Ross, A theory of the term structure of interest rates, *Econometrica* **53** (1985), 385–408.

[9] J.F. Crow and M. Kimura, *An introduction to population genetics theory*, Harper & Row Publishers, New York et al., 1970.

[10] R.A. Davis, Maximum and minimum of one-dimensional diffusions, *Stochastic Processes Appl.* **13** (1982), 1–9.

[11] P. Embrechts, C. Klüppelberg and T. Mikosh, *Modelling Extremal Events for Insurance and Finance*, Springer, Berlin et al., 1997.

[12] H.J. Engelbert and W. Schmidt, On solutions of one-dimensional stochastic differential equations without drift, *Z. Wahrscheinlichkeitstheorie verw. Gebiete* **68** (1985), 287–314.

[13] H. Föllmer, M. Schweizer, A microeconomic approach to diffusion models for stock prices, *Math. Finance* **3** (1) (1993), 1–23.

[14] J. Galambos, *Asymptotic Theory of Extreme Order Statistics*, Wiley, New York, 1978.

[15] B.V. Gnedenko, Sur la distribution limite du terme d'une serie aléatoire, *Ann. Math.* **44** (1943), 423–453.

[16] B. Grigelionis, Generalized z-distributions and related stochastic processes, Institute of Mathematics and Informatics, Preprint No 2000–22, Vilnius, 2000.

[17] B. Grigelionis, Processes of Meixner type, *Lietuvos Matem. Rink.* **39** (1) (1999), 40–51.

[18] F.B. Hanson and H.C. Tuckwell, Diffusion approximation for neural activity including synaptic reversal potentials, *J. Theoret. Neurobiol.* **2** (1983), 127–153.

[19] P.I.M. Johannesma, Diffusion models for the stochastic activity of neurons, In: *Neural Networks*, Caianiello E. R. (ed.), Springer, Berlin, 1968.

[20] G. Kallianpur and R. Wolpert, Weak convergence of stochastic neuronal models, In: *Stochastic Methods in Biology*, M. Kimura, G. Kallianpur and T. Hida (eds.), Springer, Berlin, New York, 1984.

[21] I. Karatzas and S.E. Shreve, *Brownian motion and stochastic calculus*, Springer, New York, 1998.

[22] S. Karlin and H.M. Taylor, *A second course in stochastic processes*, Academic Press, New York, 1981.

[23] M. Kessler and M. Sørensen, Estimating equations based on eigenfunctions for a discretely observed diffusion process, *Bernoulli* **5** (2) (1999), 299–314.

[24] M.R. Leadbetter, G. Lindgren and H. Rootzén, *Extremes and related properties of random sequences and processes*, Springer, Berlin, 1983.

[25] P. Mandl, *Analytical treatement of one-dimensional Markov processes*, Springer, New York, 1968.

[26] J. Meixner, Orthogonale Polynomsysteme mit einer besonderen Gestalt der erzeugenden Funktion, *J. London Math. Soc.* **9** (1934), 6–13.

[27] G.F. Newell, Asymptotic extreme value distribution for one-dimensional diffusion process, *J. Math. Mech.* **11** (1962), 481–496.

[28] R.L. Prentice, Discrimination among some parametric models, *Biometrika* **62** (1975), 607–614.

[29] S.I. Resnick, *Extreme values, regular variation and point processes*, Springer, New York, 1987.

[30] T.H. Rydberg, Generalized hyperbolic diffusions with applications in finance, *Math. Finance* **9** (2) (1999), 183–201.

[31] W. Schoutens, *Stochastic processes and orthogonal polynomials*, Lecture Notes in Statistics **146**, Springer, New York, 2000.

[32] W. Schoutens and J.L. Teugels, Lévy processes, polynomials and martingales, *Commun. Statist.-Stochastic Models* **14** (1, 2) (1998), 335–349.

[33] O.A. Vasicek, An equilibrium characterization of the term structure, *J. Financial Econ.* **5** (1977), 177–188.

[34] R. von Mises, La distribution de la plus grande de n valeurs, *Revue Mathématique de l'Union Interbalkanique* (Athens) **1** (1936), 141–160.

[35] M. Yamazoto, Unimodality of infinitely divisible distribution functions of class L, *Ann. Probab.* **6** (4) (1978), 523–531.

PORTFOLIO OPTIMISATION WITH TRANSACTION COSTS AND EXPONENTIAL UTILITY

RALF KORN and SILKE LAUE

Universität Kaiserslautern, FB Mathematik
D-67653 Kaiserslautern, Germany

Abstract The optimal portfolio problem under fixed and proportional transaction costs is examined for the case of exponential utility. Asymptotic solutions in markets with one or two stocks are given and their properties are highlighted.

Key Words Optimal portfolios, impulse control, transaction costs.

1. INTRODUCTION

The list of papers dealing with portfolio optimisation under transaction costs is still not very long. Especially the highly realistic case of transaction costs including both a fixed and a variable cost component is not dealt with very often. In Korn [4] an approach of Eastham and Hastings [2] was taken up and refined where the portfolio problem under fixed and proportional transaction costs was transformed into an impulse control problem. As an explicit example the exponential utility case in a market with just one stock and bond was considered. However, as it seems to be impossible to solve the corresponding quasi-variational inequalities explicitly, an asymptotic solution method was presented.

In the current paper we take up this asymptotic approach, simplify it and examine some of its characteristcs. Further, we consider the exponential utility portfolio problem in a market with two stocks and bonds and derive asymptotic solutions for it. Our findings will also indicate how to deal with the general n stock market setting.

171

2.   THE PROBLEM

We consider a financial market consisting of a riskless bond with constant price $P_0(t) = 1$ (i.e., we assume a zero interest rate) and a risky stock with price $P_1(t)$ given by

$$dP_1(t) = P_1(t) \left[\mu dt + \sigma dW_t \right]$$

where μ, σ are real constants with $\sigma > 0$ and where W_t is a one-dimensional standard Brownian motion with the usual right-continuous Brownian filtration $\mathcal{F}_t$. Our goal is the maximization of the expected utility from wealth $X(T)$ at the time horizon T. In this paper, we specialize to the case of the exponential utility function, i.e., we look at the problem

$$\max_{(\vartheta_i, \Delta S_i) \in Z} E(-e^{-\lambda X(T)})$$

where $\lambda > 0$ is the coefficient of risk aversion. The set Z denotes the set of all admissible impulse control strategies, i.e., of all $\{(\vartheta_i, \Delta S_i), i \in N\}$ where $\{\vartheta_i\}_{i \in N}$ is a sequence of intervention times ϑ_i and $\{\Delta S_i\}_{i \in N}$ a sequence of corresponding actions with

i) $0 \leq \vartheta_i \leq \vartheta_{i+1}$ a.s.,

ii) ϑ_i is a stopping time with respect to the filtration

$$\mathcal{G}_t := \sigma(\,(S(s-), B(s-)), s \leq t\,),$$

iii) ΔS_i is $\mathcal{G}_{\vartheta_i}$-measurable,

iv) $P(\lim_{i \to \infty} \vartheta_i \leq T) = 0$.

The times ϑ_i denote the trading times of the investor. His trades then consist of changing the amount of money invested in the stock by ΔS_i and a corresponding adjustment of the bond component. Instead of looking at the evolution of the stock and bond prices it will be more convenient to consider the evolution of the amounts of money invested in both stock and bond, which we denote by $S(t)$ and $B(t)$, respectively. Due to the above assumed price evolution the evolution of the stock and bond holdings between intervention times (i.e., times where the investor rebalances his holdings) are given by

$$\begin{aligned} dB(t) &= 0, \\ dS(t) &= S(t)\,(\mu dt + \sigma dW_t)\,. \end{aligned}$$

At the time of the i th transaction the investor has to pay transaction costs given by

$$K + k|\Delta S_i|, \quad 0 < K, \, 0 < k < 1$$

which are subtracted from the bond holdings. Here K will be the fixed cost component and $k|\Delta S_i|$ is proportional to the value of the transaction. We assume that there will be no transaction costs at the terminal time T (or equivalently: the investor is only interested in maximising his paper wealth). The analytical analogue to the Hamilton–Jacobi–Bellman equation of stochastic control in the above setting is to characterize the value function

$$v(t, b, s) := \sup_{(\vartheta_i, \Delta S_i) \in Z} E_{t,b,s}(-e^{-\lambda X(T)})$$

as a (certain kind of) solution of the so-called quasi variational inequalities (see Korn [4] for their derivation):

i) $Lv(t, b, s) := \frac{1}{2}\sigma^2 s^2 v_{ss}(t, b, s) + \mu s\, v_s(t, b, s) + v_t(t, b, s) \leq 0,$

ii) $v(t, b, s) \geq M\, v(t, b, s),$

iii) $(v(t, b, s) - Mv(t, b, s))\, Lv(t, b, s) = 0,$

iv) $v(T, b, s) = -e^{-\lambda(b+s)}$

with

$$Mv(t, b, s) := \max_{\Delta S \in R} v(t, b - K - \Delta S - k|\Delta S|, s + \Delta S).$$

As these quasi variational inequalities could not be solved explicitly an asymptotic method was proposed in Korn [4] based on a similar asymptotic method by Whalley and Wilmott [6] in the context of option pricing.

First we note that we have

$$v(t, b, s) = e^{-\lambda b} v(t, 0, s).$$

So to solve the problem, we can compute the optimal stock strategy independently of the bond holdings! We thus only have to find $v(t, 0, s)$. To this end, the function

$$q(t, p, y) = v(t, 0, yp)$$

is introduced and it is assumed that the (p, y)-space (i.e., the space of number of stocks held y and stock prices p) can be divided into three regions:

- the no-transaction region NT (i.e., the region in the (p, y)-space where it is optimal to do nothing);

- the selling region SS (i.e., the region in the (p, y)-space where it is optimal to sell shares);

- the buying region BS (i.e., the region in the (p, y)-space where it is optimal to buy shares).

As $v(t, b, s)$ should satisfy i) as an equality in NT, the relevant equality for $q(t, p, y)$ should hold. By expanding $q(t, p, y)$ in an infinite series of powers of $\varepsilon^{1/2}$ around the optimal pairs

$$(p, y^*(p)) = \left(p, \frac{\mu}{\lambda \sigma^2 p}\right)$$

for the problem without transaction costs (see Pliska [5]), putting this into the equation for q in NT, neglecting terms of higher order than ε, and using smoothness requirements at the borders between NT, SS and BS, it is shown in Korn [4] that the border points $y^*(p) + \varepsilon^{1/4} \widehat{Y}^+(p)$ between NT and SS and the optimal points $y^*(p) + \varepsilon^{1/4} Y^+(p)$ to which the investor should rebalance his stock holdings after reaching the border are given as the unique solution of

$$(Y^+(p) + \widehat{Y}^+(p)) \, \widehat{Y}^+(p) \, Y^+(p) = \frac{3 \mu^2 A}{\sigma^4 (\lambda p)^3} ,$$

$$(\widehat{Y}^+(p) - Y^+(p))^3 \, (\widehat{Y}^+(p) + Y^+(p)) = \frac{12 E \mu^2}{\lambda^3 (\sigma p)^4}$$

with $A = \frac{k}{\varepsilon^{3/4}}$ and $E = \frac{K}{\varepsilon}$. The border points $y^*(p) + \varepsilon^{1/4} \widehat{Y}^-(p)$ between NT and BS and the corresponding optimal restarting points $y^*(p) + \varepsilon^{1/4} Y^-(p)$ are given by

$$\widehat{Y}^-(p) = -\widehat{Y}^+(p), \quad Y^-(p) = -Y^+(p) .$$

To understand the properties of this solution is the main aim of the next section.

3. PROPERTIES OF THE ASYMPTOTIC SOLUTION

3.1. Constant Boundaries and Restarting Points

In the case without transaction costs Pliska's optimal volume of stock holdings is constant in p. In this part we will show that in the case of strictly positive transaction costs the (approximately) optimal boundaries and restarting points for the volume (not for the number of shares!) of the stock holdings as presented in the previous section are also constant in p.

Notation Let us denote the difference between the approximatively optimal upper restarting point and Pliska's optimal number of shares as

$$Y_\varepsilon^+(p) := \varepsilon^{1/4} Y^+(p)$$

and the difference between the approximatively optimal upper boundary of the no-transaction region and Pliska's solution as

$$\widehat{Y}_\varepsilon^+(p) := \varepsilon^{1/4} \widehat{Y}^+(p) .$$

Proposition 1 Let $k, K \neq 0$ and $Y_\varepsilon^+ > 0$. Then there exists exactly one $f > 1$ such that we have

$$\widehat{Y}_\varepsilon^+(p) = Y_\varepsilon^+(p) f \qquad \forall p > 0 .$$

Proof. Setting $y := Y_\varepsilon^+(p)$, the statement of Proposition 1 holds if the system

$$\begin{cases} (1+f) f y^3 = \frac{c}{p^3}, & c = \frac{3 \mu^2 k}{\sigma^4 \lambda^3}, \\ (f-1)^3 (f+1) y^4 = \frac{d}{p^4}, & d = \frac{12 \mu^2 K}{\sigma^4 \lambda^3} \end{cases}$$

has a solution (y, f) with $y > 0$ and $f > 1$ (since the remarks at the end of the previous section then imply that $(y, y\,f)$ is the only solution $(Y_\varepsilon^+(p), \widehat{Y}_\varepsilon^+(p))$ with $0 < Y_\varepsilon^+(p) < \widehat{Y}_\varepsilon^+(p)$). We prove the existence of a solution to the above system by computing it: Solving the first equation for y and putting this into the second one leads to

$$\frac{c^4}{d^3}(f - 1)^9 = f^4 + f^5$$

which has exactly one solution $f > 1$. From this we then obtain

$$y = \left(\frac{c}{p^3\, f\,(1 + f)}\right)^{1/3}.$$

□

Remark 1 The solution f depends only on the given and fixed parameters $k, K, \mu, \sigma, \lambda$ and is constant in p whereas $Y_\varepsilon^+(p)$ and $\widehat{Y}_\varepsilon^+(p)$ depend on both these parameters and on p. Note that $Y_\varepsilon^+(p)$ and $\widehat{Y}_\varepsilon^+(p)$ are independent of ε ! Let us therefore introduce the following notations:

$$y^+(p) := Y_\varepsilon^+(p), \quad \widehat{y}^+(p) := \widehat{Y}_\varepsilon^+(p).$$

Proposition 2 Let $k, K \geq 0$. The boundaries of the no-transaction region

$$\widehat{S}^+(p) := p\,y^* + p\,\widehat{y}^+(p)$$

and the restarting points

$$S^+(p) := p\,y^* + p\,y^+(p)$$

are constant in p. Thus, $\widehat{S}^+(p) \equiv \widehat{S}^+$ and $S^+(p) \equiv S^+$. The analogous results for $\widehat{S}^-(p)$ and $S^-(p)$ are also valid.

Proof. We consider the following four possible cases:

a) No transaction costs: $k = K = 0$. Then $\widehat{y}^+(p) = y^+(p) = 0$ and hence $p\,\widehat{y}^+(p) = p\,y^+(p) = 0$.

b) Only fixed transaction costs: $k = 0, K \neq 0$. In this case $y^+(p) = 0 = p\,y^+(p)$ and $p\,\widehat{y}^+(p) = \left(\frac{12\,\mu^2\,K}{\lambda^3\,\sigma^4}\right)^{1/4}$.

c) Only proportional transaction costs: $k \neq 0, K = 0$. This means $p\,y^+(p) = p\,\widehat{y}^+(p) = \left(\frac{3\,\mu^2\,k}{2\,\lambda^3\,\sigma^4}\right)^{1/3}$.

d) Fixed and proportional transaction costs: $k \neq 0, K \neq 0$. From the proof of Proposition 1 it follows that $p\,y^+(p) = \left(\frac{c}{f(1+f)}\right)^{1/3}$ is constant in p and, consequently, $p\,\widehat{y}^+(p) = p\,y^+(p)\,f$ is constant in p.

Since $p\,y^*$ is also constant in p Proposition 2 is proved.

□

3.2. Optimal Strategies

In the last part we found out that the boundaries of the no-transaction region and the restarting points (for the volume of the stock holdings) are constant in p. Now we want to analyse the position of the boundaries and restarting points, respectively. We will see that it depends on the combination of the parameters $k, K, \mu, \sigma, \lambda$ whether the lower boundary and restarting point, respectively, are above or below zero which then determines the strategy the investor has to follow. We will see that there are cases where the investor is only allowed to buy shares (if he starts with negative stock holdings) or to sell (if he starts with positive stock holdings) and others where both transactions can occur.

Proposition 3 a) *Let* $k = 0, K \neq 0$. *Then* $S^- \equiv S^* > 0$ *and*

$$\widehat{S}^- \begin{smallmatrix} > \\ < \end{smallmatrix} 0 \quad \Longleftrightarrow \quad \frac{\mu^2}{K\lambda\sigma^4} \begin{smallmatrix} > \\ < \end{smallmatrix} 12.$$

b) *Let* $k \neq 0, K = 0$. *Then*

$$\widehat{S}^- \equiv S^- \begin{smallmatrix} > \\ < \end{smallmatrix} 0 \quad \Longleftrightarrow \quad \frac{\mu}{k\sigma^2} \begin{smallmatrix} > \\ < \end{smallmatrix} \frac{3}{2}.$$

c) *Let* $k, K \neq 0$ *and define* $U := 3\frac{k\sigma^2}{\mu}$ *and* $V := 12\frac{K\lambda\sigma^4}{\mu^2}$. *Let* f *be given as in Proposition 1. Then*

$$\widehat{S}^- < S^- < 0 \quad \Longleftrightarrow \quad \begin{cases} U > & f(1+f), \\ V > & (f-1)^3(f+1), \end{cases}$$

$$\widehat{S}^- < 0 < S^- \quad \Longleftrightarrow \quad \begin{cases} \frac{1+f}{f^2} < U < & f(1+f), \\ \frac{(f-1)^3(f+1)}{f^4} < V < & (f-1)^3(f+1), \end{cases}$$

$$0 < \widehat{S}^- < S^- \quad \Longleftrightarrow \quad \begin{cases} U < \frac{1+f}{f^2}, \\ V < \frac{(f-1)^3(f+1)}{f^4}. \end{cases}$$

Proof. The lower boundary and restarting points are defined as $\widehat{S}^- = py^* - p\widehat{y}^+(p)$ and $S^- = py^* - py^+(p)$ with $py^* = \frac{\mu}{\lambda\sigma^2}$. In each case we replace $p\widehat{y}^+(p)$ and $py^+(p)$ by the suitable expressions given in the proofs of Proposition 1 and 2. Then the above statements follow immediately. □

Remark 2 In the following we will show that each of the above cases can occur for suitable parameter sets. While this is obvious in the first two cases where we only have fixed or proportional costs it is not in the case with fixed and proportional costs because there is no closed form solution for the factor f. Nevertheless, in this case we are able to find suitable examples numerically.

Examples 1

a)

k	K	μ	σ	λ	$S^* \equiv S^-$	$\widehat{S}^-$
0	1	0.01	0.2	0.01	25	-4.428
0	0.5	0.3	0.1	0.01	3000	2728.92

b)

k	K	μ	σ	λ	S^*	$S^- \equiv \widehat{S}^-$
0.1	0	0.05	0.6	0.001	138.89	-3.61
0.001	0	0.3	0.15	0.01	1333.33	1268.97

c)

k	K	μ	σ	λ	S^*	S^-	$\widehat{S}^-$
0.1	0.4	0.02	0.5	0.0001	800	-67.1	-316.37
0.01	0.3	0.01	0.3	0.01	11.11	9.66	-4.18
0.001	0.1	0.3	0.25	0.001	4800	4614,42	4275,49

The form of the optimal strategy a) Only fixed transaction costs ($k = 0$): If the lower boundary of the no-transaction region is below zero, the strategy of the investor is as follows: Starting with a positive number of shares (positive volume of stock holdings) the investor has to sell shares whenever he has to intervene. Starting with a negative number of shares, his first transaction will be to buy shares whereas any following transaction will be to sell shares.

If the lower boundary is above zero, the investor has to sell shares if the stock price increases sufficiently and to buy shares if the stock price decreases sufficiently.

In the case of fixed transaction costs a transaction always shifts the number of shares (volume of stock holdings) to Pliska's optimal number of shares (volume of stock holdings).

b) Only proportional transaction costs ($K = 0$): If the lower boundary of the no-transaction region is below zero the investor's strategy is to sell shares if he starts with positive stock holdings and to buy shares if he starts with negative stock holdings. In the case of the lower boundary being above zero, the investor's strategy is the same as in a).

With proportional transaction costs intervening means doing the least transaction that keeps the number of shares (volume of stock holdings) inside the no-transaction region (boundary = restarting point), i.e., the action is of local time type.

c) Fixed and proportional transaction costs: If lower boundary and lower restarting point both are below zero the investor has to sell shares if he starts with positive stock holdings and to buy shares if he starts with negative stock holdings. Starting with positive stock holdings, he also has to sell shares in the case when the lower boundary is below zero but the lower restarting point is above zero. In this case, if he starts with negative stock holdings, his first transaction will be to buy shares whereas any following transaction is to sell shares. If lower boundary and restarting point both are above zero, transactions can be selling or buying depending on the movements of the stock price.

3.3. Economic Justification

Since we computed an approximate solution to our optimisation problem, we are interested in at least a reasonable behaviour (in an economic sense) of this solution. Varying the transaction cost parameters should lead to changes of the boundaries and restarting points that match economically. For example, increasing the fixed cost parameter we expect the upper boundary of the no-transaction region to increase, too. In this part we will show that our approximate solution does react in a reasonable way to changes of the parameters.

Proposition 4 a) Let $k = 0, K \neq 0$. Then $S^+ \equiv S^- \equiv S^*$ and

$$K \uparrow \implies \widehat{S}^+ \uparrow, \ \widehat{S}^- \downarrow .$$

b) Let $k \neq 0, K = 0$. Then

$$k \uparrow \implies \widehat{S}^+ = S^+ \uparrow, \ \widehat{S}^- = S^- \downarrow .$$

c) Let $k, K \neq 0$. Then

$$k \uparrow \implies \widehat{S}^+ \uparrow, S^+ \uparrow, \qquad \widehat{S}^- \downarrow, S^- \downarrow, \qquad \widehat{S}^+ - S^+ = \widehat{S}^- - S^- \downarrow,$$

$$K \uparrow \implies \widehat{S}^+ \uparrow, S^+ \downarrow, \qquad \widehat{S}^- \downarrow, S^- \uparrow, \qquad \widehat{S}^+ - S^+ = \widehat{S}^- - S^- \uparrow .$$

Proof. Again we replace $p\widehat{y}^+(p)$ and $py^+(p)$ by the suitable expressions given in the proofs of Propositions 1 and 2. Then Proposition 4 follows via differentiation, since the proof of Proposition 1 also implies $\frac{\partial f}{\partial k} < 0$ and $\frac{\partial f}{\partial K} > 0$. □

Examples 2 Let $\mu = 0.2$, $\sigma = 0.3$, $\lambda = 0.001$.

	k	K	$\widehat{S}^-$	S^*	$\widehat{S}^+$
a)	0	0.1	1944.77	2222.22	2499.67
	0	0.2	1892.27	2222.22	2552.17

	k	K	$\widehat{S}^-$	S^*	$\widehat{S}^+$
b)	0.001	0	2027.29	2222.22	2417.16
	0.002	0	1976.62	2222.22	2467.82

	k	K	S^*	S^+	$\widehat{S}^+$	$\widehat{S}^+ - S^+$
c)	0.001	0.1	2222.22	2322.19	2560.44	238.25
	0.002	0.1	2222.22	2372.73	2597.00	224.27
	0.001	0.2	2222.22	2306.58	2601.22	294.64
	0.002	0.2	2222.22	2354.55	2633.85	279.3

Interpretation in the case of fixed and proportional transaction costs An increase of the proportional cost parameter leads to an increase of the upper boundary and restarting point and a decrease of the volume of transaction: The investor intervenes "later" than in the case of a lower proportional cost parameter in the sense that he allows his stock holdings to differ more from Pliska's optimal stock holdings without intervening. If the investor has to intervene, the volume of transaction is lower because the investor thus tries to compensate for the higher proportional cost parameter. An increase of the fixed transaction costs leads to an increase of the upper boundary of the no-transaction region, a decrease of the upper restarting point and thus to an increase of the volume of transaction: Again the investor intervenes "later" since the transaction costs are higher. If he has to intervene, his transaction has a higher volume because he wants to benefit as good as possible from the higher fixed transaction costs, which he has to pay anyway.

Remark 3 Without transaction costs the wealth process is of the following form (see Korn [3], p. 90–91):

$$X(t) = x + \vartheta^2\, t + \vartheta\, W_t\,.$$

Therefore, the wealth process always has a positive probability to attain negative values. In the case of strictly positive transaction costs we can guarantee – at least in some cases – a wealth process that will stay non-negative during the whole period of time:

a) In the case of $k = 0, K > 0$, the proof of Proposition 2 implies that the profit of selling shares $\widehat{S}^+ - S^*$ exceeds the transaction costs K if and only if $12 > \frac{\lambda^3\, \sigma^4\, K^3}{\mu^2}$.

b) Let the condition in a) be satisfied. In the case of $\widehat{S}^- < 0$ the wealth process remains non-negative for all $t \in [0, T]$ if

1) $S(0) > 0,\ B(0) > 0$ or

2) $S(0) < 0,\ B(0) > K + (\widehat{S}^+ - S^*)\,.$

Proof. (We only prove the second case since the first one is trivial.) Let

$$\tau_0 := \inf\{t : S(t) = \widehat{S}^-\}\,.$$

Then we obtain
1. $X(t) > 0$ for all $0 \le t < \tau_0$:

$$
\begin{aligned}
X(t) &= B(0) + S(t)\\
&> K + (\widehat{S}^+ - S^*) + \widehat{S}^-\\
&= K + (\widehat{S}^+ - S^*) + (2\,S^* - \widehat{S}^+)\\
&= K + S^*\\
&> 0\,.
\end{aligned}
$$

2. $X(\tau_0) > 0$ (immediately after the transaction (buying)):

$$X(\tau_0) = (B(0) - K - (\widehat{S}^+ - S^*)) + S^* > 0\,.$$

3. $X(t) > 0$ for all $t > \tau_0$ without any further transaction:

$$X(t) = B(\tau_0) + S^{S^*}(t - \tau_0) > B(\tau_0) > 0\,.$$

4. $X(t) > 0$ for $t > \tau_0$ in the case of further transactions (selling): $S(t)$ stays above zero. Since the profit exceeds the transaction costs the bond holdings increase, thus $B(t) > 0$, too.   □

Remark 4 As we proved in the first part of this section, the approximately optimal boundaries and restarting points for the stock holdings are constant in p and over time. This finding gives us the idea of trying to solve our optimisation problem by assuming that the boundaries and restarting points we are looking for are constant. This then only seems to be a four parameter static (!) problem which at first sight should be easier to solve than the original impulse control problem. However, it turns out that this problem is much more complicated than the one given by the asymptotic analysis. Even more, obtaining an explicit solution seems to be impossible.

4. MULTI-DIMENSIONAL CASE

4.1. Asymptotic Analysis

Now we will analyse the optimisation problem in the case of a market with $n \geq 1$ stocks and one bond with interest rate $r = 0$. The stock price processes are given by

$$dP_i(t) = P_i(t) \left[\mu_i \, dt + \sum_{j=1}^{n} \sigma_{ij} \, dW_t^j \right]$$

where W_t is an n-dimensional Brownian motion. Let $p := (P_1, \ldots, P_n)$ and $1/p := (1/P_1, \ldots, 1/P_n)$.

Paralleling the computations of Example 3.31 in Korn [3] we easily show that in the absence of transaction costs it is optimal to hold exactly

$$y^* = \frac{1}{\lambda} \operatorname{diag} \left(\frac{1}{p} \right) (\sigma\sigma^T)^{-1} \mu$$

shares at each time instant where $\operatorname{diag}(x)$ is defined as the diagonal matrix with entries $x_1, \ldots, x_n$ on its diagonal.

In order to find the optimal strategy in the presence of transaction costs we will now carry out the same asymptotic procedure as in the one-dimensional case.

Again we set $B = 0$ and replace $v(t, 0, y\,p)$ by $q(t, p, y)$. With

$$Lq(t, p, y) \ := \ q_t(t, p, y) + q_p(t, p, y)(\operatorname{diag}(p)\, b)$$
$$+ \frac{1}{2} \operatorname{trace}(q_{pp}(t, p, y)\, \operatorname{diag}(p)\, \sigma\sigma^T \operatorname{diag}(p)\,)$$

and

$$Mq(t, p, y) := \max_{u \in R^d} \{ e^{\lambda(K + k|u|^T p + u^T p)} q(t, p, y + u) \}$$

where $|u| = (|u_1|, \ldots, |u_n|)$, the quasi variational inequalities for q are given by

$$Lq(t, p, y) \ \leq \ 0,$$

$$q(t,p,y) \geq Mq(t,p,y),$$
$$(q(t,p,y) - Mq(t,p,y)) Lq(t,p,y) = 0,$$
$$q(T,p,y) = e^{-\lambda p^T y}.$$

We define $Q(t,p,Y) := q(t,p,y)$ where $y = y^* + \varepsilon^{1/4}Y$ and expand Q in NT in powers of $\varepsilon^{1/2}$ which again leads to

$$Q(t,p,Y) = -e^{-[\lambda p(y^* + \varepsilon^{1/4}Y) + H_0(t) + \varepsilon^{1/2}H_1(p,t) + \varepsilon H_2(p,t,Y) + \cdots]}.$$

Due to this expansion of Q the derivatives of q are

$$q_t = -Q[H_{0_t}(t) + \varepsilon^{1/2}H_{1_t}(t,p) + \varepsilon H_{2_t}(t,p,Y) + \cdots],$$

$$q_{p_i} = -Q[\lambda y_i^* + \lambda \varepsilon^{1/4}Y_i + \varepsilon^{1/2}H_{1_{p_i}}(t,p) - \varepsilon^{3/4}H_{2_{Y_i}}(t,p,Y)(y_i^*)_{p_i}$$
$$+ \varepsilon H_{2_{p_i}}(t,p,Y) + \cdots],$$

$$q_{p_ip_j} = Q\{\lambda^2 y_i^* y_j^* + \varepsilon^{1/4}\lambda^2(y_i^*Y_j + y_j^*Y_i)$$
$$+ \varepsilon^{1/2}[\lambda^2 Y_iY_j + \lambda(y_i^*H_{1_{p_j}}(t,p) + y_j^*H_{1_{p_i}}(t,p))$$
$$- H_{1_{p_ip_j}}(t,p) - H_{2_{Y_iY_j}}(t,p,Y)(y_i^*)_{p_i}(y_j^*)_{p_j}] + \varepsilon^{3/4}\cdots\}$$

which we plug into the above formula for Lq and rearrange by powers of $\varepsilon^{1/4}$. The usual neglection of higher order terms leads to

$$0 = Q\{[-H_{0_t} - \lambda \sum_{i=1}^{n} p_i \mu_i y_i^* + \frac{1}{2}\lambda^2 \sum_{i=1}^{n}\sum_{j=1}^{n}\sum_{k=1}^{n} \sigma_{ik}\sigma_{jk} p_i p_j y_i^* y_j^*]$$

$$+\varepsilon^{1/4}[-\lambda \sum_{i=1}^{n} p_i \mu_i Y_i + \frac{1}{2}\lambda^2 \sum_{i=1}^{n}\sum_{j=1}^{n}\sum_{k=1}^{n} \sigma_{ik}\sigma_{jk} p_i p_j (y_i^* Y_j + y_j^* Y_i)]$$

$$+\varepsilon^{1/2}[-H_{1_t} - \sum_{i=1}^{n} p_i \mu_i H_{1_{p_i}} + \frac{1}{2}\sum_{i=1}^{n}\sum_{j=1}^{n}\sum_{k=1}^{n} \sigma_{ik}\sigma_{jk} p_i p_j$$

$$(\lambda^2 Y_i Y_j + \lambda y_i^* H_{1_{p_j}} + \lambda y_j^* H_{1_{p_i}} - H_{1_{p_ip_j}} - H_{2_{Y_iY_j}} (y_i^*)_{p_i}(y_j^*)_{p_j})]\}$$

in NT, and we thus have to solve the following three equations:

$$O(1): \quad 0 = H_{0_t} + \lambda y^{*'}(\mathrm{diag}(p)\,\mu)$$
$$- \frac{1}{2}\lambda^2 \mathrm{trace}[(\mathrm{diag}(p)\,\sigma\sigma'\mathrm{diag}(p))\,(y^*\,y^{*'})],$$

$$O(\varepsilon^{1/4}): \quad 0 = \lambda Y'(\mathrm{diag}(p)\,\mu)$$
$$- \frac{1}{2}\lambda^2 \mathrm{trace}[(\mathrm{diag}(p)\,\sigma\sigma'\mathrm{diag}(p))\,(y^*\,Y' + Y\,y^{*'})],$$

$$O(\varepsilon^{1/2}): \quad 0 = H_{1_t} + H_{1_p}(\mathrm{diag}(p)\,\mu) - \frac{1}{2}\mathrm{trace}[(\mathrm{diag}(p)\,\sigma\sigma'\mathrm{diag}(p))$$
$$(\lambda^2 Y\,Y' + \lambda\,(y^*\,H_{1_p} + (y^*\,H_{1_p})') - H_{1_{pp}} - y_p^{*'}\,H_{2_{YY}}\,y_p^*)].$$

If we define $U := \mathrm{diag}(p)\,\sigma\sigma'\mathrm{diag}(p)$ and use $\mathrm{trace}(M\,X\,Y') = X'M'Y$ for any vectors $X, Y \in R^n$ and any matrix $M \in R^{n,n}$, equation $O(\varepsilon^{1/4})$ leads to

$$\lambda Y' \text{diag}(p) \mu = \lambda^2 Y' U y^*.$$

This is correct due to the form of y^* which implies even more:

$$\text{diag}(p) \mu = \lambda U y^*.$$

We use the last equality to solve equation $O(1)$:

$$\begin{aligned}
0 &= H_{0_t} + \lambda^2 y^{*'} U y^* - \frac{1}{2} \lambda^2 \underbrace{\text{trace}(U y^* y^{*'})}_{=y^{*'} U y^*} \\
&= H_{0_t} + \frac{1}{2} \lambda^2 y^{*'} U y^*.
\end{aligned}$$

Together with the boundary condition $H_0(T) = 0$, which ensures that Q is equal to the value function without transaction costs in the case $\varepsilon = 0$, this leads to

$$\begin{aligned}
H_0(t) &= \frac{1}{2} \lambda^2 y^{*'} U y^* (T - t) \\
&= \frac{1}{2} \mu' (\sigma \sigma')^{-1} \mu (T - t).
\end{aligned}$$

Finally we have to solve equation $O(\varepsilon^{1/2})$.

4.2. Two Uncorrelated Stocks

As this is much more complicated in the multi-stock setting we will solve the last equation and thus the whole optimisation problem in the 2-dimensional case with uncorrelated stocks. Even more, we restrict ourselves to the cases of either only fixed transaction costs or only proportional transaction costs. The more general case will be considered in future work. At first we simplify equation $O(\varepsilon^{1/2})$ in the case of two uncorrelated stocks. Using again the notation U we obtain

$$\begin{aligned}
0 &= H_{1_t} + H_{1_p}(\text{diag}(p) \mu) - \frac{1}{2} \text{trace}[(\text{diag}(p) \sigma \sigma' \text{diag}(p)) \\
&\quad (\lambda^2 Y Y' + \lambda (y^* H_{1_p} + (y^* H_{1_p})') - H_{1_{pp}} - y_p^{*'} H_{2_{YY}} y_p^*)] \\
&= H_{1_t} + \frac{1}{2} \text{trace}(U H_{1_{pp}}) - \frac{1}{2} \lambda^2 Y' U' Y + \frac{1}{2} \text{trace}(U y_p^{*'} H_{2_{YY}} y_p^*) \\
&=: F(H_1) - \frac{1}{2} \lambda^2 Y' U' Y + \frac{1}{2} \text{trace}(U y_p^{*'} H_{2_{YY}} y_p^*) \\
&= F(H_1) - \frac{1}{2} \lambda^2 (a_1 Y_1^2 + a_2 Y_2^2) + \frac{1}{2} \frac{1}{\lambda^2} \left(\frac{\mu_1^2}{a_1} H_{2_{Y_1 Y_1}} + \frac{\mu_2^2}{a_2} H_{2_{Y_2 Y_2}} \right)
\end{aligned}$$

where $a_1 := p_1^2 \sigma_1^2$ and $a_2 := p_2^2 \sigma_2^2$. To simplify our further calculations we set $H := H_2$, $x := Y_1$, $y := Y_2$ and $F_1 := F(H_1)$ which leads to

$$0 = F_1 - \frac{1}{2} \lambda^2 (a_1 x^2 + a_2 y^2) + \frac{1}{2} \frac{1}{\lambda^2} \left(\frac{\mu_1^2}{a_1} H_{xx} + \frac{\mu_2^2}{a_2} H_{yy} \right). \qquad (*)$$

We fix (p_1, p_2) and t and therefore drop these variables.

A simple integration to obtain H as in the one-stock case is no longer possible. We therefore determine H as the unique solution of an appropriate free boundary problem thereby also obtaining the optimal intervention and restarting points for our problem. As those solutions are unique (see Crank [1] for standard results on free boundary problems), we only have to construct the relevant solution which we will do below.

Let us at first look at the case of only proportional transaction costs:

4.2.1. Proportional transaction costs We will show that for fixed (p_1, p_2) in the case of $K = 0, k > 0$ the function H is of the form

$$H(x, y) = A x^4 + B x^2 + C y^4 + D y^2 + E \quad \forall (x, y) \in NT$$

where

$$NT = \{(x^*, y^*)' + \varepsilon^{1/4} (x, y)' : |x| \le \alpha, |y| \le \beta\}$$

for suitable constants $A, \ldots, E$. The boundary of NT is thus given by a rectangle and also coincides with the restarting line: $x^+ = \hat{x}^+$ and $y^+ = \hat{y}^+$. Relation (∗) and the assumed form of H and NT determine α and β as

$$\alpha = \varepsilon^{-\frac{1}{4}} \left(\frac{3}{2} \frac{k \mu_1^2}{\sigma_1^4 p_1^3 \lambda^3} \right)^{\frac{1}{3}}, \qquad \beta = \varepsilon^{-\frac{1}{4}} \left(\frac{3}{2} \frac{k \mu_2^2}{\sigma_2^4 p_2^3 \lambda^3} \right)^{\frac{1}{3}}.$$

Proof. While solving the one-dimensional case, we have derived some conditions on H which have to be satisfied in the two-dimensional case, too (see Korn [4]). Taking into consideration the form of NT these conditions read

(1) $\lambda K(x^+ - \hat{x}^+, y^+ - \hat{y}^+) = H(x^+, y^+) - H(\hat{x}^+, \hat{y}^+) = 0$,

(2) $H_x(\alpha, \hat{y}^+) = \lambda K_x(\alpha - \alpha, y^+ - \hat{y}^+) = -\lambda k \varepsilon^{-\frac{3}{4}} p_1$ for all $\hat{y}^+, y^+$ with $|\hat{y}^+| = |y^+| = \beta$,

(3) $H_x(-\alpha, \hat{y}^+) = \lambda K_x(-\alpha + \alpha, y^+ - \hat{y}^+) = \lambda k \varepsilon^{-\frac{3}{4}} p_1$ for all $\hat{y}^+, y^+$ with $|\hat{y}^+| = |y^+| = \beta$,

(4) $H_y(\hat{x}^+, \beta) = \lambda K_y(x^+ - \hat{x}^+, \beta - \beta) = -\lambda k \varepsilon^{-\frac{3}{4}} p_2$ for all $\hat{x}^+, x^+$ with $|\hat{x}^+| = |x^+| = \alpha$,

(5) $H_y(\hat{x}^+, -\beta) = \lambda K_y(x^+ - \hat{x}^+, -\beta + \beta) = \lambda k \varepsilon^{-\frac{3}{4}} p_2$ for all $\hat{x}^+, x^+$ with $|\hat{x}^+| = |x^+| = \alpha$,

(6) $H_{xx}(x^+, y^+) = 0$ for all x^+, y^+ with $|x^+| = \alpha, |y^+| = \beta$,

(7) $H_{yy}(x^+, y^+) = 0$ for all x^+, y^+ with $|x^+| = \alpha, |y^+| = \beta$.

Now we determine the constants $A, \ldots, E$ such that the above conditions as well as relation (∗) are satisfied: Since $\lambda K(0, 0) = 0$, equality (1) is satisfied. The equalities (6) and (7) read

$$H_{xx}(x^+, y^+) = 12 A \alpha^2 + 2 B = 0$$

and

$$H_{yy}(x^+, y^+) = 12\,C\,\beta^2 + 2\,D = 0$$

and thus are satisfied if and only if

$$A = -\frac{B}{6\,\alpha^2} \quad \text{and} \quad C = -\frac{D}{6\,\beta^2}\,.$$

This leads to

$$H(x,y) = -\frac{B}{6\,\alpha^2}\,x^4 + B\,x^2 - \frac{D}{6\,\beta^2}\,y^4 + D\,y^2 + E\,.$$

Relation $(*)$ is satisfied if and only if

$$\frac{1}{2}\lambda^2\,a_1 = -\frac{1}{\lambda^2}\frac{\mu_1^2}{a_1\,\alpha^2}\,B \quad \text{and} \quad \frac{1}{2}\lambda^2\,a_2 = -\frac{1}{\lambda^2}\frac{\mu_2^2}{a_2\,\beta^2}\,D$$

which together with the above relations between A and B and C and D, respectively, leads to

$$A = \frac{1}{12}\lambda^4\frac{a_1^2}{\mu_1^2}, \quad C = \frac{1}{12}\lambda^4\frac{a_2^2}{\mu_2^2}\,.$$

Now we use the remaining conditions to determine α and β: From condition (2)

$$-\lambda\,k\,\varepsilon^{-\frac{3}{4}}\,p_1 = -\frac{2}{3}\frac{B}{\alpha^2}\,\alpha^3 + 2\,B\,\alpha = \frac{4}{3}\,B\,\alpha$$

we obtain that

$$\alpha = \varepsilon^{-\frac{1}{4}}\left(\frac{3}{2}\frac{k\,\mu_1^2}{\sigma_1^4\,p_1^3\,\lambda^3}\right)^{\frac{1}{3}}\,.$$

In analogy we derive β which completes the proof. $\square$

Considering the wealth invested in the first and the second stock, respectively, we have to rebalance our portfolio whenever our wealth-combination reaches the boundary of $(S_1^*, S_2^*)' + (S_1, S_2)'$ where

$$(S_1, S_2)' \in \left\{ |S_1| \le \left(\frac{3}{2}\frac{k\,\mu_1^2}{\sigma_1^4\,\lambda^3}\right)^{\frac{1}{3}}, |S_2| \le \left(\frac{3}{2}\frac{k\,\mu_2^2}{\sigma_2^4\,\lambda^3}\right)^{\frac{1}{3}} \right\}\,.$$

It might be surprising at first sight that NT has a rectangular form. However, as both stocks are independent and as there are no fixed transaction costs there is no bargaining effect of combining actions in both stocks.

4.2.2. Fixed transaction costs We will show that for fixed (p_1, p_2) in the case of $K > 0$, $k = 0$ the function H is of the form

$$H(x,y) = A\left(\frac{x^2}{\alpha^2} + \frac{y^2}{\beta^2} - 1\right)^2 + B, \quad (x,y) \in NT,$$

where

$$NT = \left\{ (x^*, y^*)' + \varepsilon^{1/4} (x, y)' : \frac{x^2}{\alpha^2} + \frac{y^2}{\beta^2} \le 1 \right\}.$$

The boundary of NT consists of the ellipsoid

$$(x^*, y^*)' + \varepsilon^{1/4} (x, y)' \quad \text{for all} \quad \frac{x^2}{\alpha^2} + \frac{y^2}{\beta^2} = 1$$

whereas the restarting line only consists of Pliska's optimal number of shares

$$(x^*, y^*)' = \frac{1}{\lambda} \left(\frac{\mu_1}{p_1 \sigma_1^2}, \frac{\mu_2}{p_2 \sigma_2^2} \right)',$$

i.e., $(x^+, y^+)' = (0, 0)$. The relation $(*)$ and the assumed form of H and NT determine α and β as

$$\alpha = \varepsilon^{-\frac{1}{4}} \left(\frac{K}{\lambda^3} \right)^{\frac{1}{4}} \frac{\mu_1}{p_1 \sigma_1} \left(\frac{192}{17 \mu_1^2 + \mu_2^2 - \sqrt{\mu_1^4 + \mu_2^4 + 34 \mu_1^2 \mu_2^2}} \right)^{\frac{1}{4}},$$

$$\beta = \varepsilon^{-\frac{1}{4}} \left(\frac{K}{\lambda^3} \right)^{\frac{1}{4}} \frac{\mu_2}{p_2 \sigma_2} \left(\frac{192}{17 \mu_2^2 + \mu_1^2 - \sqrt{\mu_1^4 + \mu_2^4 + 34 \mu_1^2 \mu_2^2}} \right)^{\frac{1}{4}}.$$

Proof. The conditions on H now read

(1) $\lambda K(-\widehat{x}^+, -\widehat{y}^+) = H(0, 0) - H(\widehat{x}^+, \widehat{y}^+)$ for all $(\widehat{x}^+, \widehat{y}^+) \in \partial NT$,

(2) $H_x(0, 0) = \lambda K_x(-\widehat{x}^+, -\widehat{y}^+)$ for $(\widehat{x}^+, \widehat{y}^+) = (\alpha, 0)$ and $(\widehat{x}^+, \widehat{y}^+) = (-\alpha, 0)$,

(3) $H_x(\widehat{x}^+, \widehat{y}^+) = \lambda K_x(-\widehat{x}^+, -\widehat{y}^+)$ for $(\widehat{x}^+, \widehat{y}^+) = (\alpha, 0)$ and $(\widehat{x}^+, \widehat{y}^+) = (-\alpha, 0)$,

(4) $H_y(0, 0) = \lambda K_y(-\widehat{x}^+, -\widehat{y}^+)$ for $(\widehat{x}^+, \widehat{y}^+) = (0, \beta)$ and $(\widehat{x}^+, \widehat{y}^+) = (0, -\beta)$,

(5) $H_y(\widehat{x}^+, \widehat{y}^+) = \lambda K_y(-\widehat{x}^+, -\widehat{y}^+)$ for $(\widehat{x}^+, \widehat{y}^+) = (0, \beta)$ and $(\widehat{x}^+, \widehat{y}^+) = (0, -\beta)$.

We determine the constants A and B such that the conditions above as well as relation $(*)$ are satisfied: The equality (1) shows that

$$A = H(0, 0) - H(\widehat{x}^+, \widehat{y}^+) = \lambda K(-\widehat{x}^+, -\widehat{y}^+) = \frac{\lambda K}{\varepsilon}.$$

Since $k = 0$ we obtain that

$$\lambda K_x(-\widehat{x}^+, -\widehat{y}^+) = 0 = 2 A \left(\frac{x^2}{\alpha^2} + \frac{y^2}{\beta^2} - 1 \right) \frac{2x}{\alpha^2} = H_x(x, y)$$

for $(x, y) = (x^+, y^+) = (0, 0)$ as well as for $(x, y) = (\widehat{x}^+, \widehat{y}^+) = (\alpha, 0)$ and $(x, y) = (\widehat{x}^+, \widehat{y}^+) = (-\alpha, 0)$. Thus equalities (2) and (3) are satisfied and simultaneously also (4) and (5). Before we use relation $(*)$ we simplify H. We set

$$\widetilde{\alpha} := \sqrt{\frac{\lambda K}{\varepsilon} \frac{1}{\alpha^2}}, \quad \widetilde{\beta} := \sqrt{\frac{\lambda K}{\varepsilon} \frac{1}{\beta^2}}.$$

This leads to

$$H(x, y) = \left(x^2 \, \tilde{\alpha} + y^2 \, \tilde{\beta} - \sqrt{\frac{\lambda K}{\varepsilon}} \right)^2 + B \, .$$

Using the second derivatives of H with respect to x and y, respectively, we see that relation $(\ast)$ is satisfied if and only if

$$\lambda^4 \, a_1 = 12 \, \frac{\mu_1^2}{a_1} \, \tilde{\alpha}^2 + 4 \, \frac{\mu_2^2}{a_2} \, \tilde{\alpha} \, \tilde{\beta} \quad \text{and} \quad \lambda^4 \, a_2 = 12 \, \frac{\mu_2^2}{a_2} \, \tilde{\beta}^2 + 4 \, \frac{\mu_1^2}{a_1} \, \tilde{\alpha} \, \tilde{\beta} \, .$$

The first equation is equivalent to

$$\tilde{\beta} = \frac{\lambda^4 \, a_1 - 12 \, \frac{\mu_1^2}{a_1} \, \tilde{\alpha}^2}{4 \, \mu_2^2 \, \tilde{\alpha}} \, a_2$$

which we insert into the second equation. Thus we get

$$\lambda^4 \, a_2 = \frac{\mu_1^2 \, a_2}{\mu_2^2 \, a_1} \left(a_1 \, \lambda^4 - 12 \, \frac{\mu_1^2}{a_1} \, \tilde{\alpha}^2 \right) + \frac{3}{4} \, \frac{a_2}{\mu_2^2 \, \tilde{\alpha}^2} \left(a_1 \, \lambda^4 - 12 \, \frac{\mu_1^2}{a_1} \, \tilde{\alpha}^2 \right)^2$$

which is equivalent to

$$\tilde{\alpha}^4 - (17 \, \mu_1^2 + \mu_2^2) \, \frac{\lambda^4 \, a_1^2}{96 \, \mu_1^4} \, \tilde{\alpha}^2 + \frac{3}{4} \, \frac{\lambda^8 \, a_1^4}{96 \, \mu_1^4} = 0 \, .$$

Setting $\hat{\alpha} := \tilde{\alpha}^2$ we get the corresponding solutions

$$\hat{\alpha}_1 = \frac{\lambda^4 \, a_1^2}{192 \, \mu_1^4} \left(17 \, \mu_1^2 + \mu_2^2 + \sqrt{\mu_1^4 + \mu_2^4 + 34 \, \mu_1^2 \, \mu_2^2} \right)$$

and

$$\hat{\alpha}_2 = \frac{\lambda^4 \, a_1^2}{192 \, \mu_1^4} \left(17 \, \mu_1^2 + \mu_2^2 - \sqrt{\mu_1^4 + \mu_2^4 + 34 \, \mu_1^2 \, \mu_2^2} \right)$$

which are positive. Since $\tilde{\alpha}$ has to be positive, too, we find

$$\tilde{\alpha}_1 = \frac{\lambda^2 \, a_1}{\sqrt{192} \, \mu_1^2} \sqrt{17 \, \mu_1^2 + \mu_2^2 + \sqrt{\mu_1^4 + \mu_2^4 + 34 \, \mu_1^2 \, \mu_2^2}}$$

and

$$\tilde{\alpha}_2 = \frac{\lambda^2 \, a_1}{\sqrt{192} \, \mu_1^2} \sqrt{17 \, \mu_1^2 + \mu_2^2 - \sqrt{\mu_1^4 + \mu_2^4 + 34 \, \mu_1^2 \, \mu_2^2}} \, .$$

In order to decide which is the right $\tilde{\alpha}$ we look at the corresponding values of $\tilde{\beta}$:

$$\tilde{\beta}_i = \underbrace{\frac{a_2}{4 \, \tilde{\alpha}_i \, \mu_2^2}}_{> 0} \underbrace{\left(a_1 \, \lambda^4 - 12 \, \frac{\mu_1^2}{a_1} \, \tilde{\alpha}_i^2 \right)}_{=: C_i}$$

where

$$C_1 = a_1 \lambda^4 - \frac{a_1 \lambda^4}{16 \mu_1^2} \underbrace{\left(17 \mu_1^2 + \mu_2^2 + \sqrt{\mu_1^4 + \mu_2^4 + 34 \mu_1^2, \mu_2^2} \right)}_{> 17 \mu_1^2} < 0 \,,$$

$$C_2 = a_1 \lambda^4 - \frac{a_1 \lambda^4}{16 \mu_1^2} \underbrace{\left(17 \mu_1^2 + \mu_2^2 - \underbrace{\sqrt{\mu_1^4 + \mu_2^4 + 34 \mu_1^2, \mu_2^2}}_{> \mu_1^2 + \mu_2^2} \right)}_{< 16 \mu_1^2} > 0 \,.$$

Therefore, there is only one positive pair $(\tilde{\alpha}, \tilde{\beta})$ which reads

$$\tilde{\alpha} = \frac{\lambda^2 a_1}{\sqrt{192} \, \mu_1^2} \, z \,, \qquad \tilde{\beta} = \frac{a_2}{4 \, \mu_2^2 \, \tilde{\alpha}} \left(a_1 \lambda^4 - 12 \frac{\mu_1^2}{a_1} \tilde{\alpha}^2 \right)$$

with

$$z := \sqrt{17 \mu_1^2 + \mu_2^2 - \sqrt{\mu_1^4 + \mu_2^4 + 34 \mu_1^2 \mu_2^2}} \,.$$

Using

$$\alpha = \left(\frac{\lambda K}{\varepsilon} \right)^{\frac{1}{4}} \frac{1}{\sqrt{\tilde{\alpha}}} \quad \text{and} \quad \beta = \left(\frac{\lambda K}{\varepsilon} \right)^{\frac{1}{4}} \frac{1}{\sqrt{\tilde{\beta}}}$$

we obtain

$$\alpha = \left(\frac{K}{\lambda^3} \right)^{\frac{1}{4}} \varepsilon^{-\frac{1}{4}} \frac{\mu_1}{p_1 \sigma_1} \sqrt[4]{192} \frac{1}{\sqrt{z}} \,,$$

$$\beta = \left(\frac{K}{\lambda^3} \right)^{\frac{1}{4}} \varepsilon^{-\frac{1}{4}} \frac{\mu_2}{p_2 \sigma_2} \sqrt[4]{192} \frac{\sqrt{z}}{\sqrt{3} \sqrt{16 \mu_1^2 - z^2}} \,.$$

Further transformations even show (as we expected it due to symmetry)

$$\beta = \left(\frac{K}{\lambda^3} \right)^{\frac{1}{4}} \varepsilon^{-\frac{1}{4}} \frac{\mu_2}{p_2 \sigma_2} \sqrt[4]{192} \frac{1}{\sqrt{z_2}}$$

where

$$z_2 := \sqrt{17 \mu_2^2 + \mu_1^2 - \sqrt{\mu_1^4 + \mu_2^4 + 34 \mu_1^2 \mu_2^2}}$$

which completes the proof.   □

Considering again the wealth invested in the two stocks we have to rebalance our portfolio whenever our wealth-combination reaches the boundary of $(S_1^*, S_2^*)' + (S_1, S_2)'$ where $\frac{S_1^2}{a^2} + \frac{S_2^2}{b^2} \leq 1$ with

$$a := \left(\frac{K}{\lambda^3} \right)^{\frac{1}{4}} \frac{\mu_1}{\sigma_1} \left(\frac{192}{17 \mu_1^2 + \mu_2^2 - \sqrt{\mu_1^4 + \mu_2^4 + 34 \mu_1^2 \mu_2^2}} \right)^{\frac{1}{4}} \,,$$

$$b := \left(\frac{K}{\lambda^3}\right)^{\frac{1}{4}} \frac{\mu_2}{\sigma_2} \left(\frac{192}{17\,\mu_2^2 + \mu_1^2 - \sqrt{\mu_1^4 + \mu_2^4 + 34\,\mu_1^2\,\mu_2^2}}\right)^{\frac{1}{4}}.$$

In contrast to the purely proportional cost problem, we now do the biggest possible transaction towards the optimal point without transaction costs. This is clear as the transaction costs are not size dependent. On the other hand, we realize that the form of NT clearly points out that a transaction now typically involves rebalancing in both stocks.

REFERENCES

[1] J. Crank, *Free and moving boundary problems*, Oxford University Press, 1984.

[2] J. Eastham, K. Hastings, Optimal impulse control of portfolios, *Mathematics of Operations Research* **13** (1988), 588–605.

[3] R. Korn, *Optimal portfolios*, World Scientific, 1997.

[4] R. Korn, Portfolio optimisation with strictly positive transaction costs and impulse control, *Finance Stoch.* **2** (1998), 85–114.

[5] S.R. Pliska, A stochastic calculus model of continuous trading: Optimal portfolios, *Mathematics of Operations Research* **11** (1986), 371–382.

[6] E. Whalley, P. Wilmott, Optimal hedging of options with small but arbitrary cost structure, *OCIAM Working paper*, Mathematical Institute Oxford, 1994.

A SEMIMARTINGALE BACKWARD EQUATION RELATED TO THE p-OPTIMAL MARTINGALE MEASURE AND THE LOWER PRICE OF A CONTINGENT CLAIM*

MICHAEL MANIA [1], MARINA SANTACROCE [2]
and   REVAZ TEVZADZE [3]

[1] A. Razmadze Mathematical Institute
Georgian Academy of Sciences, Tbilisi, Georgia
[2] Department of Probability and Statistics
La Sapienza University, Rome, Italy
[3] Institute of Cybernetics
Georgian Academy of Sciences, Tbilisi, Georgia

Abstract We consider an incomplete financial market model, where the dynamics of asset prices is determined by an R^d-valued continuous semimartingale. Using the dynamic programming approach we give a description of the p-optimal martingale measure in terms of the value process for a suitable problem of an optimal equivalent change of measure and show that this value process uniquely solves the corresponding semimartingale backward equation. This result is applied to approximate the lower price and the corresponding hedging strategy of a contingent claim.

Key Words Semimartingale backward equation, contingent claim pricing, p-optimal martingale measure, incomplete markets, lower and upper prices.

2000 Mathematics Subject Classification 91B28, 60H30, 90C39.

*Research supported by Grant INTAS 97-30204.

1. INTRODUCTION AND THE MAIN RESULTS

Assume that the dynamics of the discounted prices of some traded assets is described by an R^d-valued continuous semimartingale $X = (X_t, t \in [0, T])$ defined on a filtered probability space $(\Omega, \mathcal{F}, F = (F_t, t \in [0, T]), P)$ satisfying the usual conditions, where $\mathcal{F} = F_T$ and $T < \infty$ is a fixed time horizon. The process X is adapted to the filtration F and admits the decomposition

$$X_t = X_0 + \Lambda_t + M_t, \tag{1.1}$$

where M is a continuous local martingale and Λ is a continuous process of finite variation. For the absence of "arbitrage" in this market it is necessary to assume that X satisfies the structure condition; this means that there exists a predictable R^d-valued process $\lambda = (\lambda_t, t \in [0, T])$ such that

$$d\Lambda_t = d\langle M \rangle_t \lambda_t \quad \text{a.s. for } t \in [0, T],$$

and $K_T = \int_0^T \lambda_s' d\langle M \rangle_s \lambda_s < \infty$ a.s., where $'$ denotes the transposition. The process K is called the mean-variance tradeoff process of X (see [29], [30] for the interpretation of the process K).

By $\mathcal{M}^{abs}$ we denote the set of measures Q absolutely continuous with respect to P on F_T such that X is a local martingale under Q. Let $\mathcal{M}^e$ be a set of equivalent martingale measures, i.e., a subset of $\mathcal{M}^{abs}$ containing probability measures which are equivalent to P. Let $Z_t(Q)$ be the density process of Q relative to the basic measure P. For any $Q \in \mathcal{M}^e$ there is a P-local martingale M^Q such that $Z^Q = \mathcal{E}(M^Q) = (\mathcal{E}_t(M^Q), t \in [0, T])$, where $\mathcal{E}(M)$ is the Doleans-Dade exponential of M. If the local martingale $\widehat{Z}_t = \mathcal{E}_t(-\lambda \cdot M), t \in [0, T])$ is a strictly positive martingale, $d\widehat{P}/dP = \widehat{Z}_T$ defines an equivalent probability measure called the minimal martingale measure for X. Here we used the notation $\lambda \cdot M$ for the stochastic integral with respect to M.

Let

$$\mathcal{M}_p^e = \{Q \in \mathcal{M}^e : E\eta \left(\frac{dQ}{dP}\right)^p < \infty\},$$

where η is a nonnegative F_T-measurable random variable.

Troughout the paper, we make the following assumptions:

A) there is an equivalent martingale measure $\tilde{Q}$ such that

$$E\eta \mathcal{E}_T^p(M^{\tilde{Q}}) < \infty;$$

B) all P-local martingales are continuous;

C) there is a constant k_1 such that $\eta \geq k_1 > 0$.

Remark Condition A) is natural and is related to some kind of no-arbitrage condition if $\eta = 1$ (see [5] for the definition of "arbitrage" and related results). We notice that, since X is continuous, the existence of an equivalent martingale measure implies that the structure condition is satisfied. Assumption B) means, in particular, the continuity of filtration F and is restrictive, but it is satisfied if the filtration F is generated by a Brownian motion, or, more generally, if F admits the integral representation property relative to some vector-valued continuous martingale. Also we notice that the main results are true if we replace condition C) by $E\eta^{\frac{1}{1-p}} < \infty$, $p > 1$.

Sometimes we replace condition A) by the stronger condition

A*) the random variable η is bounded, i.e.,

$$\eta \leq k_2 \tag{1.2}$$

for some constant $k_2 > k_1$ and there exists the minimal martingale measure satisfying the reverse Hölder inequality $R_p(P)$, i.e., there is a constant C such that

$$E(\mathcal{E}^p_{\tau,T}(-\lambda \cdot M)|F_\tau) \leq C$$

for any stopping time τ.

Here and in the following we use the notation

$$\mathcal{E}_{\tau,T}(N) = \frac{\mathcal{E}_T(N)}{\mathcal{E}_\tau(N)} = \mathcal{E}_T(N - N_{.\wedge\tau})$$

for a continuous local martingale N.

We consider the optimization problems:

$$\min_{Q\in\mathcal{M}^e_p} E\eta\mathcal{E}^p_T(M^Q), \quad p \geq 1, \tag{1.3}$$

$$\max_{Q\in\mathcal{M}^e} E\eta\mathcal{E}^p_T(M^Q), \quad 0 < p \leq 1. \tag{1.4}$$

Let

$$V_t(p) = \operatorname*{ess\,inf}_{Q\in\mathcal{M}^e_p} E(\eta\mathcal{E}^p_{tT}(M^Q)|F_t), \quad p \geq 1, \tag{1.5}$$

and

$$\overline{V}_t(p) = \operatorname*{ess\,sup}_{Q\in\mathcal{M}^e_p} E(\eta\mathcal{E}^p_{tT}(M^Q)|F_t), \quad 0 < p \leq 1. \tag{1.6}$$

be the value processes of the problems (1.3) and (1.4), respectively.

For $p = 1$, the processes $V_t(p)$ and $\overline{V}_t(p)$ represent the lower and upper prices of a contingent claim η at the moment t.

For $\eta = 1$, (1.3) is the problem of finding the p-optimal martingale measure, in particular, for $p = 2$ the solution of problem (1.3) gives the variance optimal martingale measure which plays an essential role in the mean variance hedging problem (see, e.g., [8], [16], [26], [27], [31]).

It is well known that the p-optimal martingale measure Q^* exists in the class $\mathcal{M}^{abs}$ and it was shown in [6] ([17] in case $p > 1$) that Q^* is equivalent to P if condition A) is satisfied and X is continuous. It was proved in [6] (this fact was already observed in [10], [28], [31] in various degrees of generality) that if X is a locally bounded semimartingale and if the measure Q^* is variance optimal then the corresponding density Z^* is represented as

$$Z^*_T = c + \int_0^T h'_s dX_s$$

for a constant c and an X-integrable process h, where the process $\int_0^t h'_s dX_s, t \in [0, T]$, is a Q-martingale for any $Q \in \mathcal{M}^e_2$.

We derive the corresponding fact for $p > 1$ (under assumptions A), B)) using the semimartingale backward equation for the value process. Moreover, we obtain an explicit expression of the integrand h in terms of the value process $V_t(p)$ and show that $V_t(p)$ uniquely solves a suitable semimartingale backward equation.

Now we formulate the main statement of this paper which is a combination of Theorem 1 b) and Corollary 2.

Let Y be a semimartingale with the decomposition

$$Y_t = Y_0 + B_t + L_t, \quad B \in \mathcal{A}_{\text{loc}}, \quad L \in \mathcal{M}^2_{\text{loc}}, \tag{1.7}$$

and let

$$L_t = \int_0^t \psi_s' dM_s + \tilde{L}_t, \quad \langle \tilde{L}, M \rangle = 0, \tag{1.8}$$

be the Galtchouk–Kunita–Watanabe decomposition of L with respect to the martingale M.

If conditions A*), B), and C) are satisfied, then the value process $V(p)$ is the unique solution of the semimartingale backward equation

$$Y_t = Y_0 - \underset{Q \in \mathcal{M}_p^e}{\text{ess inf}} \left[\frac{1}{2} p(p-1) \int_0^t Y_s \, d\langle M^Q \rangle_s + p \langle M^Q, L \rangle_t \right] + L_t, \quad t < T, \tag{1.9}$$

with the boundary condition

$$Y_T = \eta \tag{1.10}$$

in the class of processes Y satisfying the two-sided inequality

$$c \le Y_t \le C \quad \text{for all } t \in [0, T] \text{ a.s.}, \tag{1.11}$$

for some positive constants $c < C$.

Moreover, the martingale measure Q^* is p-optimal if and only if its density $Z^* = \mathcal{E}_T(M^{Q^*})$ is expressed as

$$\mathcal{E}_T^{(p-1)}(M^{Q^*}) = Y_0 + \int_0^T \mathcal{E}_s\left(\frac{\psi}{Y} + (1-p)\lambda\right) \cdot X\right)\left(\frac{\psi_s}{Y_s} + (1-p)\lambda_s\right)' dX_s. \tag{1.12}$$

We also show that the value process satisfies (1.9), (1.10) if we replace A*) by condition A) (Theorem 1.a), but in this case the class of processes in which this solution is unique is not explicitly described.

The same problem was studied by Laurent and Pham [21], in the case $p = 2$, $\eta = e^{-\int_0^T r_s ds}$, where the process r is the instantaneous interest rate. Using the dynamic programming approach, they obtain a characterization of the variance-optimal martingale measure in terms of the value function of a stochastic control problem (equivalent to (1.3)) in the case of Brownian filtration. In [24] the case $p = 2$, $\eta = 1$ under the present assumptions was considered, where in addition a backward equation for the value process $V(2)$ was derived.

The second question considered here is the problem of computing the upper and lower prices of a contingent claim.

It is well known ([12], [20], [15]) that the lower price $V_t(1) = V_t$ admits a decomposition

$$V_t = V_0 + \int_0^t \varphi_s dX_s + A_t \quad t \in [0, T], \tag{1.13}$$

where φ is a predictable X-integrable process and $A = (A_t, t \in [0, T])$ is an adapted increasing process of saving. This assertion is called an optional decomposition of the wealth process, since the increasing process A from (1.13) is optional, in general. In the case under consideration, the process A is predictable (since by condition B) we have no purely discontinuous martingales), but this process can be discontinuous as it is shown in example 1 of [12].

The optional decomposition (1.13) is invariant with respect to $Q \in \mathcal{M}^e$ and it gives a representation of the value process as a controlled portfolio with saving. However, V (hence, H and C) is generally difficult to compute, since it is usually impossible to represent V as a unique solution of a backward SDE (or as a unique solution of the Bellman equation in the Markov case).

By El Karoui and Quenez [12], in the context of Brownian model, the compact subclasses $(\mathcal{M}_n^e, n \geq 1)$ of martingale measures were considered, so that the essential infimum V^n taken over all elements of $\mathcal{M}_n^e$ is attained. The subclass $\mathcal{M}_n^e$ is chosen so that V_t is the limit of V_t^n (for each $t \in [0, T]$ a.s.) and V^n is determined as a unique solution of a backward equation. In [23] this result was extended, where the asset price process is driven by a locally bounded semimartingale.

Here we give an alternative way for computing V using the value processes $V(\varepsilon)$ of (1.5) type. We show (Theorem 2) that

$$\lim_{\varepsilon \downarrow 0} V_\tau(\varepsilon) = V_\tau \quad \text{a.s.,} \tag{1.14}$$

for any stopping time τ, where the process $V(\varepsilon)$ is determined by a semimartingale backward equation.

Note that we don't require here the existence of an increasing process which dominates the square characteristics $\langle M^Q \rangle$ for all $Q \in \mathcal{M}^e$, used in [12], [23] and in addition we show the convergence of the martingale parts, so that an approximation of the hedging strategies is also possible. It should be mentioned that the passage to the limit in the semimartingale backward equation for the process $V(\varepsilon)$ gives the optional decomposition (1.13), where the process A is predictable in our particular case.

The paper is organized as follows. In Section 2 it is shown that the value process related to the p-optimal martingale measure uniquely solves the corresponding backward semimartingale equation and a description of the p-optimal martingale measure in terms of this value process is given. In Section 3 we apply the results of Section 2 to approximate the lower price of a contingent claim and the corresponding hedging strategy. At the end of Sections 2 and 3 we formulate corresponding statements for the case $0 < p < 1$.

The backward SDEs were introduced by Bismut [1] for the linear case, by Chitashvili [2] and Pardoux and Peng [25] (see also [13] for references and related results) for more general generators. The semimartingale backward equation, as a stochastic version of the Bellman equation in optimal control problems, was first derived by Chitashvili [2] (see also [3], [4]).

For all unexplained notations concerning the martingale theory we refer to Jacod [18], Dellacherie and Meyer [7], and Liptzer and Shiryayev [22]. About BMO-martingales and reverse Hölder conditions see Doleans-Dade and Meyer [9] or Kazamaki [19].

2. BACKWARD SEMIMARTINGALE EQUATION FOR THE VALUE PROCESS

We say that the process B strongly dominates the process A and write $A \prec B$, if the difference $B - A \in \mathcal{A}_{\text{loc}}^+$, i.e., is a locally integrable increasing process.

Let $(A^Q, Q \in \mathcal{Q})$ be a family of processes of finite variations, zero at time zero. Denote by $\underset{Q \in \mathcal{Q}}{\text{ess inf}} (A^Q)$ the largest increasing process, zero at time zero, which is strongly dominated by the process $(A_t^Q, t \in [0, T])$ for every $Q \in \mathcal{Q}$, i.e., this is an "ess inf" of the family $(A^Q, Q \in \mathcal{Q})$ relative to the partial order $\prec$.

We recall the definition of BMO-martingales and the reverse Hölder condition.

The square integrable continuous martingale M belongs to the class BMO if there is a constant $C > 0$ such that

$$E(\langle M \rangle_T - \langle M \rangle_\tau | F_\tau) \leq C^2, \quad P\text{-a.s.}$$

for every stopping time τ. The smallest constant with this property (or $+\infty$ if this does not exist) is called the BMO norm of M and is denoted by $\|M\|_{\mathrm{BMO}}$.

A strictly positive adapted process Z satisfies the reverse Hölder inequality $R_p(P)$, where $1 < p < \infty$, if there is a constant C such that

$$E\left(\left(\frac{Z_T}{Z_\tau}\right)^p|F_\tau\right) \le C, \quad P\text{-a.s.}$$

for every stopping time τ.

We will use the following assertion proved by Delbaen and Schachermayer [6] in the case $p = 2$.

Proposition 1 *If $U = (U_t, t \in [0,T])$ is a non-negative p-integrable martingale ($p > 1$) with $U_0 > 0$, if the stopping time $\tau = \inf\{t : U_t = 0\}$ is predictable and announced by a sequence of stopping times $(\tau_n, n \ge 1)$, then*

$$E\left(\frac{U_T^p}{U_{\tau_n}^p}|F_{\tau_n}\right) \to \infty, \quad n \to \infty$$

on the F_{τ_-}-measurable set $\{U_\tau = 0\}$.

If $p < 1$ and U is a uniformly integrable martingale, then

$$E\left(\frac{U_T^p}{U_{\tau_n}^p}|F_{\tau_n}\right) \to 0, \quad n \to \infty,$$

on the set $\{U_\tau = 0\}$.

Proof. For $p > 1$ the proof is the same as in [6]. In case $0 < p < 1$ one can prove this assertion using arguments similar to [6]. Using the Hölder inequality we have

$$E\left(\frac{U_T^p}{U_{\tau_n}^p}|F_{\tau_n}\right) = E\left(\frac{U_T^p}{U_{\tau_n}^p}I_{(U_\tau \ne 0)}|F_{\tau_n}\right) \le E^{1-p}(I_{(U_\tau \ne 0)}|F_{\tau_n})$$

and the Lévy theorem implies that $E^{1-p}(I_{(U_\tau=0)}|F_{\tau_n})$ tends to zero on the set $(U_\tau = 0)$. □

Since X is continuous, any element Q of $\mathcal{M}^e$ is given by the density $Z_t(Q)$ which is expressed as an exponential martingale of the form

$$\mathcal{E}_t(-\lambda \cdot M + N)$$

where N is a local martingale strongly orthogonal to M.

By $\mathcal{N}(X)$ we denote the class of local martingales N strongly orthogonal to M such that the process $(\mathcal{E}_t(-\lambda \cdot M + N), t \in [0,T])$ is a martingale under P.

Let $\mathcal{N}_p(X)$ be the subclass of $\mathcal{N}(X)$ of local martingales N such that the process $(\mathcal{E}_t(-\lambda \cdot M + N), t \in [0,T])$ is a strictly positive P-martingale with $E\eta\mathcal{E}_T^p(-\lambda \cdot M + N) < \infty$. Then

$$\mathcal{M}_p^e = \{Q \sim P : \frac{dQ}{dP}|F_T = \mathcal{E}_T(-\lambda \cdot M + N), N \in \mathcal{N}_p(X)\}. \qquad (2.1)$$

The following assertion can be proved in a standard manner (see, e.g., [14], [12], [21]).

Proposition 2 (Optimality Principle) a) *There exists an RCLL semimartingale, still denoted by $V_t(p)$, such that for each $t \in [0,T]$*

$$V_t(p) = \operatorname*{ess\,inf}_{Q \in \mathcal{M}_p^e} E(\eta \mathcal{E}_{tT}^p(M^Q)|F_t) \quad a.s.$$

$V_t(p)$ is the largest RCLL process equal to η at time T such that $V_t(p)\mathcal{E}_t^p(M^Q)$ is a submartingale for every $Q \in \mathcal{M}_p^e$.

b) *The following properties are equivalent:*

(i) Q^* is p-optimal, i.e.,

$$V_0(p) = \inf_{Q \in \mathcal{M}_p^e} E\eta\mathcal{E}_T^p(M^Q) = E\eta\mathcal{E}_T^p(M^{Q^*}),$$

(ii) Q^* is p-optimal for all conditional criteria, i.e., $\forall t \in [0,T]$

$$V_t(p) = E(\eta\mathcal{E}_{tT}^p(M^{Q^*})|F_t) \quad a.s.$$

(iii) $V_t(p)\mathcal{E}_t^p(M^{Q^*})$ is a P-martingale.

We recall that the process X belongs to the class D if the family of random variables $X_\tau I_{(\tau \le T)}$ for all stopping times τ is uniformly integrable.

Let S (resp. S_+) be the class of semimartingales (resp. strictly positive semimartingales).

Definition 1 We say that Y belongs to the class D_p if Y is a RCLL process such that for every $Q \in \mathcal{M}_p^e$ the process $\mathcal{E}_t^p(M^Q)Y_t$ is in D.

Remark Since for every $Q \in \mathcal{M}_p^e$ the process $\mathcal{E}_t^p(M^Q)$ belongs to the class D, as a positive submartingale (see Dellacherie and Meyer 1980), then any bounded positive process Y belongs to the class D_p.

Definition 2 By $S(X)$ we denote the class of strictly positive semimartingales Y such that $Y \in D_p$ and $-\frac{1}{(p-1)Y} \cdot \tilde{L} \in \mathcal{N}(X)$, i.e., such that $(\mathcal{E}_t(-\lambda \cdot M - \frac{1}{(p-1)Y} \cdot \tilde{L}), t \in [0,T])$ is a martingale, where $\tilde{L}$ is the local martingale introduced in (1.8).

Let us consider the optimization problem (1.3). One can rewrite the value process $V(p)$ of this problem in the form

$$V_t(p) = \operatorname*{ess\,inf}_{N \in \mathcal{N}_p(X)} E(\eta\mathcal{E}_{tT}^p(-\lambda \cdot M + N)|F_t). \tag{2.2}$$

Since $\mathcal{M}^e \ne \varnothing$ the process $V(p)$ is a semimartingale with respect to the measure P and let

$$V_t(p) = m_t + A_t, \quad m \in M_{loc}^2, \quad A \in \mathcal{A}_{loc}, \tag{2.3}$$

be the canonical decomposition of $V(p)$.

Let

$$m_t = \int_0^t \varphi_s dM_s + \tilde{m}_t, \quad \langle \tilde{m}, M \rangle = 0, \tag{2.4}$$

be the Galtchouk–Kunita–Watanabe decomposition of m with respect to M.

Theorem 1 *Let conditions* A), B) *and* C) *be satisfied. Then*

a) *the value process $V(p)$ is a solution of the semimartingale backward equation*

$$Y_t = Y_0 - \operatorname*{ess\,inf}_{N \in \mathcal{N}_p(X)} \left[\frac{1}{2}p(p-1) \int_0^t Y_s \, d\langle -\lambda \cdot M + N \rangle_s + p \langle \lambda \cdot M + N, L \rangle_t\right] + L_t, \quad t < T, \tag{2.5}$$

with the boundary condition

$$Y_T = \eta. \tag{2.6}$$

This solution is unique in the class $S(X)$ of semimartingales. Moreover, the martingale measure Q^* is p-optimal if and only if it is given by the density $dQ^* = \mathcal{E}_T(M^{Q^*})dP$, where

$$M_t^{Q^*} = -\int_0^t \lambda_s' dM_s - \frac{1}{p-1}\int_0^t \frac{1}{V_s(p)} d\tilde{m}_s. \tag{2.7}$$

b) *If in addition condition* A*) *is satisfied then the value process V is the unique solution of the semimartingale backward equation* (2.5), (2.6) *in the class of semimartingales Y satisfying the two-sided inequality*

$$c \le Y_t \le C \quad \text{for all} \quad t \in [0, T] \quad a.s. \tag{2.8}$$

for some positive constants $c < C$.

Proof. a) *Existence.* According to Proposition 2, $\mathcal{E}_t^p(M^Q)V_t(p)$ is a P-submartingale for every $Q \in \mathcal{M}_p^e$. Therefore, by assumption A (since there exists $Q \in \mathcal{M}_p^e$ with $\mathcal{E}(M^Q)$ strictly positive) $V(p)$ will be a P-semimartingale with decomposition (2.3). Using the equality

$$\mathcal{E}^p(M^Q) = \mathcal{E}(pM^Q + \frac{p(p-1)}{2}\langle M^Q \rangle)$$

and the Itô formula for $\mathcal{E}_t^p(M^Q)V_t(p)$ we have that

$$\begin{aligned}
&\mathcal{E}_t^p(M^Q)V_t(p) \\
&= V_0(p) + \int_0^t \mathcal{E}_s^p(M^Q)dV_s(p) + \int_0^t V_{s-}(p)\mathcal{E}_s^p(M^Q)d(pM_s^Q + \frac{p(p-1)}{2}\langle M^Q \rangle_s) \\
&\quad + p\int_0^t \mathcal{E}_s^p(M^Q)d[V(p), M^Q]_s \\
&= V_0(p) + \int_0^t \mathcal{E}_s^p(M^Q)d(A_s + \frac{p(p-1)}{2}(V_-(p)\cdot\langle M^Q \rangle)_s + p\langle M^Q, m\rangle_s) \\
&\quad + \int_0^t \mathcal{E}_s^p(M^Q)dm_s + p\int_0^t V_{s-}(p)\mathcal{E}_s^p(M^Q)dM_s^Q. \tag{2.9}
\end{aligned}$$

Since $\mathcal{E}_t^p(M^Q)V_t(p)$ is a P-submartingale for all $Q \in \mathcal{M}_p^e$ and $\mathcal{E}_t(M^Q)$ is strictly positive, we obtain from (2.9) that

$$A_t + \frac{p(p-1)}{2}\int_0^t V_s(p)d\langle M^Q \rangle_s + p\langle M^Q, m\rangle_t \in \mathcal{A}_{loc}^+ \tag{2.10}$$

for every $Q \in \mathcal{M}_p^e$. It is well known that for convex coercive continuous functions defined on a closed convex subset of a reflexive Banach space the infimum is attained (see, Ekeland and Temam [11]). Since the set of densities $Z_T(Q)$ of absolutely continuous local martingale measures Q with $E\eta Z_T^p(Q) < \infty$ is a closed convex subset of $L^p(\eta \cdot P)$ and $\|\cdot\|_{L^p(\eta \cdot P)}^p$ is a convex coercive function then an optimal martingale measure Q^* exists. Note that the class of densities $(Z(Q), Q \in \mathcal{M}_p^e)$ is not closed in general. Therefore, we have only that $Q^* \in \mathcal{M}_p^{abs}$, where

$$\mathcal{M}_p^{abs} = \{Q \in \mathcal{M}^{abs} : E\eta\left(\frac{dQ}{dP}\right)^p < \infty\}.$$

Let us show that the existence of an equivalent martingale measure $\tilde{Q}$ with

$$E\eta\mathcal{E}_T^p(M^{\tilde{Q}}) < \infty$$

implies that Q^* is equivalent to P. We prove this fact using the idea of Delbaen and Schacher-mayer [6]. Since Q^* is optimal we have that

$$E\eta Z_T^p(Q^*) \le E\eta \mathcal{E}_T^p(M^{\tilde{Q}}), \tag{2.11}$$

where by $Z_t(Q^*)$ we denote the density process of Q^* relative to the measure P. Following [6], we define the stopping times

$$\tau_n = \inf\{t : Z_t(Q^*) \le 1/n\} \text{ and } \tau = \inf\{t : Z_t(Q^*) = 0\}. \tag{2.12}$$

The inequality (2.11) implies that for every $n \ge 1$

$$E\Big[\eta \frac{Z_T^p(Q^*)}{Z_{\tau_n}^p(Q^*)}|F_{\tau_n}\Big] \le E\Big[\eta \frac{\mathcal{E}_T^p(M^{\tilde{Q}})}{\mathcal{E}_{\tau_n}^p(M^{\tilde{Q}})}|F_{\tau_n}\Big] \quad \text{a.s.} \tag{2.13}$$

Indeed, if the measure of the set B defined by

$$B = \{\omega : E\Big[\eta \frac{Z_T^p(Q^*)}{Z_{\tau_n}^p(Q^*)}|F_{\tau_n}\Big] > E\Big[\eta \frac{\mathcal{E}_T^p(M^{\tilde{Q}})}{\mathcal{E}_{\tau_n}^p(M^{\tilde{Q}})}|F_{\tau_n}\Big]\} \tag{2.14}$$

is strictly positive, then constructing a new (absolutely continuous) martingale measure $\widehat{Q}$ by $d\widehat{Q} = \widehat{Z}_T dP$,

$$\widehat{Z}_T = I_B Z_{\tau_n}(Q^*)\frac{\mathcal{E}_T(M^{\tilde{Q}})}{\mathcal{E}_{\tau_n}(M^{\tilde{Q}})} + I_{B^c} Z_T(Q^*)$$

we have that

$$E\eta(\widehat{Z}_T)^p = E\eta Z_{\tau_n}^p(Q^*)\Big[I_B \frac{\mathcal{E}_T(M^{\tilde{Q}})}{\mathcal{E}_{\tau_n}(M^{\tilde{Q}})} + I_{B^c}\frac{Z_T(Q^*)}{Z_{\tau_n}(Q^*)}\Big]^p$$

$$= E Z_{\tau_n}^p(Q^*)\Big[I_B E(\eta \frac{\mathcal{E}_T^p(M^{\tilde{Q}})}{\mathcal{E}_{\tau_n}^p(M^{\tilde{Q}})}|F_{\tau_n}) + I_{B^c} E(\eta \frac{Z_T^p(Q^*)}{Z_{\tau_n}^p(Q^*)})|F_{\tau_n})\Big] < E\eta Z_T^p(Q^*),$$

which contradicts the optimality of Q^*. Now by Proposition 1 and condition C) the left-hand side of (2.13) tends to infinity on the set $Z_\tau(Q^*) = 0$ as $n \to \infty$. On the other hand, since the measure $\tilde{Q}$ is equivalent to P, the limit of the right-hand side of (2.13) is finite. Thus, $P(Z_\tau(Q^*) = 0) = 0$, hence Q^* is an equivalent local martingale measure.

Therefore, by the optimality principle (see Proposition 2) the process $V_t(p)\mathcal{E}_t^p(M^{Q^*})$ is a martingale and using the Itô formula (2.9) for $V_t(p)\mathcal{E}_t^p(M^{Q^*})$ we obtain that

$$A_t + \frac{p(p-1)}{2}\int_0^t V_s(p)d\langle M^{Q^*}\rangle_s + p\langle M^{Q^*}, m\rangle_t = 0. \tag{2.15}$$

The last equality, together with relation (2.10), implies that

$$A_t = -\operatorname*{ess\,inf}_{Q \in \mathcal{M}_p^e}\Big[\frac{p(p-1)}{2}\int_0^t V_s(p)d\langle M^Q\rangle_s + p\langle M^Q, m\rangle_t\Big], \tag{2.16}$$

hence the value process $V(p)$ satisfies equation (2.5) (it is evident that $V(p)$ satisfies also the boundary condition $V_t(p) = \eta$). We notice that, equality (2.15) implies that the process A_t and, hence, $V_t(p)$ is continuous.

Let us show now that the optimal martingale measure Q^* is given by (2.7) and that the value process $V(p)$ belongs to the class $S(X)$ of semimartingales. From (2.16) and (2.1) we have

$$
\begin{aligned}
A_t &= -\frac{p(p-1)}{2} \int_0^t V_s(p) d\langle \lambda \cdot M \rangle_s + p \int_0^t d\langle \lambda \cdot M, m \rangle_s \\
&\quad - \operatorname*{ess\,inf}_{N \in N_p(X)} \left(\frac{p(p-1)}{2} \int_0^t V_s(p) d\langle N \rangle_s + p\langle N, m \rangle_t \right) \\
&= -\frac{p(p-1)}{2} \int_0^t V_s(p) d\langle \lambda \cdot M \rangle_s + p \int_0^t d\langle \lambda \cdot M, m \rangle_s \\
&\quad - \operatorname*{ess\,inf}_{N \in N_p(X)} \left[\left\langle \sqrt{\frac{p(p-1)}{2}} \int_0^t \sqrt{V_s(p)} dN_s + \sqrt{\frac{p}{2(p-1)}} \int_0^t \frac{1}{\sqrt{V_s(p)}} d\tilde{m}_s \right\rangle_t \right. \\
&\qquad\qquad \left. - \frac{p}{2(p-1)} \int_0^t \frac{1}{V_s(p)} d\langle \tilde{m} \rangle_s \right] \\
&= -\frac{p(p-1)}{2} \int_0^t V_s(p) d\langle \lambda \cdot M \rangle_s + p \int_0^t d\langle \lambda \cdot M, \varphi \cdot M \rangle_s + \frac{p}{2(p-1)} \int_0^t \frac{1}{V_s(p)} d\langle \tilde{m} \rangle_s
\end{aligned}
\tag{2.17}
$$

and the infimum is attained for the martingale

$$
\tilde{N}_t = -\frac{1}{p-1} \int_0^t \frac{1}{V_s(p)} d\tilde{m}_s.
\tag{2.18}
$$

We observe that by the Jensen inequality we have, from condition C), the inequality $V_t(p) \geq k_1$, so that all integrals in (2.17) are well defined. By the optimality principle $V_t(p)\mathcal{E}_t^p(M^{Q^*})$ is a martingale. Since $V(p)$ solves equation (2.5), this implies that

$$
\begin{aligned}
\operatorname*{ess\,inf}_Q &\left[\frac{p(p-1)}{2} \int_0^t V_s(p) d\langle M^Q \rangle_s + p\langle M^Q, m \rangle_t \right] \\
&= \frac{p(p-1)}{2} \int_0^t V_s(p) d\langle M^{Q^*} \rangle_s + p\langle M^{Q^*}, m \rangle_t.
\end{aligned}
\tag{2.19}
$$

Since M^{Q^*} is represented in the form $-\lambda \cdot M + N^*$ for some $N^* \in N^p(X)$, it follows from (2.17) and (2.19) that the processes N^* and $\tilde{N}$ and, hence, the processes M^{Q^*} and $-\int_0^t \lambda_s dM_s - \frac{1}{p-1} \int_0^t \frac{1}{V_s(p)} d\tilde{m}_s$ are indistinguishable. So, the p-optimal martingale measure is unique and it admits represetation (2.7).

By definition of $V(p)$ we have that for any $Q \in \mathcal{M}_p^e$

$$
V_\tau(p)\mathcal{E}_\tau^p(M^Q) \leq E(\eta \mathcal{E}_T^p(M^Q)|F_\tau).
\tag{2.20}
$$

Therefore, for any $Q \in \mathcal{M}_p^e$ the process $V_t(p)\mathcal{E}_t^p(M^Q)$ is a submartingale of the class D, as a positive process majorized by a uniformly integrable martingale (see Dellacherie, Meyer [7]) and $V(p) \in D_p$ by Definition 1.

Finally, since $Q^* \in \mathcal{M}_p^e$ and the processes M^{Q^*} and $-\lambda \cdot M - \frac{1}{p-1}\frac{1}{V(p)} \cdot \tilde{m}$ are indistinguishable, we have that $\mathcal{E}_t(-\lambda \cdot M - \frac{1}{(p-1)V} \cdot \tilde{m})$ is a martingale, hence $V(p) \in S(X)$. $\square$

Uniqueness. Let Y be a solution of (2.5), (2.6) from the class $S(X)$. This means that Y is a semimartingale with decomposition (1.7), (1.8), such that $Y_T = \eta$,

$$
B_t = -\operatorname*{ess\,inf}_{Q \in \mathcal{M}_p^e} \left(\frac{p(p-1)}{2} \int_0^t Y_s d\langle M^Q \rangle_s + p\langle M^Q, L \rangle_t \right)
\tag{2.21}
$$

and $(\mathcal{E}_t(-\lambda \cdot M - \frac{1}{(p-1)Y} \cdot \tilde{L}), t \in [0, T])$ is a martingale.

Since (2.21) implies that

$$B_t + \frac{p(p-1)}{2} \int_0^t Y_s d\langle M^Q \rangle_s + p \langle M^Q, L \rangle_t \in \mathcal{A}_{loc}^+,$$

using decomposition (1.7) and the Itô formula for $\mathcal{E}_t^p(M^Q)Y_t$ we obtain that $\mathcal{E}_t^p(M^Q)Y_t$ is a local submartingale for all $Q \in \mathcal{M}_p^e$. Since $Y \in D_p$ we have that $\mathcal{E}_t^p(M^Q)Y_t$ is a submartingale of the class D. Therefore, it follows from the boundary condition (2.6) that for every $Q \in \mathcal{M}_p^e$

$$\mathcal{E}_t^p(M^Q)Y_t \le E[\mathcal{E}_T^p(M^Q)Y_T|F_t] = E[\eta \mathcal{E}_T^p(M^Q)|F_t].$$

Hence,

$$Y_t \le E[\eta \mathcal{E}_{tT}^p(M^Q)|F_t]$$

for all $Q \in \mathcal{M}_p^e$ and

$$Y_t \le \operatorname*{ess\,inf}_Q E[\eta \mathcal{E}_{tT}^p(M^Q)|F_t] = V_t(p). \tag{2.22}$$

Let us show the inverse inequality. Similarly to (2.17) one can show that

$$B_t = -\frac{p(p-1)}{2} \int_0^t Y_s \lambda_s' d\langle M \rangle_s \lambda_s + p \int_0^t \lambda_s' d\langle M \rangle_s \psi_s + \frac{p}{2(p-1)} \int_0^t \frac{1}{Y_s} d\langle \tilde{L} \rangle_s \tag{2.23}$$

and the infimum is attained for the martingale

$$N_t^0 = -\frac{1}{p-1} \int_0^t \frac{1}{Y_s} d\tilde{L}_s, \tag{2.24}$$

where $\tilde{L}$ is the orthogonal martingale part of L in the Kunita–Watanabe decomposition (1.8). Therefore, again using the Itô formula, one can show that $\mathcal{E}_t^p(-\lambda \cdot M + N^0)Y_t$ is a local martingale, since (2.23) and (1.8) imply that

$$\mathcal{E}_t^p(-\lambda \cdot M + N^0)Y_t$$
$$= Y_0 + \int_0^t \mathcal{E}_s^p(-\lambda \cdot M + N^0)(\psi_s - pY_s\lambda_s)' dM_s - \frac{1}{p-1} \int_0^t \mathcal{E}_s^p(-\lambda \cdot M + N^0) d\tilde{L}_s.$$

By Definition 2 of the class $S(X)$ we have that the process $\mathcal{E}_t(M^{Q^0})$ is a martingale, where $M^{Q^0} = -\lambda \cdot M + N^0$. Hence $dQ^0 = \mathcal{E}_T(M^{Q^0})dP$ is an absolutely continuous local martingale measure. Let us show that $Q^0 \in \mathcal{M}_p^e$. To show that $\mathcal{E}_T(M^{Q^0})$ is strictly positive we use again the Lemma of Delbaen–Schachermayer (see Proposition 1). Let τ_n and τ be stopping times defined by (2.12) for the process $\mathcal{E}_t(M^{Q^0})$. From inequality (2.22) we have that for any stopping time σ

$$Y_\sigma \le V_\sigma(p) = \operatorname*{ess\,inf}_{Q \in \mathcal{M}_p^e} E(\eta \mathcal{E}_{\tau,T}^p(M^Q)|F_\sigma) \le E(\eta \mathcal{E}_{\sigma,T}^p(M^{\tilde{Q}})|F_\sigma) \tag{2.25}$$

for any $\tilde{Q} \in \mathcal{M}_p^e$. Since any positive local martingale is a supermartingale we have that

$$\mathcal{E}_\sigma^p(M^{Q^0})Y_\sigma \ge E(Y_T \mathcal{E}_T^p(M^{Q^0})|F_\sigma) \tag{2.26}$$

and from the boundary condition (2.6), replacing σ by τ_n, we obtain

$$Y_{\tau_n} \ge E\left[\eta \frac{\mathcal{E}_T^p(M^{Q^0})}{\mathcal{E}_{\tau_n}^p(M^{Q^0})}\Big|F_{\tau_n}\right]. \tag{2.27}$$

Therefore, (2.25) and (2.27) imply the inequality

$$E\big[\eta \frac{\mathcal{E}_T^p(M^{Q^0})}{\mathcal{E}_{\tau_n}^p(M^{Q^0})}|F_{\tau_n}\big] \le E\big[\eta \frac{\mathcal{E}_T^p(M^{\tilde{Q}})}{\mathcal{E}_{\tau_n}^p(M^{\tilde{Q}})}|F_{\tau_n}\big]. \tag{2.28}$$

Now (2.28), Proposition 1 and condition C) imply that Q^0 is an equivalent local martingale measure. On the other hand using inequalities (2.25) and (2.26) for $\sigma = 0$ we have that

$$E\eta\mathcal{E}_T^p(M^{Q^0}) \le Y_0 \le V_0(p) \le E\eta\mathcal{E}_T^p(M^{\tilde{Q}}) < \infty$$

and $\eta\mathcal{E}_T^p(M^{Q^0})$ is integrable.

Thus, $Q^0 \in \mathcal{M}_p^e$ and since $Y \in D_p$ the process $Y_t\mathcal{E}_t^p(M^{Q^0})$ is from the class D and hence it is a uniformly integrable martingale. Now, the martingale property and the boundary condition imply that

$$Y_t = E(\eta\mathcal{E}_{t,T}^p(-\lambda \cdot M + N^0)|F_t). \tag{2.29}$$

Therefore, (2.22) and (2.29) imply the equality $Y_t = V_t(p)$ a.s. for all $t \in [0, T]$, hence the solution of equation (2.5), (2.6) is unique in the class $S(X)$.

b) It is easy to see that the value process satisfies the two-sided inequality

$$k_1 \le V_t(p) \le Ck_2 \quad \text{a.s.} \tag{2.30}$$

for all $t \in [0, T]$. By Jensen's inequality

$$V_t(p) = \operatorname*{ess\,inf}_{Q\in\mathcal{M}_p^e} E(\eta\mathcal{E}_{t,T}^p(M^Q)|F_t) \ge k_1 \operatorname*{ess\,inf}_{Q\in\mathcal{M}_p^e} E^p(\mathcal{E}_{t,T}(M^Q)|F_t) = k_1.$$

On the other hand, if there exists a martingale measure $\tilde{Q}$ satisfying the reverse Hölder inequality, we have that V is bounded from above, since

$$V_t(p) = \operatorname*{ess\,inf}_{Q\in\mathcal{M}_p^e} E(\eta\mathcal{E}_{t,T}^p(M^Q)|F_t) \le E(\eta\mathcal{E}_{t,T}^p(M^{\tilde{Q}})|F_t) \le Ck_2.$$

To prove this part of the theorem we should show that any solution Y which satisfies the two-sided inequality (1.11) belongs to the class $S(X)$. Since any bounded positive process belongs to the class D_p (see Remark after Definition 1), we should show that the process

$$(\mathcal{E}_t(-\lambda \cdot M - \frac{1}{(p-1)Y} \cdot \tilde{L}), t \in [0, T])$$

is a martingale. According to Theorem 2.3 from [19] it is sufficient to prove that the process $-\lambda \cdot M - \frac{1}{(p-1)Y} \cdot \tilde{L}$ belongs to the class BMO. Since the minimal martingale measure satisfies the reverse Hölder condition, Proposition 6 of [9] implies $-\lambda \cdot M \in$ BMO. On the other hand since $Y \ge k_1$ and $\langle \tilde{L} \rangle \prec \langle L \rangle$, it is sufficient to show that $L \in$ BMO.

Let us now show that, if the random variable η is bounded and if there is an equivalent local martingale measure Q satisfying the reverse Hölder condition, or, if the associated local martingale M^Q belongs to BMO, then the martingale part L of any bounded solution Y of (2.5), (2.6) belongs to the class BMO.

By the Itô formula

$$Y_t^2 = Y_0^2 + 2\int_0^t Y_s dY_s + \langle L \rangle_t. \tag{2.31}$$

Since $Y_T = \eta$ and $Y_\tau \geq c$ from (2.31) we have that

$$\langle L \rangle_T - \langle L \rangle_\tau + 2 \int_\tau^T Y_s d(B_s + L_s) = \eta^2 - Y_\tau^2 \leq k_2^2. \tag{2.32}$$

Since Y satisfies (2.5), the process

$$B_t + \frac{p(p-1)}{2} \int_0^t Y_s d\langle M^Q \rangle_s + p \langle M^Q, L \rangle_t$$

is increasing and (2.32) implies that

$$\langle L \rangle_T - \langle L \rangle_\tau + 2 \int_\tau^T Y_s dL_s - p(p-1) \int_\tau^T Y_s^2 d\langle M^Q \rangle_s - 2p \int_\tau^T Y_s d\langle M^Q, L \rangle_s) \leq k_2^2. \tag{2.33}$$

Without loss of generality we may assume that L is a square integrable martingale, otherwise one can use the localization arguments. Therefore, if we take conditional expectations, having inequality $Y_t \leq C$ in mind, we obtain

$$E(\langle L \rangle_T - \langle L \rangle_\tau | F_\tau) \tag{2.34}$$

$$\leq C^2 p(p-1) E(\langle M^Q \rangle_T - \langle M^Q \rangle_\tau | F_\tau) + k_2^2 + 2pCE(\int_\tau^T |d\langle M^Q, L \rangle_s| | F_\tau).$$

Now using the Kunita–Watanabe inequality

$$E(\int_\tau^T |d\langle M^Q, L \rangle_s| | F_\tau) \leq E^{1/2}(\langle M^Q \rangle_T - \langle M^Q \rangle_\tau | F_\tau) E^{1/2}(\langle L \rangle_T - \langle L \rangle_\tau | F_\tau) \tag{2.35}$$

and that $M^Q \in$ BMO we obtain from (2.34) that

$$E(\langle L \rangle_T - \langle L \rangle_\tau | F_\tau) \leq c_1 + c_2 E^{1/2}(\langle L \rangle_T - \langle L \rangle_\tau | F_\tau) \tag{2.36}$$

for some positive constants c_1 and c_2 which do not depend on τ. The last inequality implies that $E(\langle L \rangle_T - \langle L \rangle_\tau | F_\tau)$ is bounded for every stopping time τ by one and the same constant, hence $L \in$ BMO. $\qquad \square$

Remark 1 In particular if $M^Q \in M^2$ and η is square integrable, the same arguments imply that m is a square integrable martingale.

Remark 2 If Condition A*) is satisfied then the p-optimal martingale measure satisfies the reverse Hölder inequality $R_p(P)$, since for any stopping time τ

$$E(\mathcal{E}_{\tau,T}^p(M^{Q^*}) | F_\tau) \leq \frac{1}{k_1} E(\eta \mathcal{E}_{\tau,T}^p(M^{Q^*}) | F_\tau) = \frac{1}{k_1} \operatorname*{ess\,inf}_{Q \in \mathcal{M}_p^e} E(\eta \mathcal{E}_{\tau,T}^p(M^Q) | F_\tau)$$

$$\leq \frac{1}{k_1} E(\eta \mathcal{E}_{\tau,T}^p(M^{\tilde{Q}}) | F_\tau) \leq C \frac{k_2}{k_1}.$$

Proposition 3 *Equation (2.5), (2.6) is equivalent to the equation*

$$\frac{\mathcal{E}_T((\bar{\psi} - p\lambda) \cdot M)}{\mathcal{E}_T^{p-1}(\bar{L})} = \bar{c}\eta \mathcal{E}_T^p(-\lambda \cdot M), \tag{2.37}$$

i.e., if Y is a solution of (2.5), (2.6) then the triple $(\bar{c}, \bar{\psi}, \bar{L})$, where

$$\bar{c} = 1/Y_0, \quad \bar{\psi} = \psi/Y, \quad \bar{L} = -\frac{1}{p-1} \int_0^t \frac{1}{Y_s} d\tilde{L}_s$$

will be a solution of (2.37). Conversely, if $(\bar{c}, \bar{\psi}, \bar{L})$ solves (2.37), then Y defined by

$$Y_t = \frac{1}{\bar{c}} \mathcal{E}_t((\bar{\psi} - p\lambda) \cdot M) \mathcal{E}_t^{1-p}(\bar{L}) \mathcal{E}_t^{-p}(-\lambda \cdot M) \tag{2.38}$$

satisfies (2.5), (2.6).

Proof. Let Y, which admits decomposition (1.7), (1.8), be a solution of (2.5), (2.6). It follows from (2.23) that

$$Y_t = Y_0 - \frac{p(p-1)}{2} \int_0^t Y_s \lambda'_s d\langle M \rangle_s \lambda_s + p \int_0^t \lambda'_s d\langle M \rangle_s \psi_s$$
$$+ \frac{p}{2(p-1)} \int_0^t \frac{1}{Y_s} d\langle \tilde{L} \rangle_s + \int_0^t \psi'_s dM_s + \tilde{L}_t. \tag{2.39}$$

We introduce

$$\overline{\psi}_t = \frac{\psi_t}{Y_t} \quad \text{and} \quad \overline{L}_t = -\frac{1}{p-1} \int_0^t \frac{1}{Y_s} d\tilde{L}_s.$$

Then

$$\psi_t = \overline{\psi}_t Y_t, \quad \tilde{L}_t = -(p-1) \int_0^t Y_s d\overline{L}_s$$

and from (2.39) we have

$$dY_t = Y_t \Big[-\frac{p(p-1)}{2} \lambda'_t d\langle M \rangle_t \lambda_t + p\lambda'_t d\langle M \rangle_t \overline{\psi}_t \tag{2.40}$$
$$+ \frac{p(p-1)}{2} d\langle \overline{L} \rangle_t + \overline{\psi}_t dM_t - (p-1) d\overline{L}_t \Big], \quad Y_T = \eta.$$

Solving this linear equation with respect to Y we obtain

$$Y_t = Y_0 \exp \Big[-\frac{p(p-1)}{2} \int_0^t \lambda'_s d\langle M \rangle_s \lambda_s + p \int_0^t \lambda'_s d\langle M \rangle_s \overline{\psi}_s \tag{2.41}$$
$$+ \frac{p(p-1)}{2} \langle \overline{L} \rangle_t - \frac{1}{2} \int_0^t \overline{\psi}'_s d\langle M \rangle_s \overline{\psi}_s - \frac{(p-1)^2}{2} \langle \overline{L} \rangle_t + \int_0^t \overline{\psi}_s dM_s - (p-1)\overline{L}_t \Big]$$

which can be expressed by means of Doleans-Dade exponents

$$Y_t = Y_0 \mathcal{E}_t((\bar{\psi} - p\lambda) \cdot M) \mathcal{E}_t^{1-p}(\overline{L}) \mathcal{E}_t^{-p}(-\lambda \cdot M). \tag{2.42}$$

Now, using the boundary condition $Y_T = \eta$ we see that (2.37) is satisfied for $\bar{c} = 1/Y_0$.

Conversely, if a triple $(\bar{c}, \overline{\psi}, \overline{L})$ satisfies (2.37) then it is also evident that Y defined by (2.38) is a solution of (2.5), (2.6). □

Corollary 1 *The semimartingale Bellman equation (2.5), (2.6) coincides with the equation*

$$V_t(p) = V_0(p) - \int_0^t \Big(\frac{p(p-1)}{2} V_s(p) \lambda'_s d\langle M \rangle_s \lambda_s - p\lambda'_s d\langle M \rangle_s \varphi_s \Big) \tag{2.43}$$

$$+ \frac{p}{2(p-1)} \int_0^t \frac{1}{V_s(p)} d\langle \tilde{m} \rangle_s + \int_0^t \varphi_s' dM_s + \tilde{m}_t, \quad V_T(p) = \eta,$$

that is the same as (2.39) written for $V(p)$ instead of Y. The equation

$$R_t = R_0 - \int_0^t \frac{1}{2}(\bar{\varphi}_s - p\lambda_s)' d\langle M \rangle_s (\bar{\varphi}_s - p\lambda_s) + \int_0^t \frac{p}{2}\lambda_s' d\langle M \rangle_s \lambda_s + \frac{p-1}{2}\langle \tilde{m} \rangle_t$$

$$+ \int_0^t p\lambda_s' dM_s + \int_0^t (\bar{\varphi}_s - p\lambda_s)' dM_s - (p-1)\tilde{m}_t,$$

$$R_T = \ln \eta \tag{2.44}$$

with respect to $(R, \bar{\varphi}, \tilde{m})$, which admits a unique solution in the class $\mathcal{S}_+ \times L^2_{loc}(\langle M \rangle) \times \mathcal{N}(X)$, is also equivalent to (2.5), (2.6).

Corollary 2 *A martingale measure Q^* is p-optimal if and only if*

$$\eta \mathcal{E}_T^{p-1}(M^{Q^*}) = c + \int_0^T h_s' dX_s \tag{2.45}$$

for a constant c and an X-integrable predictable process h such that $(\int_0^t h_s' dX_s, t \in [0,T])$ is a Q-martingale for every $Q \in \mathcal{M}_p^e$.

Proof. Let Q^* be a p-optimal martingale measure. According to Theorem 1 M^{Q^*} admits representation (2.7), hence

$$\mathcal{E}_T^{p-1}(M^{Q^*}) = \mathcal{E}_T^{p-1}(-\lambda \cdot M)\mathcal{E}_T^{p-1}\left(-\frac{1}{p-1}\frac{1}{V(p)} \cdot \tilde{m}\right). \tag{2.46}$$

Therefore, using (2.38) and the equality

$$\frac{\mathcal{E}(X)}{\mathcal{E}(Y)} = \mathcal{E}(X - Y - \langle X - Y, Y \rangle),$$

valid for continuous semimartingales X and Y, we obtain

$$V_t(p)\mathcal{E}_t^{p-1}(M^{Q^*}) = V_0(p)\frac{\mathcal{E}_t((\bar{\varphi} - p\lambda) \cdot M)}{\mathcal{E}_t(-\lambda \cdot M)} = V_0(p)\mathcal{E}_t((\bar{\varphi} + (1-p)\lambda) \cdot X).$$

Thus, the boundary condition $V_T(p) = \eta$ implies that $\eta \mathcal{E}_T^{p-1}(M^Q)$ is of the form (2.45) with

$$h_s = (\bar{\varphi}_s + (1-p)\lambda_s)\mathcal{E}_s((\bar{\varphi} + (1-p)\lambda) \cdot X), \quad s \in [0,T]. \tag{2.47}$$

Besides it follows from (2.47) that

$$V_0(p) + \int_0^t h_s' dX_s = V_t(p)\mathcal{E}_t^{p-1}(M^{Q^*})$$

and, hence, $\int_0^t h_s' dX_s$ is a Q^*-martingale by the optimality principle. The latter equality implies that $\int_0^t h_s' dX_s \geq -V_0(p)$. Since $\int_0^t h_s' dX_s$ is a Q-local martingale, then it is also a supermartingale and $E^Q \int_0^t h_s' dX_s \leq 0$ (for any $Q \in \mathcal{M}_p^e$). On the other hand, since Q^* is optimal, from Proposition A of Appendix we have

$$E^Q \int_0^T h_s' dX_s = E^Q \eta \mathcal{E}_T^{p-1}(M^{Q^*}) - V_0(p)$$

$$= E\eta\mathcal{E}_T^{p-1}(M^{Q^*})\big(\mathcal{E}_T(M^Q) - \mathcal{E}_T(M^{Q^*})\big) \geq 0,$$

which implies that $E^Q \int_0^T h_s' dX_s = 0$, hence $\int_0^t h_s' dX_s$ is a martingale for all $Q \in \mathcal{M}_p^e$.

Conversely, if Q^0 is a martingale measure satisfying the relation (2.45) and the process $(\int_0^t h_s' dX_s, t \in [0,T])$ is a Q-martingale for every $Q \in \mathcal{M}_p^e$ then

$$E^Q \eta \mathcal{E}_T^{p-1}(M^{Q^0}) = E^{Q^0} \eta \mathcal{E}_T^{p-1}(M^{Q^0})$$

for any Q which implies that Q^0 is optimal by Proposition A of Appendix.   □

Corollary 3 *The minimal martingale measure is p-optimal if and only if*

$$\eta\mathcal{E}_T^p(-\lambda \cdot M) = c + \int_0^T g_s' dM_s \tag{2.48}$$

for some M-integrable predictable g and the process $(\int_0^t g_s' dX_s, t \in [0,T])$ is a P-martingale.

The proof is similar to the proof of Corollary 2.

Remark It is evident that if $\langle \lambda \cdot M \rangle$ is deterministic and $\eta = $ const, then

$$\mathcal{E}_T^p(-\lambda \cdot M) = \mathcal{E}_T(-p\lambda \cdot M)\exp\{\frac{p(p-1)}{2}\langle \lambda \cdot M \rangle_T\}$$

$$= \exp\{\frac{p(p-1)}{2}\langle \lambda \cdot M \rangle_T\}(1 - p\int_0^T \mathcal{E}_s(-p\lambda \cdot M)\lambda_s' dM_s)$$

and (2.48) is satisfied.

The semimartingale backward equation for the value process $\bar{V}_t(p)$ defined by (1.6) can be derived in a similar way. Here we give only the corresponding theorem and remark some differences. Assume that the following conditions are satisfied:

A') $\mathcal{M}^e \neq \emptyset$;

B) all P-local martingales are continuous;

C') η is a strictly positive F_T-measurable random variable such that

$$E\eta^{\frac{1}{1-p}} < \infty.$$

Theorem 1' *Let $0 < p < 1$ and let conditions A'), B) and C') be satisfied. Then*

a) *the value process V is a solution of the semimartingale backward equation*

$$Y_t = Y_0 - \operatorname*{ess\,sup}_{N \in \mathcal{N}(X)}\big[\frac{1}{2}p(p-1)\int_0^t Y_s d\langle -\lambda \cdot M + N \rangle_s + p\langle \lambda \cdot M + N, L \rangle_t\big] + L_t, \quad t < T, \tag{2.49}$$

with the boundary condition

$$Y_T = \eta. \tag{2.50}$$

This solution is unique in the class $S(X)$ of semimartingales. Moreover, the martingale measure Q^* is p-optimal if and only if it is given by the density $dQ^* = \mathcal{E}_T(M^{Q^*})dP$, where

$$M_t^{Q^*} = -\int_0^t \lambda_s' dM_s + \frac{1}{1-p}\int_0^t \frac{1}{V_s(p)}d\tilde{m}_s.$$

b) *If in addition, the conditions $k_1 \leq \eta \leq k_2$, $\lambda \cdot M \in BMO$ are satisfied and there is a constant c_1 such that*

$$E(\mathcal{E}^p_{\tau,T}(-\lambda \cdot M)|F_\tau) \geq c_1 \tag{2.51}$$

for any stopping time τ, then the value process $\bar{V}(p)$ is the unique solution of the semimartingale backward equation (2.49), (2.50) in the class of semimartingales Y satisfying the two-sided inequality

$$c \leq Y_t \leq C \quad \text{for all } t \in [0,T] \text{ a.s..}$$

for some constants $0 < c < C$.

The proof is essentially similar to the proof of Theorem 1. In this case $\bar{V}_t(p)\mathcal{E}^p_t(M^Q)$ is a P-supermartingale for all $Q \in \mathcal{M}^e$ and the classes D_p and $S(X)$ are defined similarly. From condition C') and the Hölder inequality we have that $\sup_Q E\eta\mathcal{E}^p_T(M^Q) < \infty$ and the existence of an optimal martingale measure Q^* in the class $\mathcal{M}^{abs}$ follows from the same arguments. We only show that conditions A') – C') imply that Q^* is equivalent to P. Since Q^* is optimal, for the optimal density Z^{Q^*} and the stopping times τ_n defined by (2.14) we have the inequality

$$E\left[\eta\frac{Z^p_T(Q^*)}{Z^p_{\tau_n}(Q^*)}|F_{\tau_n}\right] \geq E\left[\eta\frac{\mathcal{E}^p_T(M^{\tilde{Q}})}{\mathcal{E}^p_{\tau_n}(M^{\tilde{Q}})}|F_{\tau_n}\right] \quad \text{a.s.} \tag{2.52}$$

By the Hölder inequality

$$E\left[\eta\frac{Z^p_T(Q^*)}{Z^p_{\tau_n}(Q^*)}|F_{\tau_n}\right] = E\left[\frac{Z^p_T(Q^*)}{Z^p_{\tau_n}(Q^*)}\eta I_{(Z_\tau(Q^*)\neq 0)}|F_{\tau_n}\right]$$

$$\leq E^{1-p}(\eta^{\frac{1}{1-p}}I_{(Z_\tau(Q^*)\neq 0)}|F_{\tau_n}). \tag{2.53}$$

Condition C') and the Lévy theorem imply that

$$E^{1-p}(\eta^{\frac{1}{1-p}}I_{(Z_\tau(Q^*)=0)}|F_{\tau_n})$$

tends to zero on the set $(Z_\tau(Q^*) = 0)$, hence the left-hand side of (2.53) tends to zero on the same set. On the other hand,

$$P\left(\sup_{t\leq T}\mathcal{E}^p_t(M^{\tilde{Q}}) \geq N\right) \leq \frac{1}{N}$$

by Doob's inequality for the supermartingale $\mathcal{E}^p_t(M^{\tilde{Q}})$ and

$$P\left(\inf_{t\leq T} E(\eta\mathcal{E}^p_T(M^{\tilde{Q}})|F_t) > 0\right) = 1,$$

since $\eta\mathcal{E}^p_T(M^{\tilde{Q}}) > 0$. Therefore the limit of the right-hand side of (2.53) is strictly positive, which implies that $P(Z_\tau(Q^*) = 0) = 0$ and Q^* is equivalent to P.

Note that it follows from (2.51) that the value process $\bar{V}(p)$ is bounded from below, but this condition (unlike to reverse Hölder condition $R_p(P)$ for $p > 1$) does not imply that $\lambda \cdot M \in BMO$. Therefore we assume in the part b) that $\lambda \cdot M \in BMO$ in order to guarantee

$$E\mathcal{E}_T\left(-\lambda \cdot M - \frac{1}{(p-1)\bar{Y}} \cdot \tilde{L}\right) = 1.$$

3. APPROXIMATION OF LOWER AND UPPER PRICES OF A CONTINGENT CLAIM

In the following we assume that the minimal martingale measure Q^λ exists and we will use Q^λ as a reference probability measure. The lower price of a contingent claim η can be written in the form (see [12])

$$V_t = \operatorname*{ess\,inf}_{N \in \mathcal{N}(X)} E^\lambda(\eta \mathcal{E}_{tT}(N)|F_t)$$

where E^λ stands for mathematical expectation relative to the measure Q^λ and $\mathcal{N}(X)$ is a set of local martingales $(N_t, t \in [0, T])$ with $N_0 = 0$ such that
  i) N is strongly orthogonal to M,
  ii) $(\mathcal{E}_t(N), t \in [0, T])$ is a strictly positive martingale under Q^λ.
  By $\mathcal{N}_\varepsilon(X)$ we denote the set of elements $N \in \mathcal{N}(X)$ such that $E^\lambda(\eta \mathcal{E}_T^{1+\varepsilon}(N)) < \infty$. For each $0 < \varepsilon < 1$ let us introduce the value process

$$V_t(\varepsilon) = \operatorname*{ess\,inf}_{N \in \mathcal{N}_\varepsilon(X)} E^\lambda((\eta + \varepsilon)\mathcal{E}_{tT}^{1+\varepsilon}(N)|F_t). \tag{3.1}$$

Let condition B) be satisfied and let

$$E^\lambda \eta^2 < \infty. \tag{3.2}$$

This implies that all conditions of Theorem 1 a) are satisfied, since in this case $P = Q^\lambda$ and for $N = 0$ we have $E^\lambda(\eta \mathcal{E}_T^{1+\varepsilon}(N)) = E^\lambda \eta < \infty$. Besides $\eta + \varepsilon \geq \varepsilon > 0$. According to Corollary 1 of Theorem 1, $V_t(\varepsilon)$ satisfies the equation

$$V_t(\varepsilon) = V_0(\varepsilon) + \frac{1+\varepsilon}{2\varepsilon} \int_0^t \frac{1}{V_s(\varepsilon)} d\langle \tilde{m}(\varepsilon) \rangle_s + m_t(\varepsilon), \tag{3.3}$$

$$V_T(\varepsilon) = \eta + \varepsilon \tag{3.4}$$

where $m(\varepsilon)$ is the martingale part in the canonical decomposition

$$V_t(\varepsilon) = V_0(\varepsilon) + A_t(\varepsilon) + m_t(\varepsilon) \tag{3.5}$$

of $V_t(\varepsilon)$ with respect to the minimal martingale measure and $\tilde{m}(\varepsilon)$ is the martingale orthogonal to X in the Galtchouk–Kunita–Watanabe decomposition

$$m_t(\varepsilon) = \int_0^t \varphi_s(\varepsilon) dX_s + \tilde{m}_t(\varepsilon), \quad \langle \tilde{m}(\varepsilon), X \rangle = 0. \tag{3.6}$$

Similarly, let

$$V_t = V_0 + m_t + A_t, \quad m_t = \int_0^t \varphi_s dX_s + \tilde{m}_t$$

be the canonical and the Galtchouk–Kunita–Watanabe decomposition of the lower price V with respect to the measure Q^λ.

Theorem 2 Let $E^\lambda \eta^2 < \infty$ and condition B) be satisfied. Then for any stopping time τ

$$\lim_{\varepsilon \downarrow 0} V_\tau(\varepsilon) = V_\tau \quad a.s. \quad and \ in \ L^2 \tag{3.7}$$

where $V_t(\varepsilon)$ is solution of equation (3.3), (3.4) for any $\varepsilon > 0$. Moreover,

$$E^\lambda \int_0^T (\varphi_s(\varepsilon) - \varphi_s)^2 d\langle X \rangle_s \to 0 \quad as \ \varepsilon \downarrow 0 \tag{3.8}$$

and

$$E^\lambda \langle \tilde{m}(\varepsilon) \rangle_T \to 0 \quad as \quad \varepsilon \downarrow 0. \tag{3.9}$$

Proof. Let us first show that $V_\tau(\varepsilon)$ is decreasing as $\varepsilon \downarrow 0$. For any $0 \leq \varepsilon_1 \leq \varepsilon_2$, in fact, we have

$$
\begin{aligned}
V_\tau(\varepsilon_2) &= \operatorname*{ess\,inf}_{N \in \mathcal{N}_{\varepsilon_2}(X)} E^\lambda [(\eta + \varepsilon_2) \mathcal{E}_{\tau T}^{1+\varepsilon_2}(N) | F_\tau] \\
&= \operatorname*{ess\,inf}_{N \in \mathcal{N}_{\varepsilon_2}(X)} E^\lambda [(\eta + \varepsilon_2) e^{(\varepsilon_2 - \varepsilon_1)\frac{1+\varepsilon_2}{1+\varepsilon_1}(\langle N \rangle_T - \langle N \rangle_\tau)} \mathcal{E}_{\tau T}^{1+\varepsilon_1} (\frac{1+\varepsilon_2}{1+\varepsilon_1} N) | F_\tau] \\
&\geq \operatorname*{ess\,inf}_{N \in \mathcal{N}_{\varepsilon_2}(X)} E^\lambda [(\eta + \varepsilon_1) \mathcal{E}_{\tau T}^{1+\varepsilon_1} (\frac{1+\varepsilon_2}{1+\varepsilon_1} N) | F_\tau] \\
&\geq \operatorname*{ess\,inf}_{N \in \mathcal{N}_{\varepsilon_1}(X)} E^\lambda [(\eta + \varepsilon_1) \mathcal{E}_{\tau T}^{1+\varepsilon_1} (\frac{1+\varepsilon_2}{1+\varepsilon_1} N) | F_\tau] = V_\tau(\varepsilon_1).
\end{aligned}
$$

In particular,

$$V_\tau(\varepsilon) \geq V_\tau \quad a.s. \tag{3.10}$$

for any $\varepsilon > 0$. Thus for any stopping time τ

$$\lim_{\varepsilon \downarrow 0} V_\tau(\varepsilon) \geq V_\tau \quad a.s. \tag{3.11}$$

On the other hand, for any $N \in \mathcal{N}_\varepsilon(X)$

$$V_\tau(\varepsilon) \leq E^\lambda((\eta + \varepsilon) \mathcal{E}_{\tau T}^{1+\varepsilon}(N) | F_\tau) \tag{3.12}$$

and for any $N \in \mathcal{N}_\varepsilon(X)$ with the bounded characteristic $\langle N \rangle$ we have that

$$\lim_{\varepsilon \downarrow 0} V_\tau(\varepsilon) \leq \lim_{\varepsilon \downarrow 0} E^\lambda((\eta + \varepsilon) \mathcal{E}_{\tau T}^{1+\varepsilon}(N) | F_\tau) = E^\lambda(\eta \mathcal{E}_{\tau T}(N) | F_\tau). \tag{3.13}$$

Therefore from (3.12) we obtain

$$\lim_{\varepsilon \downarrow 0} V_\tau(\varepsilon) \leq \operatorname*{ess\,inf}_{N \in \mathcal{N}(X):\langle N \rangle_T \in L^\infty} E^\lambda(\eta \mathcal{E}_{\tau T}(N) | F_\tau) = \operatorname*{ess\,inf}_{N \in \mathcal{N}(X)} E^\lambda(\eta \mathcal{E}_{\tau T}(N) | F_\tau) = V_\tau. \tag{3.14}$$

Thus, relations (3.11) and (3.14) imply the almost sure convergence (3.7) of the value processes. Since

$$V_\tau(\varepsilon) \leq E^\lambda(\eta | F_\tau) + 1, \tag{3.15}$$

$$V_\tau \leq E^\lambda(\eta | F_\tau) \tag{3.16}$$

the Lebesgue theorem implies the convergence of the value processes in L^2.

Let us show now that the martingale part of $V(\varepsilon)$ converges in L^2 to the martingale part of V. By the Ito formula for $(V(\varepsilon) - V)^2$, taking into account the boundary conditions $V_T(\varepsilon) = \eta + \varepsilon$ and $V_T = \eta$ we have that

$$\langle m(\varepsilon) - m \rangle_T + 2 \int_0^T (V_s(\varepsilon) - V_s) d(V_s(\varepsilon) - V_s) \leq \varepsilon^2. \tag{3.17}$$

Since $m(\varepsilon) \in \mathcal{M}^2$ for any $\varepsilon \geq 0$ (see Remark 1 after Theorem 1), from (3.15) and (3.16), using the Doob and Hölder inequalities we have

$$E^\lambda \Big(\int_0^T (V_s(\varepsilon) - V_s)^2 d\langle m(\varepsilon) \rangle_s \Big)^{\frac{1}{2}} \leq 2E^\lambda \Big(\int_0^T E^2(\eta | F_s) d\langle m(\varepsilon) \rangle_s \Big)^{\frac{1}{2}}$$

$$\leq 2E^\lambda \langle m(\varepsilon)\rangle_T^{\frac{1}{2}} \sup_{s\leq T} E^\lambda(\eta|F_s)$$

$$\leq \mathrm{const}(E^\lambda \eta^2 E^\lambda \langle m(\varepsilon)\rangle_T)^{\frac{1}{2}} < \infty.$$

Therefore the stochastic integrals with respect to $m(\varepsilon)$ and m in (3.17) are martingales, taking expectations we obtain

$$E^\lambda[\langle m(\varepsilon) - m\rangle_T] + 2E^\lambda[\int_0^T (V_s(\varepsilon) - V_s)d(A_s(\varepsilon) - A_s)] \leq \varepsilon^2. \tag{3.18}$$

Since $A(\varepsilon)$ is an increasing process, it follows from (3.18) and (3.10) that

$$E^\lambda[\langle m(\varepsilon) - m\rangle_T] \leq 2E^\lambda[\int_0^T (V_s(\varepsilon) - V_s)d(A_s)] + \varepsilon^2. \tag{3.19}$$

It is easy to see that $E^\lambda \eta^2 < \infty$ implies $E^\lambda A_T^2 < \infty$. Since

$$V_s(\varepsilon) - V_s \leq V_s(\varepsilon) - 1 \leq E^\lambda(\eta|F_t)$$

and

$$E^\lambda[\int_0^T E^\lambda(\eta|F_s)dA_s] \leq E^\lambda[\sup_{s\leq T} E^\lambda(\eta|F_s)A_T] \leq \mathrm{const}(E^\lambda \eta^2 E^\lambda A_T^2)^{\frac{1}{2}} < \infty, \tag{3.20}$$

by the Lebesgue dominated convergence theorem it follows from (3.19) and (3.20) that

$$\lim_{\varepsilon\downarrow 0} E^\lambda[\langle m(\varepsilon) - m\rangle_T] = 0. \tag{3.21}$$

From the orthogonality of $\tilde{m}(\varepsilon)$ and $\tilde{m}$ to X it follows that (3.8) holds and

$$\lim_{\varepsilon\downarrow 0} E^\lambda[\langle \tilde{m}(\varepsilon) - \tilde{m}\rangle_T] = 0. \tag{3.22}$$

By the optional decomposition (1.13) we have $\tilde{m} = 0$ and therefore (3.9) follows from (3.22). But we will prove the convergence (3.9) and then the validity of the decomposition (1.13) directly using the properties of the semimartingale backward SDE (3.3).

Thus, from (3.7) and (3.21) we have that

$$A_\tau(\varepsilon) \to A_\tau \text{ in } L^2 \text{ as } \varepsilon \downarrow 0, \tag{3.23}$$

hence in L^2 from (3.23) thanks to (3.3) it follows

$$\frac{\varepsilon + 1}{2\varepsilon} \int_0^\tau \frac{1}{V_s(\varepsilon)} d\langle \tilde{m}(\varepsilon)\rangle_s \to A_\tau \text{ as } \varepsilon \downarrow 0, \tag{3.24}$$

which implies that

$$E^\lambda(\int_0^T \frac{1}{V_s(\varepsilon)} d\langle \tilde{m}(\varepsilon)\rangle_s) \to 0,$$

as $\varepsilon \downarrow 0$. But since

$$E^\lambda \frac{\langle \tilde{m}(\varepsilon)\rangle_\tau}{\sup_{s\leq T} E^\lambda(\eta|F_s) + 1} \leq E^\lambda(\int_0^T \frac{1}{V_s(\varepsilon)} d\langle m(\varepsilon)\rangle_s)$$

and

$$Q^\lambda(\sup_{s \leq T} E^\lambda(\eta|F_s) \geq k) \leq \frac{1}{k} E^\lambda \eta^2 \to 0, \quad k \to \infty$$

(by the Doob inequality) we obtain that $\langle \tilde{m}(\varepsilon) \rangle_T \to 0$ in probability as $\varepsilon \downarrow 0$. This, together with (3.22) implies that (3.9) holds, $\tilde{m} = 0$ and

$$V_t = V_0 + \int_0^t \varphi_s dX_s + A_t, \tag{3.25}$$

which finishes the proof of Theorem 2. □

Let us now formulate a version of Theorem 2 for the upper price. For each $0 < \varepsilon$ let us introduce the value process

$$\overline{V}_t(\varepsilon) = \operatorname*{ess\,sup}_{N \in \mathcal{N}(X)} E^\lambda(\eta(\varepsilon)\mathcal{E}_{tT}^{1-\varepsilon}(N)|F_t) \tag{3.26}$$

and let

$$\overline{V}_t = \overline{V}_t(0) = \operatorname*{ess\,sup}_{N \in \mathcal{N}(X)} E^\lambda(\eta\mathcal{E}_{tT}(N)|F_t) \tag{3.27}$$

be the upper price of η at the moment t. Assume that

$$E^\lambda \eta^2 < \infty \quad \text{and} \quad \sup_{Q \in \mathcal{M}^e} E^Q \eta < \infty. \tag{3.28}$$

Let $(\eta(\varepsilon), \varepsilon > 0)$ be a family of F_T-measurable strictly positive random variables such that: for every $\varepsilon > 0$ it holds $\eta(\varepsilon) \leq C(\varepsilon)$ for a constant $C(\varepsilon)$, which depends only on ε, such that

$$E^\lambda(\eta(\varepsilon) - \eta)^2 \to 0, \quad \varepsilon \to 0.$$

Theorem 2′ *Let condition B) and (3.28) be satisfied. Then for any stopping time τ*

$$\lim_{\varepsilon \downarrow 0} \overline{V}_\tau(\varepsilon) = \overline{V}_\tau \quad a.s.,$$

where $\overline{V}_t(\varepsilon)$ satisfies equation

$$\overline{V}_t(\varepsilon) = \overline{V}_0(\varepsilon) - \frac{1-\varepsilon}{2\varepsilon} \int_0^t \frac{1}{\overline{V}_s(\varepsilon)} d\langle \tilde{m}(\varepsilon) \rangle_s + m_t(\varepsilon),$$

$$\overline{V}_t(\varepsilon) = \eta(\varepsilon)$$

for any $\varepsilon > 0$. Moreover,

$$E^\lambda \int_0^T (\varphi_s(\varepsilon) - \varphi_s)^2 d\langle X \rangle_s \to 0 \quad and \quad E^\lambda \langle \tilde{m}(\varepsilon) \rangle_T \to 0$$

as $\varepsilon \downarrow 0$.

APPENDIX

The proof of the following assertion for the case $p = 2$ can be found in [31]. For all cases we give the proof for $p > 1$.

Proposition A $\tilde{Z}_T \in \mathcal{M}_p^{abs}$ is p-optimal if and only if

$$E\eta(Z_T - \tilde{Z}_T)\tilde{Z}_T^{p-1} \geq 0 \tag{A.1}$$

for all $Z \in \mathcal{M}_p^{abs}$.

Proof. Let (A.1) be satisfied. We consider the function

$$f(x) = E\eta(xZ_T + (1 - x)\tilde{Z}_T)^p.$$

It is evident that f is convex and continuously differentiable, since the derivative

$$pE\eta(Z_T - \tilde{Z}_T)(\bar{x}Z_T + (1 - \bar{x})\tilde{Z}_T)^{p-1}$$

of the function $\eta(xZ_T + (1 - x)\tilde{Z}_T)^p$ is majorized by the integrable random variable

$$2^{p-1}\eta(Z_T + \tilde{Z}_T)(Z_T^{p-1} + \tilde{Z}_T^{p-1}).$$

According to (A.1), $f'(0) \geq 0$. It follows from the convexity of f that

$$f(\varepsilon) - f(0) \leq f(x) - f(x - \varepsilon)$$

for all ε, x such that $0 < \varepsilon < x \leq 1$, which implies that $f'(x) \geq f'(0) \geq 0$. Hence f is a non-decreasing function and

$$E\eta Z_T^p = f(1) \geq f(0) = E\eta \tilde{Z}_T^p.$$

Thus $\tilde{Z}_T$ is p-optimal. Conversely, if $\tilde{Z}$ is p-optimal then it is evident that $f'(0) \geq 0$ for any $Z \in \mathcal{M}^{abs}$ which gives (A.1). □

REFERENCES

[1] J.M. Bismut, Conjugate convex functions in optimal stochastic control, *J. Math. Anal. Appl.* **44** (1973), 384–404.

[2] R. Chitashvili, Martingale ideology in the theory of controlled stochastic processes, *Lecture Notes in Math.* **1021**, 73–92, Springer-Verlag, New York, 1983.

[3] R. Chitashvili and M. Mania, Optimal locally absolutely continuous change of measure: finite set of decisions, *Stochastics Stochastics Rep.* **21** (1987), 131–185 (part 1), 187–229 (part 2).

[4] R. Chitashvili and M. Mania, Generalized Ito's formula and derivation of Bellman's equation, *Stochastic Processes and Related Topics*, Stochastics Monographs **10**, 1–21, Gordon & Breach Science Publishers, London, 1996.

[5] F. Delbaen and W. Schachermayer, A general version of the fundamental theorem of asset pricing, *Math. Ann.* **300**, 463–520, Springer-Verlag, Berlin, 1994.

[6] F. Delbaen and W. Schachermayer, Variance-optimal martingale measure for continuous processes, *Bernoulli* **2** 1 (1996), 81–105.

[7] C. Dellacherie and P.A. Meyer, *Probabilités et potentiel II*, Hermann, Paris, 1980.

[8] F. Delbaen, P. Monat, W. Schachermayer, W. Schweizer and C. Stricker, Weighted norm inequalities and hedging in incomplete markets, *Finance Stoch.* **1** (1997), 181–227.

[9] C. Doleans-Dade and P.A. Meyer, Inegalités de normes avec poids, *Séminaire de Probabilités XIII*, Lecture Notes in Math. **721**, 313–331, Springer-Verlag, New York, 1979.

[10] D. Duffie and H.R. Richardson, Mean-Variance hedging in continuous time, *Ann. Appl. Probab.* **1** (1991), 1–15.

[11] I. Ekeland and R. Temam, *Convex analysis and variational problems*, Oxford, 1976.

[12] N. El Karoui and M.C. Quenez, Dynamic programming and pricing of contingent claims in an incomplete market, *SIAM J. Control Optim.* **33** 1 (1995), 29–66.

[13] N. El Karoui, S. Peng and M.C. Quenez, Backward stochastic differential equations in finance, *Math. Finance* **7** 1 (1997), 1–71.

[14] R.J. Elliott, *Stochastic calculus and applications*, Springer, New York, 1982.

[15] H. Föllmer and Ju.M. Kabanov, Optional decomposition and Lagrange multipliers, *Finance Stoch.* **2** 1 (1998), 69–81.

[16] C. Gourieroux, J.P. Laurent and H. Pham, Mean-Variance hedging and numeraire, *Math. Finance* **8** 3 (1998), 179–200.

[17] P. Grandits and L. Krawczyk, Closeness of some spaces of stochastic integrals, *Séminaire de Probabilités XXXII*, Lecture Notes in Math. **1686**, 75–87, Springer-Verlag, New York, 1998.

[18] J. Jacod, *Calcul stochastique et problèmes de martingales*, Lecture Notes in Math. **714**, Springer-Verlag, New York, 1979.

[19] N. Kazamaki, *Continuous exponential martingales and* BMO, Lecture Notes in Math. **1579**, Springer, New York, 1994.

[20] D.O. Kramkov, Optional decomposition of supermartingales and hedging contingent claims in incomplete security markets, *Probab. Theory Relat. Fields* **105** (1996), 459–479.

[21] J.P. Laurent and H. Pham, Dynamic programming and mean-variance hedging, *Finance Stoch.* **3** (1999), 83–110.

[22] R.Sh. Liptzer and A.N. Shiryayev, *Martingale theory*, Nauka, Moscow, 1986.

[23] M. Mania, A general problem of an optimal equivalent change of measure and contingent claim pricing in an incomplete market, *Stochastic Processes Appl.* **90** (2000), 19–42.

[24] M. Mania and R. Tevzadze, A semimartingale Bellman equation and the variance-optimal martingale measure, *Georgian Math. J.* **7** 4 (2000), 765–792.

[25] E. Pardoux and S.G. Peng, Adapted solution of a backward stochastic differential equation, *Systems Control Lett.* **14** (1990), 55–61.

[26] H. Pham, T. Rheinländer and M. Schweizer, Mean-variance hedging for continuous processes: New proofs and examples, *Finance Stoch.* **2** (1998), 173–198.

[27] T. Rheinländer, Optimal martingale measures and their applications in mathematical finance, *Dissertation* zur Erlangung des Akademischen Grades eines Doktors der Naturwissenschaften, 1999.

[28] M. Schäl, On quadratic cost criteria for option hedging, *Math. Oper. Res.* **19** (1994), 121–131.

[29] M. Schweizer, Mean-variance hedging for general claims, *Ann. Appl. Probab.* **2** (1992), 171–179.

[30] M. Schweizer, Approximating random variables by stochastic integrals, *Ann. Probab.* **22** 3 (1994), 1536–1575.

[31] M. Schweizer, Approximation pricing and the variance optimal martingale measure, *Ann. Probab.* **24** 1 (1996), 206–236.

SUBORDINATORS RELATED TO THE EXPONENTIAL FUNCTIONALS OF BROWNIAN BRIDGES AND EXPLICIT FORMULAE FOR THE SEMIGROUPS OF HYPERBOLIC BROWNIAN MOTIONS

HIROYUKI MATSUMOTO[1],
LAURENT NGUYEN[2] and MARC YOR[2]

[1] Nagoya University
School of Informatics and Sciences
Chikusa-ku, Nagoya 464-8601, Japan

[2] Université Pierre et Marie Curie
Laboratoire de Probabilités
F-75252 Paris, France

Abstract We prove that, if $\{b_t(s), 0 \leqq s \leqq t\}$ denotes the standard Brownian bridge of length t and $a_t = \int_0^t \exp(2b_t(s))ds$, then the subordinator $\{\mathcal{K}_s, s \geqq 0\}$ with no drift and with Lévy measure $K_0(x)e^{-x}x^{-1}dx$ satisfies $(\mathcal{K}_s)^{-1} \overset{(\text{law})}{=} a_{1/s}$ for fixed s. Variants and extensions of this result, in particular to general Brownian bridges and their exponential functionals are also discussed.

Key Words Subordinator, exponential functional, Brownian bridge, Brownian motion.

1. MOTIVATIONS AND MAIN RESULTS

1.1. In his paper [14], Gruet obtained a formula for the heat kernel $p_t^n(r)$ of the semigroup generated by the Laplacian on the real hyperbolic space $\mathbf{H}^n$, where r denotes the hyperbolic

213

distance between two points in $\mathbf{H}^n$. His formula holds for any dimension n and is quite different from the well known classical formulae (see, e.g., [3], p.178). If we concentrate on the two-dimenisonal case, Gruet's formula is

$$p_t^2(r) = \frac{\exp(-t/8)}{4\pi^{3/2}t^{1/2}} \int_0^\infty \frac{\exp((\pi^2 - \varrho^2)/2t)\sinh(\varrho)\sin(\pi\varrho/t)}{(\cosh(\varrho) + \cosh(r))^{3/2}} \, d\varrho,$$

whereas the classical formula is

$$p_t^2(r) = \frac{\sqrt{2}\exp(-t/8)}{(2\pi t)^{3/2}} \int_r^\infty \frac{\varrho\exp(-\varrho^2/2t)}{(\cosh(\varrho) - \cosh(r))^{1/2}} \, d\varrho.$$

Since Gruet's result is very convenient to use in different set-ups (see [1], [15], [17]), it is of interest to understand what lies behind the identity between these two expressions.

Gruet himself started this investigation in [15] and extended the identity, but we found that our recent works on exponential functionals of Brownian motion and, more specifically, Brownian bridges lead us quite naturally to further these inquiries.

Let us be more precise. As remarked by Gruet [15], the identity between the two expressions for the hyperbolic semigroups boils down to the equality between

$$G_1(t) = \int_0^\infty \frac{\exp((\pi^2 - \varrho^2)/2t)\sinh(\varrho)\sin(\pi\varrho/t)}{(\cosh(\varrho) + \cosh(r))^{\beta+1}} \, d\varrho$$

and

$$G_2(t) = \frac{\sin(\pi\beta)}{\beta t} \int_r^\infty \frac{\varrho\exp(-\varrho^2/2t)}{(\cosh(\varrho) - \cosh(r))^\beta} \, d\varrho$$

for $0 < \beta < 1$. In fact, the expression for $G_2(t)$ found in [15] has $\sinh(\varrho)$ instead of ϱ in the numerator and $(\cosh(\varrho) - \cosh(r))^{\beta+1}$ in the denominator. But this is obviously incorrect and it should be corrected as indicated above.

We shall show in this paper that the equality $G_1 = G_2$ is "equivalent" to the following particular case of the Lipschitz–Hankel formula for the modified Bessel function (Watson [25], p. 388 and Gradshteyn–Ryzhik [13], p. 694):

$$\nu \int_0^\infty e^{-ax} I_\nu(x) \frac{dx}{x} = (a + \sqrt{a^2 - 1})^{-\nu}, \qquad a \geq 1, \tag{1.1}$$

which will play an important role in this paper. By two "equivalent" identities, we mean that one can be deduced from the other in only a few steps. In fact, the identity (1.1) may be interpreted in the following probabilistic manner (cf. Feller [10], which develops [12], p. 336, and Sato [23], p. 234, Ex. 34.2).

Theorem 1.1 (i) *There exists a subordinator* $\{\mathcal{J}_\nu, \nu \geq 0\}$ *with no drift and Lévy measure*

$$\ell_{\mathcal{J}}(dx) = I_0(x)e^{-x} \frac{dx}{x}. \tag{1.2}$$

(ii) *For every* $\nu > 0$, *the law of* $\mathcal{J}_\nu$ *is given by*

$$P(\mathcal{J}_\nu \in dx) = \nu I_\nu(x)e^{-x} \frac{dx}{x}$$

and its Laplace transform is

$$E[\exp(-\lambda \mathcal{J}_\nu)] = ((1+\lambda) + \sqrt{(1+\lambda)^2 - 1}\,)^{-\nu}$$
$$= \exp(-\nu a_c(1+\lambda)),$$

where $a_c(x) = \mathrm{Argcosh}(x)$, $x \geq 1$.

In the present paper, we discuss the following Theorem 1.3, which is clearly an analogue of Theorem 1.1 and in which the role of the modified Bessel function I_0 is taken by the other modified Bessel (Macdonald) function K_0 of index 0. However, we first need to recall the following fact about the (first) Hartman–Watson law (cf. Hartman–Watson [16], Yor [26]).

Proposition 1.2 *Fix* $v > 0$. *Then the function* $\lambda \mapsto I_{\sqrt{2\lambda}}(v)$ *on* $\mathbf{R}_+$ *is the Laplace transform of a positive function* $\vartheta_v(t)$, $t > 0$, *that is,*

$$I_{\sqrt{2\lambda}}(v) = \int_0^\infty e^{-\lambda t} \vartheta_v(t)\, dt \tag{1.3}$$

and $\vartheta_v(t)$ *is given by*

$$\vartheta_v(t) = \frac{v}{(2\pi^3 t)^{1/2}} \int_0^\infty e^{(\pi^2 - \xi^2)/2t} e^{-v\cosh(\xi)} \sinh(\xi) \sin\left(\frac{\pi\xi}{t}\right) d\xi. \tag{1.4}$$

Note. We keep the notation $\vartheta_v(t)$ from the previous papers ([26], [28]); this should not create any confusion with the Jacobi theta function $\Theta(v)$ which is given by (6.5) and appears in our discussions in Section 6 below.

We may now state our first result.

Theorem 1.3 (i) *There exists a subordinator* $\{\mathcal{K}_t, t \geq 0\}$ *with no drift and Lévy measure*

$$\ell_\mathcal{K}(dv) \equiv k(v)dv = K_0(v)e^{-v}\,\frac{dv}{v}.$$

(ii) *For every* $t \geq 0$, *the Laplace transform of* $\mathcal{K}_t$ *is given by*

$$E[\exp(-\lambda \mathcal{K}_t)] = \exp\left\{-\frac{t}{2}\left(a_c(1+\lambda)\right)^2\right\} \tag{1.5}$$

and the distribution of $\mathcal{K}_t$ *is*

$$P(\mathcal{K}_t \in dx) = \sqrt{\frac{2\pi}{t}} \vartheta_x\left(\frac{1}{t}\right) e^{-x}\,\frac{dx}{x}. \tag{1.6}$$

Corollary 1.4 *The subordinator* $\mathcal{J} = \{\mathcal{J}_\nu, \nu \geq 0\}$ *may be obtained by the Bochner subordination from* $\mathcal{K} = \{\mathcal{K}_t, t \geq 0\}$ *and the stable* $(1/2)$-*subordinator* $\sigma = \{\sigma_\nu = \inf\{t; B_t \geq \nu\}, \nu \geq 0\}$ *for a one-dimensional Brownian motion* $\{B_t, t \geq 0\}$ *starting from* 0:

$$\{\mathcal{J}_\nu, \nu \geq 0\} \overset{\text{(law)}}{=} \{\mathcal{K}_{\sigma_\nu}, \nu \geq 0\},$$

where, on the right hand side, $\mathcal{K}$ *and* σ *are assumed independent.*

The subordination relation obtained in the corollary gives another check on the formula (1.2) presented in Theorem 1.1. Indeed, from the subordination relation linking $\mathcal{J}$ and $\mathcal{K}$, we deduce (this is a very particular instance of Lemma 6.2 below)

$$\ell_{\mathcal{J}}(dx) = \left(\int_0^\infty \frac{1}{\sqrt{2\pi u^3}} \varkappa_u(x) \, du \right) dx, \tag{1.7}$$

where $\varkappa_u(x)$ is the density of $\mathcal{K}_u$ presented in Theorem 1.3. Thus we obtain

$$\ell_{\mathcal{J}}(dx) = \left(\int_0^\infty \vartheta_x\left(\frac{1}{u}\right) \frac{du}{u^2} \right) e^{-x} \frac{dx}{x} = \left(\int_0^\infty \vartheta_x(v) \, dv \right) e^{-x} \frac{dx}{x}.$$

Therefore we find, in agreement with formula (1.2),

$$\ell_{\mathcal{J}}(dx) = I_0(x)e^{-x} \frac{dx}{x}$$

from the definition of the function ϑ in Proposition 1.2.

In fact, our main result in this paper is the following (partial!) realization of the subordinator $\mathcal{K}$ in terms of the Brownian bridges.

Theorem 1.5 *Let $\{b_u(s), 0 \leqq s \leqq u\}$ denote the Brownian bridge of length u and define*

$$A(b_u) = \int_0^u \exp(2b_u(s)) \, ds.$$

Then, for any fixed $t > 0$, one has

$$\mathcal{K}_t \overset{\text{(law)}}{=} \left(A(b_{1/t}) \right)^{-1} \overset{\text{(law)}}{=} t\left(\int_0^1 \exp\left(\frac{2}{\sqrt{t}}b(u)\right) \, du \right)^{-1}, \tag{1.8}$$

where $\{b(u), 0 \leqq u \leqq 1\}$ denotes the standard Brownian bridge.

It is worth discussing our result (1.8) in the following manner: For any $u > 0$, the probability law of $(A(b_u))^{-1}$ is infinitely divisible, more precisely, since $\mathcal{K}_s + \mathcal{K}_t \overset{\text{(law)}}{=} \mathcal{K}_{t+s}$, where on the left hand side $\mathcal{K}_s$ and $\mathcal{K}_t$ are independent, one has

$$\frac{1}{A(b_{1/t})} + \frac{1}{A(\tilde{b}_{1/s})} \overset{\text{(law)}}{=} \frac{1}{A(b_{1/(t+s)})},$$

where $b_{1/t}$ and $\tilde{b}_{1/s}$ are independent.

1.2. We now describe how the rest of this paper is organized. In Section 2, we give some useful formulae involving Bessel functions; in particular, the Lipschitz–Hankel formulae and their variants. In Section 3, we prove the equivalence between the identity $G_1 = G_2$ and the Lipschitz–Hankel formula (1.1). In Section 4, we prove Theorems 1.1, 1.3 and 1.5 and give some probabilistic interpretation of the Lipschitz–Hankel formula.

From Section 5 onwards, we consider extensions and applications of our main results. In Section 5, we extend Theorem 1.5, which deals with the standard Brownian bridge starting and ending at 0, to a general Brownian bridge. In Section 6, we discuss the Bochner subordination of our process $\mathcal{K}$ by some hyperbolic subordinators. In Section 7, we discuss about the subordinators whose Lévy measures are given in terms of the general modified Bessel functions I_ν and K_ν. Some related questions are discussed in the final Section 8. We found it convenient for the reader to gather our main results and notations for the various subordinators involved, in a short Appendix and in the forms of the Tables.

1.3. We conclude this Introduction with several remarks, which may give the reader some perspective on the present paper.

(i) The identity in law (1.8) between the one-dimensional marginals of the subordinator $\{\mathcal{K}_t\}$ and those of the exponential functionals of Brownian bridge is reminiscent of the well known identities: for any $t > 0$,

$$T_{\sqrt{t}} \stackrel{\text{(law)}}{=} \frac{1}{(S_{1/t})^2} \stackrel{\text{(law)}}{=} \frac{1}{(B_{1/t})^2},$$

where

$$T_a = \inf\{u; B_u \geq a\} \tag{1.9}$$

for a Brownian motion $B = \{B_u, u \geq 0\}$ starting from 0 and $S_u = \sup_{s \leq u} B_s$.

(ii) Given the complexity of the joint law of $A_t^{(\mu)} \equiv \int_0^t \exp(2(B_s + \mu s))ds$ of the exponential functional of Brownian motion with constant drift (see [28]), it is interesting to look for some (independent) random times T such that $A_T^{(\mu)}$ has a simple enough distribution; this was achieved in the note [27].

Again, the present paper aims at some analogous results for the exponential functionals of Brownian bridges, taking advantage of the identity (1.8) and some descriptions of $\mathcal{K}_T$, as in Section 6, say.

(iii) For completeness, we mention that some description of the law of $A(\pm r_t)$, where $\{r_t(u), u \leq t\}$ denotes a three-dimensional Bessel bridge with duration t, is found in [4]. As is well known, the processes $\{r_t\}_{t \geq 0}$ play a most important role in the theory of Brownian excursions.

2. PRELIMINARIES: INTEGRALS INVOLVING BESSEL FUNCTIONS

2.1. Given the important roles played in this paper by the inverse functions of the hyperbolic functions, it may be convenient to introduce and to recall the following elementary formulae:

$$a_c(x) \equiv \text{Argcosh}(x) = \log(x + \sqrt{x^2 - 1}), \qquad x \geq 1;$$
$$a_c'(x) = (x^2 - 1)^{-1/2}, \qquad x > 1.$$
$$a_s(x) \equiv \text{Argsinh}(x) = \log(x + \sqrt{x^2 + 1}), \qquad x \geq 0;$$
$$a_s'(x) = (x^2 + 1)^{-1/2}, \qquad x \geq 0.$$

2.2. Likewise, the modified Bessel functions I_ν and K_ν are crucial in this paper. We refer the reader to, e.g., Gradshteyn–Ryzhik [13] and Watson [25] for the main definitions and properties of these functions. For convenience, we recall

$$K_\nu(z) = \frac{\pi}{2\sin(\nu\pi)}(I_{-\nu}(z) - I_\nu(z)) \qquad (\nu \neq 0) \tag{2.1}$$

$$= \frac{1}{2}\int_0^\infty \exp\left(-\frac{z}{2}\left(y + \frac{1}{y}\right)\right)y^{\nu-1}\, dy \tag{2.2}$$

and

$$K_0(z) = -\frac{d}{d\nu}I_\nu(z)\big|_{\nu=0}. \tag{2.3}$$

2.3. A number of our results may be deduced from the following particular case of the Lipschitz–Hankel formulae (cf.[13], [25]):

$$\int_0^\infty e^{-u\cosh(r)} I_\nu(u)\, du = \frac{e^{-\nu r}}{\sinh(r)}, \qquad \nu > -1,\ r > 0. \tag{2.4}$$

As a consequence, we obtain

$$\int_0^\infty e^{-u\cosh(r)}(I_{-\nu}(u) - I_\nu(u))\, du = \frac{2\sinh(\nu r)}{\sinh(r)},\ 0 < \nu < 1,\ r > 0 \tag{2.5}$$

and, hence, from (2.1),

$$\int_0^\infty e^{-u\cosh(r)} K_\nu(u)\, du = \frac{\pi}{\sin(\pi\nu)} \frac{\sinh(\nu r)}{\sinh(r)}, \qquad 0 < |\nu| < 1. \tag{2.6}$$

Moreover, letting $\nu \to 0$, we also obtain

$$\int_0^\infty e^{-u\cosh(r)} K_0(u)\, du = \frac{r}{\sinh(r)}, \qquad r > 0. \tag{2.7}$$

Going back to (2.4), bringing $\sinh(r)$ on the left hand side and integrating with respect to r, we arrive at the following variant

$$\int_0^\infty e^{-u\cosh(r)} I_\nu(u)\, \frac{du}{u} = \frac{1}{\nu} e^{-\nu r}, \qquad \nu > 0. \tag{2.8}$$

We should note the following general formula, which will also be useful in the sequel:

$$\int_0^\infty e^{-u\cosh(r)} I_\nu(u) u^{\mu-1}\, du = \sqrt{\frac{2}{\pi}} e^{-(\mu-1/2)\pi i} \frac{Q_{\nu-1/2}^{\mu-1/2}(\cosh(r))}{(\sinh(r))^{\mu-1/2}}, \tag{2.9}$$

where, on the right hand side, Q_ν^μ is the associated Legendre function. For details, see [13], p. 694, and [25], p. 388. But the reader should be aware of the difference of conventions between [13] and [25]. We have followed the notation in [13].

2.4. In fact, in the sequel, it will be more convenient to work with the argument $x = \cosh(r)$ and we shall then use, e.g., the formulae (2.4) and (2.7), in the form

$$\int_0^\infty e^{-ux} I_\nu(u)\, du = a_c'(x) a_c(x) \exp(-\nu a_c(x)), \qquad x > 1,\ \nu > -1, \tag{2.10}$$

and

$$\frac{d}{dx}\left(\frac{1}{2} a_c(x)^2\right) = a_c'(x) a_c(x) = \int_0^\infty e^{-xu} K_0(u)\, du, \qquad x > 1. \tag{2.11}$$

3. ON THE EQUIVALENCE BETWEEN $\{G_1 = G_2\}$ AND (1.1)

We shall first show that the identity $G_1(t) = G_2(t), t > 0$ (in the sequel, we shall write $\{G_1 = G_2\}$ for short) is "equivalent" to

$$\Gamma(1-\beta) \int_0^\infty I_\nu(x) e^{-u\cosh(r)} u^{\beta-1}\, du = \int_r^\infty \frac{e^{-\nu\varrho}}{(\cosh(\varrho) - \cosh(r))^\beta}\, d\varrho \tag{3.1}$$

for all $\beta \in [0, 1)$ and then it is "equivalent" to (3.1) in the case $\beta = 0$, which is precisely (1.1). Throughout the proofs of these equivalences, Proposition 1.2 will play an important role and is taken for granted.

To show the "equivalence" between $\{G_1 = G_2\}$ and (3.1), we first begin by transforming $G_1(t)$ into a double integral with the help of the identity

$$c^{-m} = \frac{1}{\Gamma(m)} \int_0^\infty u^{m-1} e^{-cu}\, du, \qquad c > 0,\ m > 0.$$

Then we may write

$$G_1(t) = \frac{1}{\Gamma(\beta+1)} \int_0^\infty u^\beta e^{-u\cosh(r)}\, du \int_0^\infty e^{(\pi^2 - \varrho^2)/2t} e^{-u\cosh(\varrho)} \sinh(\varrho) \sin\left(\frac{\pi\varrho}{t}\right)\, d\varrho.$$

On the other hand, making use of the two equalities in Proposition 1.2, we obtain that the equality $\{G_1 = G_2\}$ is "equivalent" to $\{\mathcal{G}_1 = \mathcal{G}_2\}$ for $0 < \beta < 1$, where

$$\mathcal{G}_1(t) = \int_0^\infty u^{\beta-1} e^{-u\cosh(r)} \vartheta_u(t)\, du$$

and

$$\mathcal{G}_2(t) = \frac{\sin(\pi\beta)}{\pi} \Gamma(\beta+1) \int_r^\infty \frac{\varrho e^{-\varrho^2/2t}}{\sqrt{2\pi t^3}(\cosh(\varrho) - \cosh(r))^\beta}\, d\varrho.$$

Now the equality $\{\mathcal{G}_1 = \mathcal{G}_2\}$ holds if and only if the identity between their Laplace transforms with respect to t holds, which leads to (3.1) as an "equivalence" to $\{G_1 = G_2\}$. In fact, by using the Lipschitz–Hankel formula (2.9), we obtain for $\lambda > 0$

$$\mathcal{L}_1(\lambda) \equiv \int_0^\infty e^{-\lambda t} \mathcal{G}_1(t)\, dt = \int_0^\infty u^{\beta-1} e^{-u\cosh(r)} I_{\sqrt{2\lambda}}(u)\, du$$

$$= \sqrt{\frac{2}{\pi}} e^{-(\beta-1/2)\pi i} \frac{Q^{\beta-1/2}_{\sqrt{2\lambda}-1/2}(\cosh(r))}{(\sinh(r))^{\beta-1/2}}.$$

On the other hand, using the classical identity

$$\int_0^\infty \varrho \exp\left(-\left(\frac{\varrho^2}{2t} + \frac{\nu^2 t}{2}\right)\right) \frac{dt}{\sqrt{2\pi t^3}} = e^{-\nu\varrho}, \qquad \nu,\ \varrho > 0, \tag{3.2}$$

which expresses the Laplace transform of $P(T_\varrho \in dt)$ with T_ϱ as in (1.9), we have

$$\mathcal{L}_2(\lambda) \equiv \int_0^\infty e^{-\lambda t} \mathcal{G}_2(t)\, dt = \frac{\sin(\pi\beta)}{\pi} \Gamma(\beta+1) \int_r^\infty \frac{e^{-\sqrt{2\lambda}\varrho}}{(\cosh(\varrho) - \cosh(r))^\beta}\, d\varrho.$$

Then, recalling an integral representation

$$Q^\mu_\nu(\cosh(r)) = \frac{\sqrt{\pi} e^{\mu\pi\sqrt{-1}}(\sinh(r))^\mu}{\sqrt{2}\Gamma(\frac{1}{2} - \mu)} \int_r^\infty \frac{e^{-(\nu+1/2)\varrho}}{(\cosh(\varrho) - \cosh(r))^{\mu+1/2}}\, d\varrho$$

of the associated Legendre function (cf. [13] p. 952) and using the identity $\Gamma(z)\Gamma(1 - z) = \pi/\sin(\pi z)$, we obtain the equality between the Laplace transforms $\mathcal{L}_1$ and $\mathcal{L}_2$, hence the identity $\mathcal{G}_1 = \mathcal{G}_2$.

We next show that (3.1) in general form is "equivalent" to (3.1) when $\beta = 0$. By writing

$$\Gamma(1 - \beta) = \int_0^\infty t^{-\beta} e^{-t} \, dt, \quad 0 < \beta < 1,$$

we see that the former is the equality of the moments of order β between two measures. Thus, (3.1) holds for every $\beta < 1$ if and only if, for every Borel function $f : \mathbf{R}_+ \to \mathbf{R}_+$, one has

$$\int_0^\infty e^{-t} \, dt \int_0^\infty f\left(\frac{u}{t}\right) e^{-u \cosh(r)} I_\nu(u) \, \frac{du}{u} = \int_r^\infty e^{-\nu \varrho} f\left(\frac{1}{\cosh(\varrho) - \cosh(r)}\right) d\varrho. \quad (3.3)$$

We transform the left hand side of (3.3) by first using Fubini's theorem and then making change of variables by $t = ux$. Then we obtain

$$\int_0^\infty e^{-t} \, dt \int_0^\infty f\left(\frac{u}{t}\right) e^{-u \cosh(r)} I_\nu(u) \, \frac{du}{u} = \int_0^\infty f\left(\frac{1}{x}\right) dx \int_0^\infty e^{-(x + \cosh(r))u} I_\nu(u) \, du.$$

Concerning the right hand side of (3.3), we make the change of variables by $y = \cosh(\varrho)$ or, equivalently, $\varrho = a_c(y) = \log(y + \sqrt{y^2 - 1})$. Then we get

$$\begin{aligned} \int_r^\infty e^{-\nu \varrho} f\left(\frac{1}{\cosh(\varrho) - \cosh(r)}\right) d\varrho &= \int_{\cosh(r)}^\infty e^{-\nu a_c(y)} f\left(\frac{1}{y - \cosh(r)}\right) a_c'(y) \, dy \\ &= \int_0^\infty e^{-\nu a_c(x + \cosh(r))} f\left(\frac{1}{x}\right) a_c'(x + \cosh(r)) \, dx. \end{aligned}$$

Hence (3.3) holds if and only if

$$\int_0^\infty e^{-ux} I_\nu(u) \, du = a_c'(x) e^{-\nu a_c(x)}, \quad x > 1,$$

which is nothing else but the identity (2.4).

4. PROOFS OF THE THEOREMS

4.1. *Proof of Theorem 1.1.* As indicated in the references preceding Theorem 1.1, this theorem is not new. An elementary "check" consists in replacing in formula (1.1) the variable $a \geq 1$ by $1 + \lambda$ with $\lambda \geq 0$, so that the right hand side of (1.1) now appears as the Laplace transform in λ of the probability distribution $\iota_\nu(dx) = \nu I_\nu(x) e^{-x} x^{-1} dx$. It is obvious that $\iota_\nu * \iota_{\nu'} = \iota_{\nu + \nu'}$ for every $\nu, \nu' \geq 0$ and hence we obtain the existence of the subordinator $\{\mathcal{J}_\nu, \nu \geq 0\}$.

4.2. *Proof of Theorem 1.3.* At first we recall (2.4). Then, by using Proposition 1.2 and formula (3.2), we see that both hand sides of (2.4) are the Laplace transforms in $\nu^2/2$ of positive measures on $\mathbf{R}_+$ and obtain

$$\sinh(r) \int_0^\infty e^{-u \cosh(r)} \vartheta_u(t) \, du = \frac{1}{\sqrt{2\pi t^3}} r e^{-r^2/2t}.$$

Integrating both hand sides with respect to r yields

$$\int_0^\infty e^{-u \cosh(r)} \vartheta_u(t) \, \frac{du}{u} = \frac{1}{\sqrt{2\pi t}} e^{-r^2/2t}. \quad (4.1)$$

Now let $\varkappa_t(dv)$ be the probability measure on $\mathbf{R}_+$ given by

$$\varkappa_t(dv) = \sqrt{\frac{2\pi}{t}}\,\vartheta_v\Big(\frac{1}{t}\Big)e^{-v}\,\frac{dv}{v}, \qquad v > 0.$$

Then, we obtain from (4.1)

$$\int_0^\infty e^{-\lambda v}\varkappa_t(dv) = \exp\Big(-\frac{t}{2}\big(a_c(1+\lambda)\big)^2\Big). \qquad (4.2)$$

Thus, the family of probability measures $\{\varkappa_t, t > 0\}$ satisfies $\varkappa_t * \varkappa_s = \varkappa_{t+s}$ for $t, s > 0$ and the second part of the theorem is proved.

The first part of the theorem follows from identity (2.11).

4.3. Proof of Corollary 1.4. By Theorem 1.1 we have

$$E[\exp(-\lambda \mathcal{J}_\nu)] = \exp(-\nu a_c(1+\lambda)).$$

On the other hand, in Theorem 1.3 we have shown

$$E[\exp(-\lambda \mathcal{K}_{\sigma_\nu})] = E\Big[\exp\Big(-\frac{\big(a_c(1+\lambda)\big)^2}{2}\sigma_\nu\Big)\Big].$$

The rest of the proof is easy.

4.4. Proof of Theorem 1.5. Given Theorem 1.3, we can easily prove the theorem if we recall the following explicit formula for the probability distribution of $A(b_t)$ (cf. Yor [28]):

$$P(A(b_t) \in du) \equiv P(A_t \in du|B_t = 0) = \sqrt{2\pi t}\vartheta_{1/u}(t)e^{-1/u}\,\frac{du}{u}.$$

5. FROM (GENERAL) BROWNIAN BRIDGE TO BROWNIAN MOTION

5.1. Letting $B = \{B_t, t \geq 0\}$ denote a standard Brownian motion starting from 0 and setting

$$A_t \equiv A_t(B) = \int_0^t \exp(2B_u)\,du,$$

we first recall the following description (cf.[18]) of the joint probability law, for fixed t, of (B_t, A_t):

$$(ee^{-B_t}A_t(B), B_t) \overset{(\text{law})}{=} (\cosh(|B_t| + L_t) - \cosh(B_t), B_t), \qquad (5.1)$$

where e denotes a standard exponential random variable, independent of B, and $\{L_t\}$ is the local time of B at 0.

We shall now see that the previous result (1.8) and the further results in this section for general Brownian bridges are easily deduced from (5.1), after we condition with respect to $B_t = x$.

For simplicity, let us take $x = 0$ and consider

$$E\Big[\exp\Big(-\frac{\lambda}{A(b_t)}\Big)\Big] = P(eA(b_t) > \lambda).$$

Then, from (5.1), this is equal to

$$P(\cosh(L_t) > 1 + \lambda | B_t = 0) = P(L_t > a_c(1 + \lambda) | B_t = 0) = P(te > \frac{1}{2}(a_c(1 + \lambda))^2)$$

since, conditioning on $B_t = 0$, L_t is distributed as $\sqrt{2te}$. Proposition 5.1 in the next subsection may be proved in the same manner.

5.2. Letting b_t^x be the Brownian bridge of length t which starts from 0 and ends at x, we consider the probability law of $A(b_t^x)$.

Proposition 5.1 *Let $x \in \mathbf{R}$ and $t > 0$. Then, for every non-negative Borel function f on $\mathbf{R}_+$, one has*

$$E[f(\frac{1}{A(b_t^x)})] = E[f(e^{-x}\mathcal{K}_{1/t}) \exp(-(\cosh(x) - 1)\mathcal{K}_{1/t} + \frac{x^2}{2t})] = E[f(\mathcal{K}_{1/t}^{(x)})], \qquad (5.2)$$

where $\mathcal{K}^{(x)} = \{\mathcal{K}_u^{(x)}, u \geq 0\}$ is a subordinator with no drift and Lévy measure

$$k^{(x)}(v)dv = K_0(e^x v) \exp(-\frac{e^{2x} + 1}{2}v) \frac{dv}{v}.$$

Remark 5.1 Formula (5.2) shows that $\mathcal{K}^{(x)}$ is obtained from $\mathcal{K}$ via two operations, the Esscher transform ([9] and [24], Section 3c) by the martingale density

$$\{\exp(-(\cosh(x) - 1)\mathcal{K}_u + \frac{x^2 u}{2}), u \geq 0\}$$

and the simple multiplication by e^{-x}.

Proof. In [7], we showed

$$E[\exp(-\frac{\lambda}{A(b_t^x)})] \equiv E[\exp(-\frac{\lambda}{A_t}) | B_t = x] = \exp(-\frac{\varphi_x(\lambda)^2 - x^2}{2t})$$

for $\lambda > 0$ and $t > 0$, where $\varphi_x(\lambda) = a_c(\lambda e^{-x} + \cosh(x))$. By (2.11), we have

$$\frac{d}{d\lambda}(\frac{1}{2}\varphi_x(\lambda)^2) = e^{-x} \int_0^\infty \exp(-(\lambda e^{-x} + \cosh(x))y) K_0(y) \, dy.$$

Therefore, integrating both hand sides with respect to λ, we obtain

$$-\frac{\varphi_x(\lambda)^2 - x^2}{2t} = \frac{1}{t} \int_0^\infty (e^{-\lambda e^{-x} y} - 1) e^{-y \cosh(x)} K_0(y) \frac{dy}{y}$$

$$= \frac{1}{t} \int_0^\infty (e^{-\lambda v} - 1) \exp(-\frac{e^{2x} + 1}{2} v) K_0(e^x v) \frac{dv}{v}$$

and $E[f(1/A(b_t^x))] = E[f(\mathcal{K}_{1/t}^{(x)})]$.
    The first identity in (5.2) is easily shown from (1.6) and (4.1). $\qquad \square$

5.3. As a variant of Proposition 5.1, we present the joint law of (A_t, B_t) in the following form. Formula (5.4) below is a key in [19] and [20] for a proof of an analogue of Pitman's $2M - X$

theorem, which asserts that the stochastic process $\{Z_t = \exp(-B_t)A_t, t \geqq 0\}$ is a diffusion process.

Proposition 5.2 *Let $t > 0$. Then the joint law of (Z_t, B_t) may be expressed in terms of the law of $\mathcal{K}_{1/t}$ via the following integral formula: for every non-negative Borel function f on $\mathbf{R}_+^2$,*

$$E[f\left(\frac{1}{Z_t}, \exp(B_t)\right)] = \frac{1}{\sqrt{2\pi t}} E[e^{\mathcal{K}_{1/t}} \int_0^\infty f(\mathcal{K}_{1/t}, y) \exp\left(-\frac{1}{2}(y + y^{-1})\mathcal{K}_{1/t}\right) \frac{dy}{y}]. \quad (5.3)$$

In particular, one has

$$P(B_t \in dx | Z_t = z) = \frac{1}{2K_0(1/z)} \exp\left(-\frac{\cosh(x)}{z}\right) dx \quad (5.4)$$

and

$$E[\varphi\left(\frac{1}{Z_t}\right)] = \sqrt{\frac{2}{\pi t}} E[e^{\mathcal{K}_{1/t}} \varphi(\mathcal{K}_{1/t}) K_0(\mathcal{K}_{1/t})]. \quad (5.5)$$

Proof. By formula (5.2), we deduce

$$E[f\left(\frac{1}{Z_t}, \exp(B_t)\right)] = \int_{-\infty}^\infty E[f\left(\frac{e^x}{A(b_t^x)}, e^x\right)] \frac{1}{\sqrt{2\pi t}} e^{-x^2/2t} \, dx$$

$$= \frac{1}{\sqrt{2\pi t}} E[e^{\mathcal{K}_{1/t}} \int_{-\infty}^\infty f(\mathcal{K}_{1/t}, e^x) e^{-\mathcal{K}_{1/t} \cosh(x)} \, dx], \quad (5.6)$$

from which we easily obtain (5.3).

Moreover, (5.6) implies

$$P(B_t \in dx | Z_t = z) = C \exp\left(-\frac{\cosh(x)}{z}\right) dx$$

for a normalizing constant C. Now, recalling the integral representation (2.2) for K_0, we easily obtain $C = (2K_0(1/z))^{-1}$. Formula (5.5) is also an easy consequence of (5.6). $\qquad\square$

6. RELATIONS TO SOME HYPERBOLIC SUBORDINATORS

Let $C = \{C_t\}, S = \{S_t\}, T = \{T_t\}$ be the subordinators defined by

$$E[\exp\left(-\frac{\lambda^2}{2}C_t\right)] = \frac{1}{(\cosh(\lambda))^t}, \qquad E[\exp\left(-\frac{\lambda^2}{2}S_t\right)] = \left(\frac{\lambda}{\sinh(\lambda)}\right)^t,$$

$$E[\exp\left(-\frac{\lambda^2}{2}T_t\right)] = \left(\frac{\tanh(\lambda)}{\lambda}\right)^t,$$

$\lambda > 0$, respectively. Note that, if S and T are independent,

$$\{C_t, t \geqq 0\} \overset{\text{(law)}}{=} \{S_t + T_t, t \geqq 0\}.$$

These subordinators are studied in Pitman–Yor [22], motivated by the importance of the random variables C_1, C_2, S_1, S_2 in relation to the Riemann zeta function and the Dirichlet L function. See Biane–Pitman–Yor [2]. It is quite interesting to subordinate our process $\mathcal{K} = \{\mathcal{K}_t\}$ with either of these three subordinators. The explicit formulae of the corresponding Lévy measures are obtained in [2] and [22] and we give them in the following table.

Table 1

Subordinator	Density of Lévy measure
C	$\ell_C(u) = u^{-1} \sum_{n=1}^{\infty} \exp(-\pi^2 (2n-1)^2 u/8)$
S	$\ell_S(u) = u^{-1} \sum_{n=1}^{\infty} \exp(-\pi^2 n^2 u/2)$
T	$\ell_T(u) = \ell_C(u) - \ell_S(u)$

The following well-known series (cf. [13], p. 924) will help us to simplify formulae for the densities of the Lévy measures of our subordinated processes.

Lemma 6.1 *One has*

$$I_0(x) + 2\sum_{n=1}^{\infty} (-1)^n I_{2n}(x) = 1 \tag{6.1}$$

and

$$I_0(x) + 2\sum_{n=1}^{\infty} I_{2n}(x) = \cosh(x). \tag{6.2}$$

6.1. *Subordination with C*

Proposition 6.1 *Let $\{\Gamma_t, t \geq 0\}$ be a gamma process. Then, one has*

$$\{\mathcal{K}_{C_t}, t \geq 0\} \overset{\text{(law)}}{=} \{\Gamma_t, t \geq 0\}.$$

Proof. By using Theorem 1.3 (ii), it is easy to show

$$E[\exp(-\lambda \mathcal{K}_{C_t})] = E\big[\exp\big(-\frac{t}{2}\big(a_c(1+\lambda)\big)^2\big)\big] = (1+\lambda)^{-t},$$

which is equal to $E[\exp(-\lambda \Gamma_t)]$. □

It is interesting to check how this identity in law is reflected at the level of the corresponding Lévy measures. To do this, we recall a well known relation between Lévy measures under a subordination scheme. We give it under some restricting assumptions. The general result is found in Sato [23], p. 197, Theorem 30.1.

Lemma 6.2 *Let X and Y be two independent subordinators with no drift. Assume that the Lévy measure of Y has a density $\ell_Y(u)$ and that the probability distribution of X_t has a density $p_t^X(x)$. Then, the Lévy measure of the subordinated process $\{X_{Y_t}, t \geq 0\}$ has a density $\ell_{XY}(x)$ given by*

$$\ell_{XY}(x) = \int_0^{\infty} p_u^X(x) \ell_Y(u) \, du. \tag{6.3}$$

We show that the relation (6.3) for $X = \mathcal{K}$ and $Y = C$ is nothing else but the identity (6.1). In fact, by using the explicit form of the density ℓ_C of the Lévy measure of the subordinator C and the definition (1.3) of the function $\vartheta_x(t)$, we easily obtain

$$e^{-x}\frac{1}{x} = \int_0^{\infty} \frac{1}{u}\sum_{n=1}^{\infty}\exp\big(-\frac{\pi^2}{2}\big(n-\frac{1}{2}\big)^2 u\big) \cdot \sqrt{\frac{2\pi}{u}}\,\vartheta_x\big(\frac{1}{u}\big)e^{-x}\frac{1}{x}\,du$$

or

$$\int_0^\infty \sqrt{\frac{2\pi}{v}} \sum_{n=1}^\infty \exp\left(-\frac{\pi^2}{2}\left(n - \frac{1}{2}\right)^2 u\right) \vartheta_x(v) \, dv = 1.$$

Moreover, particular cases of the Poisson summation formula yield the Jacobi identity

$$\sqrt{u}\,\Theta(u) = \Theta\left(\frac{1}{u}\right), \qquad (6.4)$$

where

$$\Theta(u) = \sum_{n=-\infty}^\infty e^{-n^2 \pi u}, \qquad (6.5)$$

as well as

$$\sum_{n=1}^\infty \exp\left(-\frac{\pi^2}{2}\left(n - \frac{1}{2}\right)^2 \frac{1}{v}\right) = \sqrt{\frac{2\pi}{v}}\left(\frac{1}{2} + \sum_{n=1}^\infty (-1)^n e^{-2n^2 v}\right). \qquad (6.6)$$

Therefore, recalling the definition (1.3) of the function $\vartheta_x(v)$, we obtain (6.1).

6.2. *Subordination with S* We denote the subordinated process by $Q = \{Q_t\}$, $Q_t = \mathcal{K}_{S_t}$. To motivate some interest in this process, we first remark that:

$$\mathcal{K}_{S_1} \overset{\text{(law)}}{=} \Lambda, \qquad (6.7)$$

where (the distribution of) Λ is given by $P(\Lambda \in d\xi) = e^{-\xi} K_0(\xi) d\xi$. For a more detailed discussion of Λ and its relation to the process $\{\mathcal{K}_t\}$, we refer the reader to Lemma 8.1 below. Identity (6.7) follows from (2.7) and (1.5). In fact, by using these identities, we easily obtain

$$E[e^{-\lambda\Lambda}] = \frac{a_c(1+\lambda)}{\sinh(a_c(1+\lambda))} = E\left[\exp\left(-\frac{(a_c(1+\lambda))^2}{2} S_1\right)\right] = E[\exp(-\lambda \mathcal{K}_{S_1})]. \qquad (6.8)$$

Proposition 6.2 (i) *It holds that*

$$E[\exp(-\lambda Q_t)] = \left(\frac{a_c(1+\lambda)}{\sqrt{(1+\lambda)^2 - 1}}\right)^t.$$

(ii) *The density $\ell_Q(x)$ of the Lévy measure of Q is*

$$\ell_Q(x) = \frac{e^{-x}}{2x}\left(\cosh(x) - \int_0^\infty I_\nu(x)\,d\nu\right).$$

Proof. The assertion of (i) is proved in a similar way to that of (6.8) above and we omit it. For a proof of (ii), we recall Lemma 6.2:

$$\ell_Q(x) = \int_0^\infty \ell_S(u) p_u^{\mathcal{K}}(x)\,du.$$

We know from Table 1:

$$\ell_S(u) = \frac{1}{u}\sum_{n=1}^\infty e^{-\pi^2 n^2/2} = \frac{1}{2u}(\Theta(u) - 1).$$

Then, using the Jacobi identity (6.4) again, we obtain

$$
\begin{aligned}
\ell_Q(x) &= \frac{e^{-x}}{x} \int_0^\infty \sqrt{\frac{\pi}{2u}} \vartheta_x\left(\frac{1}{u}\right) \cdot \frac{1}{2u}(\Theta(u) - 1)\, du \\
&= \frac{e^{-x}}{2x} \int_0^\infty \vartheta_x(v)\left(\Theta\left(\frac{2v}{\pi}\right) - \sqrt{\frac{\pi}{2v}}\right) dv.
\end{aligned}
$$

By the definition (1.3) of the function $\vartheta_x(v)$ and by the identity (6.2), we easily obtain

$$
\int_0^\infty \vartheta_x(v)\Theta\left(\frac{2v}{\pi}\right) dv = \sum_{n=-\infty}^\infty I_{|2n|}(x) = \cosh(x). \tag{6.9}
$$

Therefore, the proof of the proposition is completed by the following lemma.   □

Lemma 6.3 *Letting $\vartheta_x(v)$, $x > 0$, $v > 0$, be the function given in Proposition 1.2, one has*

$$
\int_0^\infty \vartheta_x(v) \frac{dv}{\sqrt{v}} = \sqrt{\frac{2}{\pi}} \int_0^\infty I_\nu(x)\, d\nu.
$$

The proof follows from (1.3) together with Fubini's theorem.

6.3. *Subordination with T* If the subordinators C and S are assumed independent, we have $C \stackrel{\text{(law)}}{=} S + T$. Moreover, thanks to the independence of increments of $\mathcal{K}$, we obtain the identity in law between processes:

$$
\mathcal{K}_{C_t} \stackrel{\text{(law)}}{=} \mathcal{K}_{S_t} + \widetilde{\mathcal{K}}_{T_t},
$$

where $\widetilde{\mathcal{K}} = \{\widetilde{\mathcal{K}}_t, t \geq 0\}$ is a copy of $\mathcal{K}$, independent of S and T.

Therefore, by using the results given above, we easily obtain the following.

Proposition 6.3 *Let $R = \{R_t\}$ be the subordinated process of $\mathcal{K}$ by T. Then the density ℓ_R of the Lévy measure of R is given by*

$$
\ell_R(x) = \frac{e^{-x}}{2x}\left(2 - \cosh(x) + \int_0^\infty I_h(x)\, dh\right)
$$

and one has

$$
E[\exp(-\lambda R_t)] = \left(\frac{\sqrt{(1 + \lambda)^2 - 1}}{(1 + \lambda)a_c(1 + \lambda)}\right)^t.
$$

7.   REPLACING I_0 BY I_ν AND K_0 BY K_ν

It is natural to look for some generalizations of the existence of the subordinators $\{\mathcal{J}_t, t \geq 0\}$ and $\{\mathcal{K}_t, t \geq 0\}$, when replacing in the expression of the densities of the Lévy measures the functions I_0 and K_0 by I_ν and K_ν, respectively.

7.1. We first develop the replacement of I_0 by I_ν, starting with the following remark: for $\nu > 0$,

$$
I_\nu(x) \sim \frac{x^\nu}{2^\nu \Gamma(1 + \nu)} \quad \text{as } x \downarrow 0 \quad \text{and} \quad I_\nu(x) \sim \frac{e^x}{\sqrt{2\pi x}} \quad \text{as } x \to \infty.
$$

Hence, we find that the measure $\tilde{i}_\nu(dx)$ on $\mathbf{R}_+$ defined by

$$\tilde{i}_\nu(dx) = e^{-x} I_\nu(x) \frac{dx}{x}$$

is a finite measure. In fact, from Theorem 1.1, $\nu \cdot \tilde{i}_\nu(dx) = \iota_\nu(dx) = P(\mathcal{J}_\nu \in dx)$ is a probability measure. A fortiori, $\tilde{i}_\nu(dx)$ is a Lévy measure.

If $-1 < \nu \le 0$, the same arguments yield that $\tilde{i}_\nu(dx)$ is a Lévy measure on $\mathbf{R}_+$ because $\int_0^\infty (x \wedge 1) \tilde{i}_\nu(dx) < \infty$. But, in this case, $\tilde{i}_\nu$ is no longer a finite measure.

In any case, let us denote by $\mathcal{J}^{(\nu)} = \{\mathcal{J}_t^{(\nu)}, t \ge 0\}$ a subordinator with no drift and Lévy measure $\tilde{i}_\nu$. In particular, we have $\mathcal{J}^{(0)} = \mathcal{J}$.

We may now give a description of $\mathcal{J}^{(\nu)}$ as a compound Poisson process.

Proposition 7.1 *Let $\nu > 0$. Then, the Lévy measure $\tilde{i}_\nu(dx)$, $x \ge 0$, of $\mathcal{J}^{(\nu)}$ satisfies*

$$\int_0^\infty (1 - e^{-\lambda x}) \tilde{i}_\nu(dx) = \frac{1}{\nu}(1 - E[\exp(-\lambda \mathcal{J}_\nu)]) = \frac{1}{\nu}(1 - e^{-\nu a_c(1+\lambda)}). \tag{7.1}$$

Consequently, one has

$$\{\mathcal{J}_t^{(\nu)}, t \ge 0\} \overset{(law)}{=} \{\mathcal{J}_{\nu N_{t/\nu}}, t \ge 0\}, \tag{7.2}$$

where $\{N_u, u \ge 0\}$ is a standard Poisson process, independent of $\mathcal{J}$.

Proof. The identities in (7.1) are easy consequences of the Lipschitz–Hankel formula (1.1) and Theorem 1.1. For a proof of (7.2), we need only to show the coincidence of the Lévy–Khintchine representations of both hand sides by using (7.1). □

We now turn our attention to $\mathcal{J}^{(\nu)}$ for $-1 < \nu < 0$; the simplest case to look at is certainly the case of $\nu = -1/2$. As is well known, we have

$$I_{-1/2}(x) = \sqrt{\frac{2}{\pi x}} \cosh(x) \tag{7.3}$$

and, therefore, we may describe $\mathcal{J}^{(-1/2)}$ in the following way without using the Lipschitz–Hankel formula.

At first we note the elementary formula related to the Lévy exponent of the stable $(1/2)$-subordinator

$$\int_0^\infty (1 - e^{-\lambda x}) \frac{e^{-\mu x}}{\sqrt{2\pi x^3}} \, dx = \sqrt{2(\lambda + \mu)} - \sqrt{2\mu}, \qquad \mu \ge 0. \tag{7.4}$$

Therefore we have

$$\int_0^\infty (1 - e^{-\lambda x}) \tilde{i}_{-1/2}(dx) = \sqrt{2\lambda} + \sqrt{2(\lambda + 2)} - 2. \tag{7.5}$$

Now we recall that, letting $B = \{B_t, t \ge 0\}$ be a one-dimensional standard Brownian motion starting from 0 and setting $T_a^{(\mu)} = \inf\{t; B_t + \mu t = a\}$, one has

$$E[\exp(-\lambda T_a^{(\mu)})] = \exp(-a(\sqrt{2\lambda + \mu^2} - \mu)).$$

Then, we obtain the following.

Proposition 7.2 *If $\nu = -1/2$, one has*

$$\{\mathcal{J}_a^{(-1/2)}, a \geq 0\} \overset{\text{(law)}}{=} \{T_a^{(0)} + T_a^{(2)}, a \geq 0\},$$

where $\{T_a^{(0)}\}$ and $\{T_a^{(2)}\}$ are assumed independent on the right hand side.

For the general case, we recall the identity (2.5). Then, setting $r = a_c(1 + \lambda)$, we have

$$\int_0^\infty e^{-(1+\lambda)x} I_{-\nu}(x)\, dx = \frac{2\cosh(\nu a_c(1+\lambda))}{\sinh(a_c(1+\lambda))} - \int_0^\infty e^{-(1+\lambda)x} I_\nu(x)\, dx$$

or

$$\frac{d}{d\lambda}\left(\int_0^\infty (1 - e^{-\lambda x})\tilde{i}_{-\nu}(dx)\right) = 2\cosh(\nu a_c(1+\lambda))\frac{d}{d\lambda}a_c(1+\lambda) - \frac{d}{d\lambda}\left(\int_0^\infty (1 - e^{-\lambda x})\tilde{i}_\nu(dx)\right)$$

when $0 < \nu < 1$. Therefore, for the exponent of the Lévy–Khintchine representation, we obtain

$$\int_0^\infty (1 - e^{-\lambda x})\tilde{i}_{-\nu}(dx) = \frac{1}{\nu}(e^{\nu a_c(1+\lambda)} - 1), \qquad 0 < \nu < 1.$$

It is easy to see that, when $\nu = 1/2$, this coincides with the right hand side of (7.5).

7.2. Next we develop a similar discussion for subordinators $\mathcal{K}^{(\nu)} = \{\mathcal{K}_t^{(\nu)}, t \geq 0\}$ with no drift and Lévy measure

$$\tilde{k}_\nu(dx) = e^{-x} K_\nu(x)\frac{dx}{x}, \qquad \nu > 0.$$

Note that, for $\nu > 0$,

$$K_\nu(x) \sim 2^{\nu-1}\Gamma(\nu)x^{-\nu} \text{ as } x \downarrow 0 \quad \text{and} \quad K_\nu(x) \sim \sqrt{\frac{\pi}{2x}}e^{-x} \text{ as } x \to \infty$$

and, therefore, that the measure $\tilde{k}_\nu(dx)$ satisfies $\int_0^\infty (x \wedge 1)\tilde{k}_\nu(dx) < \infty$ if and only if $\nu < 1$, which we assume in the following.

From formula (2.1), we deduce

$$c_\nu \int_0^\infty (1 - e^{-\lambda x})\tilde{k}_\nu(dx) = \int_0^\infty (1 - e^{-\lambda x})\tilde{i}_{-\nu}(dx) - \int_0^\infty (1 - e^{-\lambda x})\tilde{i}_\nu(dx),$$

where $c_\nu = 2\pi^{-1}\sin(\nu\pi)$. Therefore we obtain the following proposition, which shows the problem of description for $\mathcal{K}^{(\nu)}$ is closely related to that for $\mathcal{J}^{(-\nu)}$ when $-1 < \nu < 0$.

Proposition 7.3 *For $0 < \nu < 1$, one has*

$$\{\mathcal{J}_t^{(-\nu)}, t \geq 0\} \overset{\text{(law)}}{=} \{\mathcal{J}_t^{(\nu)} + \mathcal{K}_{c_\nu t}^{(\nu)}, t \geq 0\},$$

where, on the right hand side, $\{\mathcal{J}_t^{(\nu)}\}$ and $\{\mathcal{K}_t^{(\nu)}\}$ are assumed independent.

We again consider the simplest case of $\nu = 1/2$. Recall

$$K_{1/2}(x) = \sqrt{\frac{\pi}{2x}}e^{-x}.$$

Then, by using (7.4), we obtain

$$\int_0^\infty (1 - e^{-\lambda x}) \widetilde{k}_{1/2}(dx) = \pi(\sqrt{2(\lambda + 2)} - 2)$$

and $\{\mathcal{K}_a^{(1/2)}, a \geq 0\} \overset{(\text{law})}{=} \{T_{\pi a}^{(2)}, a \geq 0\}$.

In the general case, we can compute the Lévy–Khintchine representation by using formula (2.6). Replacing r by $a_c(1 + \lambda)$, we have

$$\int_0^\infty e^{-(1+\lambda)x} K_\nu(x)\, dx = \frac{\pi}{\sin(\nu\pi)} \frac{\sinh(\nu a_c(1 + \lambda))}{\sinh(a_c(1 + \lambda))}, \tag{7.6}$$

which yields

$$\int_0^\infty (1 - e^{-\lambda x}) \widetilde{k}_\nu(dx) = \frac{\cosh(\nu a_c(1 + \lambda)) - 1}{s_\nu},$$

where $s_\nu = \pi^{-1}\nu \sin(\nu\pi)$. Therefore, we obtain

$$E[\exp(-\lambda \mathcal{K}_t^{(\nu)})] = \exp\left(-t\, \frac{\cosh(\nu a_c(1 + \lambda)) - 1}{s_\nu}\right). \tag{7.7}$$

We now go back to formula (7.6), which we exploit to obtain the following identity:

$$P(\mathcal{K}_{T_{(\nu,1)}^{(3)}} \in dx) = \frac{\sin(\pi\nu)}{\pi\nu} e^{-x} K_\nu(x)\, dx, \qquad 0 < \nu < 1, \tag{7.8}$$

where $T_{(\nu,1)}^{(3)} = \inf\{t; R_\nu^{(3)}(t) = 1\}$ denotes the first hitting time of 1 by a three-dimensional Bessel process $\{R_\nu^{(3)}(t), t \geq 0\}$ starting from ν and independent of $\{\mathcal{K}_t, t \geq 0\}$. It is well known that

$$E[\exp\left(-\frac{\alpha^2}{2} T_{(\nu,1)}^{(3)}\right)] = \frac{\sinh(\nu\alpha)}{\nu \sinh(\alpha)},$$

which, together with (7.6), easily implies the identity (7.8).

8. DISCUSSIONS AND EXTENSIONS OF THE RESULTS

8.1. *Other occurrence of $\mathcal{J}_\nu$* In their study of infinitely divisible laws related to Bessel processes, Pitman–Yor [21] found some random variables which are distributed as $\mathcal{J}_\nu$ for fixed ν, using the following arguments. Recall that, for $\nu > 0$,

$$p_t^{(\nu)}(x, y) = t^{-1}\left(\frac{y}{x}\right)^\nu I_\nu\left(\frac{xy}{t}\right) \exp\left(-\frac{x^2 + y^2}{2t}\right) y\, dy \tag{8.1}$$

is the transition probability density with respect to the Lebesgue measure of the Bessel process with index ν. Consequently, as is proved in [21] for general transient diffusion processes, for fixed x, y, $p_t^{(\nu)}(x, y)dt$ is (up to a multiplicative constant) the distribution of the last passage time $L_y^{(\nu)}$ at y for the Bessel process $\{R_x^{(\nu)}(t), t \geq 0\}$ of index ν starting at x.

A time inversion argument yields that $\sigma_y^{(\nu,x)} \equiv \inf\{t; R_t^{(\nu,x)} = ty\}$, where $\{R_t^{(\nu,x)}, t \geq 0\}$ denotes a Bessel process with index ν and drift x starting from 0, is distributed as $1/L_y^{(\nu)}$.

Finally, from formula (8.1), we find

$$\sigma_1^{(\nu,1)} \overset{(\text{law})}{=} \mathcal{J}_\nu. \tag{8.2}$$

Hence we obtain

$$P(\sigma_y^{(\nu,x)} \in dt) = \nu\left(\frac{y}{x}\right)^\nu I_\nu(xyt) \exp\left(-\frac{(x^2+y^2)t}{2}\right) \frac{dt}{t}. \tag{8.3}$$

The infinitely divisible distribution given by (8.3) was also encountered by Feller [11] (see also [12]) in the study of the first passage times for a continuous time random walk: for positive integers ν, the probability law (8.3) is the distribution of the first passage time to ν of a compound Poisson process starting from 0 with jumps $+1$ at rate $2(x^2+y^2)^{-1}x^2$ and with jumps -1 at rate $2(x^2+y^2)^{-1}y^2$.

8.2. *Around Bougerol's identity* Clearly, in Section 2, the function

$$a_c(x) = \log(x + \sqrt{x^2-1}), \quad x \geq 0,$$

plays an essential role. This is reminiscent of Bougerol's identity: for any fixed $t > 0$,

$$\sinh(B_t) \overset{(\text{law})}{=} \gamma_{A_t}, \tag{8.4}$$

where $\{\gamma_s, s \geq 0\}$ is a Brownian motion starting from 0 independent of $\{B_s, s \geq 0\}$ and

$$A_t = \int_0^t \exp(2B_s)\, ds.$$

As remarked in Yor [28], formula (1.e), the identity (8.4) is "equivalent" to

$$E[\sqrt{\frac{t}{A_t}} \exp\left(-\frac{u}{2A_t}\right)] = \frac{1}{\sqrt{1+u}} \exp\left(-\frac{a_s(\sqrt{u})^2}{2t}\right). \tag{8.5}$$

See the discussion in Dufresne [8], p. 414.

Now, from the elementary duplication formula,

$$\sinh(2x) = 2\sinh(x)\cosh(x),$$

one deduces

$$a_c(1+2u) = 2a_s(\sqrt{u}),$$

so that we may write (8.5) as

$$E[\sqrt{\frac{t}{A_t}} \exp\left(-\frac{u}{2A_t}\right)] = \frac{1}{\sqrt{1+u}} \exp\left(-\frac{a_c(1+2u)^2}{8t}\right).$$

Therefore, from Theorems 1.3 and 1.5, we deduce

$$E[\sqrt{\frac{t}{A_t}} \exp\left(-\frac{u}{2A_t}\right)] = \frac{1}{\sqrt{1+u}} E[\exp\left(-\frac{2u}{A(b_{4t})}\right)]$$

$$= E[\exp\left(-u\left(\frac{N^2}{2} + \frac{2}{A(b_{4t})}\right)\right)]$$

$$= E[\exp(-\frac{u}{2}(\frac{1}{T} + \frac{4}{A(b_{4t})}))],$$

where N is a standard normal random variable independent of the Brownian bridge b_{4t} of length $4t$ and $T \overset{(law)}{=} N^{-2}$ is the stable $(1/2)$-variable. Thus we have obtained, for every non-negative Borel function on $\mathbf{R}_+$,

$$E[\sqrt{\frac{t}{A_t}}f(\frac{1}{A_t})] = E[f(\frac{1}{T} + \frac{4}{A(b_{4t})})], \tag{8.6}$$

which should be compared with our previous result in [6]

$$E[\sqrt{\frac{t}{A_t}}f(\frac{1}{A_t})] = E[f(\frac{1}{T})\exp(\frac{1}{A(b_t)} - \frac{T}{2(A(b_t))^2})].$$

8.3. *Negative moments of* $A(b_t)$ We may exploit the relationship (8.6) to present some formulae for the negative moments of $A(b_t)$, which we denote simply by a_t in the following. For the related results, see [7] and [8]. One finds a number of explicit results for the negative moments of A_t in [5] and [6].

At first we recall some results in [6] and [8]: for every $n \in \mathbf{N}$, there exists a polynomial P_n of degree n such that

$$m_{0,n+1/2}(t) \equiv E[(A_t)^{-(n+1/2)}] = \frac{1}{t^{1/2}}P_n(\frac{1}{t})$$

and the sequence $\{P_n\}$ satisfies

$$P_{n+1}(x) = (2n+1)P_n(x) + \frac{1}{2n+1}(xP_n(x) + x^2 P_n'(x)).$$

Moreover, in [6] Proposition 5.2, one finds an integral representation of P_n:

$$P_n(x) = \frac{2^{n+1/2}}{\Gamma(n+1/2)}\int_0^\infty e^{-u^2/2}F_n(\sqrt{x}u)\,du, \tag{8.7}$$

where

$$F_n(y) = \prod_{j=0}^{n-1}((j+\frac{1}{2})^2 + \frac{y^2}{4}).$$

We have also shown in Proposition 4.2 in [6] that there exists a polynomial $\widehat{Q}_n$ of degree n such that

$$m_{2,n+3/2}(t) \equiv E[(A_t)^{-(n+3/2)}\exp(2B_t)] = t^{-3/2}\widehat{Q}_n(\frac{1}{t}).$$

It is elementary that

$$(m_{0,n+1/2})'(t) = -(n+\frac{1}{2})m_{2,n+3/2}(t)$$

and, hence,

$$\frac{1}{2}P_n(x) + xP_n'(x) = (n+\frac{1}{2})\widehat{Q}_n(x).$$

We now come back to the relation (8.6), from which we deduce the following.

Proposition 8.1 *For each* $k \in \mathbf{N}$, *there exists a polynomial* Π_k *of order* k *such that*

$$E[(a_t)^{-k}] = \Pi_k(\frac{1}{t})$$

and it holds that

$$P_n(x) = \sum_{k=0}^{n} 4^k \binom{n}{k} E[N^{2(n-k)}]\Pi_k\left(\frac{x}{4}\right), \tag{8.8}$$

where N is a standard normal random variable.

Proof. Formula (8.8) is an immediate consequence of (8.6). Since P_n is a polynomial, some inductive arguments ensures that Π_k is also a polynomial. $\qquad\square$

Before proceeding further, we comment on the identity (8.8). Assuming that we know the sequence of the polynomials $\{P_n\}$, which is indeed the case (see (8.7) above), we may consider (8.8) as a recurrence relation giving Π_n in terms of Π_k, $k < n$.

The main result of this article, that is, for fixed t, $\mathcal{K}_t \stackrel{(law)}{=} (a_{1/t})^{-1}$ and the fact that $\Pi_n(t) = E[(\mathcal{K}_t)^n]$ yield naturally another recurrence relationship which expresses Π_n in terms of $\{\Pi_k, \; k < n\}$.

Before mentioning the result, we show the following.

Lemma 8.1 (i) *Let N_1 and N_2 be two independent standard normal random variables. Then we have*

$$P(|N_1 N_2| \in d\xi) = \frac{2}{\pi} K_0(\xi) \, d\xi \tag{8.9}$$

and

$$E[\exp(-|N_1 N_2|)] = \frac{2}{\pi}. \tag{8.10}$$

(ii) *Let Λ be a random variable such that*

$$P(\Lambda \in d\xi) = \frac{\pi}{2} e^{-\xi} P(|N_1 N_2| \in d\xi) = e^{-\xi} K_0(\xi) d\xi. \tag{8.11}$$

Then, for any $\alpha \geqq 0$, one has

$$E[\Lambda^\alpha] = \frac{2^\alpha (\Gamma(1+\alpha))^3}{\Gamma(2+2\alpha)}.$$

(iii) *One has $E[\exp((2-\varepsilon)\Lambda)] < \infty$ for any $\varepsilon > 0$ and $E[\exp(2\Lambda)] = \infty$.*

Proof. (i) Identity (8.9) is easily obtained from the integral representation (2.2) of K_0. Identity (8.10) follows from the formula $\int_0^\infty e^{-x} K_0(x) dx = 1$, which is itself a consequence of (2.11).

(ii) Using the representation (2.2) again, we obtain

$$E[\Lambda^\alpha] = \frac{1}{2} \int_0^\infty \frac{du}{u} \int_0^\infty x^\alpha \exp\left(-\frac{1}{2}\left(\sqrt{u} + \frac{1}{\sqrt{u}}x\right)\right) dx$$

$$= 2^\alpha \Gamma(1+\alpha) \int_0^\infty \left(\sqrt{u} + \frac{1}{\sqrt{u}}\right)^{-2(1+\alpha)} \frac{du}{u}.$$

The rest of the proof is elementary.

(iii) By the trivial inequality $ab \leqq (a^2 + b^2)/2$, we have

$$E[\exp((1-\varepsilon)|N_1 N_2|)] < \infty$$

for any $\varepsilon > 0$. Moreover, we have

$$E[\exp(|N_1 N_2|)] \geqq E[\exp(N_1|N_2|)] = E[\exp(N^2/2)] = \infty.$$

Therefore the assertion follows from (i) and the definition of Λ. $\qquad\square$

Corollary 8.2 *One has*

$$E[\exp((2-\varepsilon)\mathcal{K}_t)] < \infty \text{ for every } \varepsilon > 0, \text{ and } E[\exp(2\mathcal{K}_t)] = \infty.$$

Remark 8.1 The results in the corollary also follow from (1.6) if we note an elementary estimate for the function $\vartheta_v(t)$:

$$\vartheta_v(t) \leqq (2\pi^3 t)^{-1/2} \exp\left(-v + \frac{\pi^2}{2t}\right).$$

Now we show another recurrence relation for the sequence of polynomials $\{\Pi_k\}_{k=0}^{\infty}$.

Proposition 8.3 *Let Λ be a random variable whose distribution is given by (8.11) and set $q_k = E[\Lambda^k]$. Then it holds that*

$$\Pi_k(t) = \sum_{j=0}^{k-1} \binom{k}{j} q_{k-j-1} \int_0^t \Pi_j(s)\,ds \tag{8.12}$$

and, therefore, letting U be a uniform random variable on $[0,1]$ and assuming that $\mathcal{K}_s, \Lambda$ and U are independent, one has

$$\Pi_k(t) = \int_0^t E[(\mathcal{K}_s + U\Lambda)^{k-1}]\,ds.$$

Proof. Formula (8.12) follows from the chain of equalities:

$$\Pi_k(t) = E[(\mathcal{K}_t)^k] = \int_0^t E\left[\int_0^\infty ((\mathcal{K}_s + x)^k - (\mathcal{K}_s)^k)\ell_{\mathcal{K}}(dx)\right] ds$$

$$= \sum_{j=0}^{k-1} \binom{k}{j} q_{k-j-1} \int_0^t \Pi_j(s)\,ds. \qquad \square$$

APPENDIX

It may be convenient to present the following tables, which gather some of the results in this paper. The results mentioned in Section 1 are included in the following.

Table 2

Subordinator Σ	$\mathcal{J}$	$\mathcal{K}$
$-\frac{1}{t}\log(E[e^{-\lambda\Sigma_t}])$	$a_c(1+\lambda)$	$\frac{1}{2}(a_c(1+\lambda))^2$
Lévy density	$x^{-1}I_0(x)e^{-x}$	$x^{-1}K_0(x)e^{-x}$
$\frac{d}{dx}P(\Sigma_t \leq x)$	$tI_t(x)x^{-1}e^{-x}$	$\sqrt{\frac{2\pi}{t}}\vartheta_x\left(\frac{1}{t}\right)x^{-1}e^{-x}$
Distribution of Σ_T	$\mathcal{J}_t^{(\nu)}$ for $T = \nu N_{t/\nu}$	Γ_t for $T = C_t$

Here $a_c(x) = \text{Argcosh}(x)$, $\{N_t\}$ is the standard Poisson process, $\{\Gamma_t\}$ is the standard Gamma process and $\{C_t\}$ is a hyperbolic subordinator defined in Section 6.

The results shown in Section 7 are gathered in the following:

Table 3

Subordinator Σ	$\mathcal{J}^{(\nu)}$	$\mathcal{K}^{(\nu)}$
$-\dfrac{1}{t}\log(E[e^{-\lambda\Sigma_t}])$	$\dfrac{1-\exp(-\nu a_c(1+\lambda))}{\nu}$	$\dfrac{\cosh(\nu a_c(1+\lambda))-1}{s_\nu}$
Lévy density	$x^{-1}I_\nu(x)e^{-x}$	$x^{-1}K_\nu(x)e^{-x}$

Here $0 < \nu < 1$ and $s_\nu = \pi^{-1}\nu\sin(\nu\pi)$.

REFERENCES

[1] L. Alili, H. Matsumoto and T. Shiraishi, On a triplet of exponential Brownian functionals, to appear in *Sém. Probab.* **XXXV**, Lecture Notes in Math. **1755**, Springer-Verlag, Berlin, 2001.

[2] P. Biane, J. Pitman and M. Yor, Probability laws related to the Jacobi theta and Riemann zeta functions, and Brownian excursions, to appear in *Bulletin AMS*.

[3] E.B. Davies, *Heat kernels and spectral theory*, Cambridge University Press, Cambridge, 1989.

[4] C. Donati-Martin and M. Yor, Some Brownian functionals and their laws, *Ann. Probab.* **25** (1997), 1011–1058.

[5] C. Donati-Martin, H. Matsumoto and M. Yor, On positive and negative moments of the integrals of geometric Brownian motions, *Stat. Probab. Lett.* **49** (2000), 45–52.

[6] C. Donati-Martin, H. Matsumoto and M. Yor, On striking identities about the exponential functionals of the Brownian bridge and Brownian motion, *Periodica Math. Hung.* **41** (2000), 103–109.

[7] C. Donati-Martin, H. Matsumoto and M. Yor, The law of geometric Brownian motion and its integral, revisited; application to conditional moments, to appear in *Proc. First Bachelier Conference*, Springer-Verlag, Berlin, 2001.

[8] D. Dufresne, Laguerre series for Asian and other options, *Math. Finance* **10** (2000), 407–428.

[9] F. Esscher, On the probability function in the collective theory of risk, *Skandinavisk Aktuarietidskrift* **15** (1932), 175–195.

[10] W. Feller, Infinitely divisible distributions and Bessel functions associated with random walks, *J. Soc. Indust. Appl. Math.* **14** (1966), 864–875.

[11] W. Feller, *An introduction to probability theory and its applications*, Vol. 1, 3rd ed., Wiley, New York, 1968.

[12] W. Feller, *An introduction to probability theory and its applications*, Vol. 2, 2nd ed., Wiley, New York, 1971.

[13] I.S. Gradshteyn and I.M. Ryzhik (ed. by A. Jeffrey), *Tables of integrals, series and products*, 6th ed., Academic Press, New York, 2000.

[14] J.-C. Gruet, Semi-groupe du mouvement Brownien hyperbolique, *Stochastics Stochastics Rep.* **56** (1996), 53–61.

[15] J.-C. Gruet, Windings of hyperbolic Brownian motion, in [29].

[16] P. Hartman and G.S. Watson, "Normal" distribution functions on spheres and the modified Bessel functions, *Ann. Probab.* **2** (1974), 593–607.

[17] H. Matsumoto, Closed form formulae for the heat kernels and Green functions on symmetric spaces of rank one, Preprint.

[18] H. Matsumoto and M. Yor, On Bougerol and Dufresne identities for exponential Brownian functionals, *Proc. Japan Acad.* **74**, Ser. A (1998), 152–155.

[19] H. Matsumoto and M. Yor, An analogue of Pitman's $2M - X$ theorem for exponential Brownian functionals, Part I: A time inversion approach, *Nagoya Math. J.* **159** (2000), 125–166.

[20] H. Matsumoto and M. Yor, An analogue of Pitman's $2M - X$ theorem for exponential Brownian functionals, Part II: The role of the generalized inverse Gaussian laws, to appear in *Nagoya Math. J.* **162** (2001).

[21] J.W. Pitman and M. Yor, Bessel processes and infinitely divisible laws, *Stochastic Integrals*, ed. by D.Williams, Lecture Notes in Math. **851**, 285–370, Springer-Verlag, Berlin, 1981.

[22] J. Pitman and M. Yor, Infinitely divisible laws associated with hyperbolic functions, Preprint, University of California, Berkeley.

[23] K. Sato, *Lévy processes and infinitely divisible distributions*, Cambridge University Press, Cambridge, 1999.

[24] A.N. Shiryaev, *Essentials of stochastic finance*(English translation by N. Kruzhilin), World Scientific, Singapore, 1999.

[25] G.N. Watson, *A treatise on the theory of Bessel functions*, Cambridge University Press, 1966.

[26] M. Yor, Loi de l'indice du lacet Brownien, et distribution de Hartman-Watson, *Z. Wahrscheinlichkeitstheorie verw. Gebiete* **53** (1980), 71–95.

[27] M. Yor, Sur les lois des fonctionnelles exponentielles du mouvement brownien, considérées en certains instants aléatoires, *C.R. Acad. Sci. Paris Sér. I* **314** (1992), 951–956.

[28] M. Yor, On some exponential functionals of Brownian motion, *Adv. Appl. Probab.* **24** (1992), 509–531.

[29] M. Yor (Ed.), *Exponential functionals and principal values related to Brownian motion*, A collection of research papers, Biblioteca de la Revista Matemática Iberoamericana, Madrid, 1997.

FIRST PASSAGE TIME
STRUCTURAL MODELS
WITH INTEREST RATE RISK

MAREK RUTKOWSKI

Warsaw University of Technology
Faculty of Mathematics and Information Science
00-661 Warszawa, Poland

and

The University of New South Wales
School of Mathematics
2052 Sydney, Australia

Abstract The structural approach (also known as the firm's value approach) to the valuation of corporate debt was initiated in the pathbreaking papers by Black and Scholes [2] and Merton [9] in the 1970s. In recent years, the so-called first passage time approach, first put forward by Black and Cox [3], has been studied and developed by several authors. The aim of this note is to derive the closed-form solution for the value of a zero-coupon defaultable bond within the framework of the first passage time model with stochastic term structure of interest rates. It should be stressed that even in the Gaussian Heath–Jarrow–Merton framework, the availability of the closed-from solution for the price of a corporate debt hinges on a judicious choice of a default triggering barrier reflecting the safety covenants.

Key Words Credit risk, structural models, first passage time, term structure models.

1. INTRODUCTION

The *structural approach* to the valuation of financial instruments sensitive to the default risk was pioneered by Black and Scholes [2] and Merton [9]. It focuses primarily on the valuation of the firm's liabilities; for this reason, it is also frequently referred to either as the *firm's value approach* or as the *option theoretic approach*. Indeed, in the structural approach the firm's liabilities are seen as contingent claims issued against the firm's assets. Default (bankruptcy) event is specified in terms of the evolution of the total unleveraged value process of the firm's assets, denoted by V in what follows, as well as in terms of some default triggering barrier. Put another way, the ability of the company to meet its liabilities – that is, the firm's solvency level – is assumed to be completely determined by the process V in conjunction with some specified bankruptcy covenants. It should be stressed that the original Merton approach does not allow for a premature default, in the sense that the default may only occur at the maturity of the firm's debt.

In their paper devoted to the *first passage time approach* to the valuation of corporate debt, Black and Cox [3] extend Merton's research in several directions. In particular, they make account for the specific features of debt contracts as: safety covenants, debt subordination, and restrictions on the sale of assets. They assume that the firm's shareholders receive a continuous dividend payment, proportional to the current value of the firm. Consequently, the SDE which governs the dynamics of the firm's value takes the following form, under the (spot) martingale measure $\mathbb{P}^*$,

$$dV_t = V_t\big((r - \varkappa)\, dt + \sigma\, dW_t^*\big), \tag{1}$$

where $\sigma > 0$ is the constant volatility, the constant $\varkappa \geq 0$ represents the dividend rate (or payout ratio), and W^* follows a standard Brownian motion under $\mathbb{P}^*$ with respect to the underlying filtration $\mathbb{F}$. It is apparent that the interest rate is assumed to be constant here; $r_t = r$, for some $r \in \mathbb{R}$. This means that the existence of the interest rate risk is not accounted for in the original Black and Cox [3] framework.

Let us first focus on the *safety covenants* in the firm's indenture provisions. Generally speaking, safety covenants provide the firm's bondholders with the right to force the firm to bankruptcy or reorganization if the firm is doing poorly according to a set standard. The standard for a poor performance is set in Black and Cox [3] in terms of a time-dependent deterministic barrier: $\bar{v}(t) = Ke^{-\gamma(T-t)}$ for $t \in [0,T)$, where $K > 0$ and $\gamma \in \mathbb{R}$ are constants. If the value of firm's assets falls below this level, the bondholders take over the firm from the shareholders. Otherwise, default takes place at the debt's maturity T or not depending on whether $V_T \geq L$ or not, where L is the face value of the total firm's debt. Let us set

$$v_t = \begin{cases} \bar{v}(t) = Ke^{-\gamma(T-t)}, & \text{for } t < T, \\ L, & \text{for } t = T. \end{cases} \tag{2}$$

The *default time* τ is the first moment in the interval $[0,T]$ when the firm's value V_t falls below the time-varying level v_t; otherwise the default event does not occur at all. Formally, we set

$$\tau = \inf\{t \in [0,T] : V_t < v_t\},$$

with the convention: $\inf \varnothing = +\infty$. Formally, we deal with the defaultable contingent claim $(X, Z, \tilde{X}, \tau)$ which settles at time T, where

$$X = L, \quad Z_t = \beta_2 V_t, \quad \tilde{X} = \beta_1 V_T.$$

Let us now explain the meaning of each term. First, the random variable X represents the firm's liabilities to be redeemed at time T – that is, the *promised claim*. If default does not occur prior to or at time T, the promised claim X is paid in full at time T. Otherwise, either:

(i) default occurs at time $t < T$, and the holder of the defaultable claim receives the recovery payoff Z_t at time t, or:

(ii) default occurs at the debt's maturity T, and the recovery payoff $\tilde{X}$ is received by the claimholder at time T.

The *recovery process* Z is assumed to be proportional to the firm's value process: $Z_t = \beta_2 V_t$ for some constant β_2. Similarly, the *recovery payoff* at maturity equals $\tilde{X} = \beta_1 V_T$ for some constant β_2. It is natural to assume that the coefficients β_1 and β_2, which represent the bankruptcy costs, are constants from the interval $[0, 1]$ (Black and Cox [3] examine the case of $\beta_1 = \beta_2 = 1$). Notice that the default time $\tau = \bar{\tau} \wedge \hat{\tau}$, where $\bar{\tau}$ is the first passage time of the firm's value process V to the deterministic barrier $\bar{v}$, specifically:

$$\bar{\tau} := \inf\left\{ t \in [0, T) : V_t \leq \bar{v}(t) \right\} = \inf\left\{ t \in [0, T) : V_t < \bar{v}(t) \right\}$$

and $\hat{\tau}$ stands for Merton's default time, i.e., $\hat{\tau} := T \mathbb{1}_{\{V_T < L\}} + \infty \mathbb{1}_{\{V_T \geq L\}}$. We assume throughout that for any $t \in [0, T)$ we have $\bar{v}(t) \leq LB(t, T)$, where $B(t, T)$ is the price at time t of a default-free zero-coupon bond maturing at T. More explicitly, we postulate that for every $t \in [0, T)$

$$Ke^{-\gamma(T-t)} \leq Le^{-r(T-t)} = LB(t, T).$$

Suppose that the default occurs at some instant t prior to the bond's maturity. The last condition ensures that the recovery payoff does not exceed the face value of debt L, discounted at a risk-free rate r.

Let us fix $t < T$. Then the price of a zero-coupon defaultable bond with face value L and maturity T, denoted by $D(t, T)$, admits the following probabilistic representation before default – that is, on the set $\{\tau > t\} = \{\bar{\tau} > t\}$,

$$\begin{aligned}
D(t, T) &= \mathbb{E}_{\mathbb{P}^*}\left(\beta_1 V_T e^{-r(T-t)} \mathbb{1}_{\{\bar{\tau} \geq T, V_T < L\}} \,\Big|\, \mathcal{F}_t \right) \\
&\quad + \mathbb{E}_{\mathbb{P}^*}\left(Le^{-r(T-t)} \mathbb{1}_{\{\bar{\tau} \geq T, V_T \geq L\}} \,\Big|\, \mathcal{F}_t \right) \\
&\quad + \mathbb{E}_{\mathbb{P}^*}\left(K\beta_2 e^{-\gamma(T-\bar{\tau})} e^{-r(\bar{\tau}-t)} \mathbb{1}_{\{t < \bar{\tau} < T\}} \,\Big|\, \mathcal{F}_t \right).
\end{aligned}$$

After default – that is, on the set $\{\tau \leq t\} = \{\bar{\tau} \leq t\}$ – we clearly have

$$D(t, T) = \beta_2 v_{\bar{\tau}} B^{-1}(\bar{\tau}, T) = K\beta_2 e^{-\gamma(T-\bar{\tau})} e^{r(T-\bar{\tau})}.$$

We find it convenient to introduce the following notation:

$$\nu = r - \varkappa - \tfrac{1}{2}\sigma^2, \quad \tilde{\nu} = \nu - \gamma = r - \varkappa - \gamma - \tfrac{1}{2}\sigma^2, \quad \tilde{a} = \tilde{\nu}\sigma^{-2}.$$

For the reader's convenience, we quote the following result, which provides the closed-form expression for the value of a defaultable zero-coupon bond in the Black and Cox [3] framework. It should be acknowledged that the valuation formula of Proposition 1 is slightly different from the expression given in the original paper by Black and Cox [3]. For the proof of Proposition 1, we refer to Bielecki and Rutkowski [1].

Proposition 1 *Assume that $\tilde{\nu}^2 + 2\sigma^2(r - \gamma) > 0$. Then the value $D(t, T)$ of a zero-coupon defaultable bond with the face value L and the maturity date T equals, on the set $\{\bar{\tau} > t\}$,*

$$\begin{aligned}
D(t, T) &= LB(t, T)\big(N\big(h_1(V_t, T-t)\big) - R_t^{2\tilde{a}} N\big(h_2(V_t, T-t)\big)\big) \\
&\quad + \beta_1 V_t e^{-\varkappa(T-t)}\big(N\big(h_3(V_t, T-t)\big) - N\big(h_4(V_t, T-t)\big)\big) \\
&\quad + \beta_1 V_t e^{-\varkappa(T-t)} R_t^{2\tilde{a}+2}\big(N\big(h_5(V_t, T-t)\big) - N\big(h_6(V_t, T-t)\big)\big) \\
&\quad + \beta_2 V_t \big(R_t^{\vartheta+\zeta} N\big(h_7(V_t, T-t)\big) + R_t^{\vartheta-\zeta} N\big(h_8(V_t, T-t)\big)\big),
\end{aligned}$$

where N is the standard normal cumulative distribution function, $R_t = \bar{v}(t)/V_t$,

$$\vartheta = \tilde{a} + 1, \ \zeta = \sigma^{-2}\sqrt{\tilde{\nu}^2 + 2\sigma^2(r - \gamma)},$$

and

$$h_1(V_t, T - t) = \frac{\ln(V_t/L) + \nu(T - t)}{\sigma\sqrt{T - t}},$$

$$h_2(V_t, T - t) = \frac{\ln\bar{v}^2(t) - \ln(LV_t) + \nu(T - t)}{\sigma\sqrt{T - t}},$$

$$h_3(V_t, T - t) = \frac{\ln(L/V_t) - (\nu + \sigma^2)(T - t)}{\sigma\sqrt{T - t}},$$

$$h_4(V_t, T - t) = \frac{\ln(K/V_t) - (\nu + \sigma^2)(T - t)}{\sigma\sqrt{T - t}},$$

$$h_5(V_t, T - t) = \frac{\ln\bar{v}^2(t) - \ln(LV_t) + (\nu + \sigma^2)(T - t)}{\sigma\sqrt{T - t}},$$

$$h_6(V_t, T - t) = \frac{\ln\bar{v}^2(t) - \ln(KV_t) + (\nu + \sigma^2)(T - t)}{\sigma\sqrt{T - t}},$$

$$h_7(V_t, T - t) = \frac{\ln(\bar{v}(t)/V_t) + \zeta\sigma^2(T - t)}{\sigma\sqrt{T - t}},$$

$$h_8(V_t, T - t) = \frac{\ln(\bar{v}(t)/V_t) - \zeta\sigma^2(T - t)}{\sigma\sqrt{T - t}}.$$

2. MAIN RESULT

We shall examine a natural generalization of the Black and Cox [3] approach, which takes into account both the credit and interest rate risk. Formally, our goal is to extend the valuation formula of Proposition 1 to the case of stochastic term structure of interest rates, as specified by the Heath et al. [6] approach. We make the following standing assumptions:

(i) the default triggering barrier $\bar{v}$ equals $\bar{v}(t) = KB(t, T)f(t)$ for some constant K, and some function $f : [0, T) \to \mathbb{R}_+$,

(ii) the volatility of the forward value of the firm follows a deterministic function.

To guarantee the existence of a closed-form solution for the value of a defaultable bond, the function f in (ii) needs to be chosen in a judicious way (see expression (5) below). On the other hand, to satisfy the second requirement above, we find it convenient to place ourselves in the Gaussian Heath–Jarrow–Morton setup. More specifically, we assume that the bond price volatility is a deterministic function.

We assume that the underlying probability space $(\Omega, \mathcal{F}, \mathbb{P})$, endowed with the filtration $\mathbb{F} = (\mathcal{F}_t)_{t \geq 0}$, is rich enough to support the short-term interest rate process r and the value process V. Let us fix a finite time horizon $T > 0$. The dynamics under the spot martingale measure $\mathbb{P}^*$ of the firm's value and of the price of a default-free zero-coupon bond $B(t, T)$ are

$$dV_t = V_t\big((r_t - \varkappa(t))\,dt + \sigma(t)\,dW_t^*\big), \tag{3}$$

and

$$dB(t, T) = B(t, T)\big(r_t\,dt + b(t, T)\,dW_t^*\big), \tag{4}$$

respectively, where W^* is a d-dimensional standard Brownian motion. Furthermore, the functions $\varkappa : [0, T] \to \mathbb{R}$, $\sigma : [0, T] \to \mathbb{R}^d$ and $b(\cdot, T) : [0, T] \to \mathbb{R}^d$ are assumed to be bounded. In view of (3)-(4), the *forward value* $F_V(t, T) := V_t / B(t, T)$ of the firm satisfies under the forward martingale measure $\mathbb{P}_T$ (see Section 13.2 in Musiela and Rutkowski [10])

$$dF_V(t, T) = -\varkappa(t) F_V(t, T)\, dt + F_V(t, T)\big(\sigma(t) - b(t, T)\big)\, dW_t^T,$$

where the process W^T, given by the formula

$$W_t^T = W_t^* - \int_0^t b(u, T)\, du, \quad \forall \tilde{\in} [0, T],$$

is known to follow a d-dimensional standard Brownian motion under $\mathbb{P}_T$. Let us introduce an auxiliary process $F_V^{\varkappa}(t, T)$ by setting, for $t \in [0, T]$,

$$F_V^{\varkappa}(t, T) = F_V(t, T) e^{-\int_t^T \varkappa(u)\, du}.$$

It is clear that $F_V^{\varkappa}(t, T)$ follows a lognormally distributed martingale under $\mathbb{P}_T$, specifically,

$$dF_V^{\varkappa}(t, T) = F_V^{\varkappa}(t, T)\big(\sigma(t) - b(t, T)\big)\, dW_t^T.$$

Furthermore, it is apparent that $F_V^{\varkappa}(T, T) = F_V(T, T) = V_T$. We consider the following modification of the Black and Cox approach:

$$X = L, \quad Z_t = \beta_2 V_t, \quad \tilde{X} = \beta_1 V_T, \quad \tau = \inf\{t \in [0, T] : V_t < v_t\},$$

where $\beta_2, \beta_1 \in [0, 1]$ are constants, and the barrier v is given by the formula

$$v_t := \begin{cases} KB(t, T) e^{\int_t^T \varkappa(u)\, du}, & \text{for } t < T, \\ L, & \text{for } t = T, \end{cases} \tag{5}$$

where the constant K satisfies $0 < K \le L$. Let us denote, for any $t \le T$,

$$\varkappa(t, T) = \int_t^T \varkappa(u)\, du, \quad \sigma^2(t, T) = \int_t^T \|\sigma(u) - b(u, T)\|^2\, du,$$

where $\| \cdot \|$ is the Euclidean norm in $\mathbb{R}^d$. We write briefly $F_t = F_V(t, T)$, and we denote

$$\eta_+(t, T) = \varkappa(t, T) + \tfrac{1}{2}\sigma^2(t, T), \quad \eta_-(t, T) = \varkappa(t, T) - \tfrac{1}{2}\sigma^2(t, T).$$

Proposition 2 *Let the barrier process v be given by (5). For any $t < T$, the forward price $F_D(t, T) = D(t, T)/B(t, T)$ of a defaultable bond with the face value L and the maturity date T equals, on the set $\{\tau > t\} = \{\tilde{\tau} > t\}$,*

$$\begin{aligned} F_D(t, T) = \; & L\big(N\big(\hat{h}_1(F_t, t, T)\big) - (F_t/K) e^{-\varkappa(t, T)} N\big(\hat{h}_2(F_t, t, T)\big)\big) \\ & + \beta_1 F_t e^{-\varkappa(t, T)}\big(N\big(\hat{h}_3(F_t, t, T)\big) - N\big(\hat{h}_4(F_t, t, T)\big)\big) \\ & + \beta_1 K\big(N\big(\hat{h}_5(F_t, t, T)\big) - N\big(\hat{h}_6(F_t, t, T)\big)\big) \\ & + \beta_2 K J_1(F_t, t, T) + \beta_2 F_t e^{-\varkappa(t, T)} J_2(F_t, t, T), \end{aligned}$$

where

$$\hat{h}_1(F_t, t, T) = \frac{\ln(F_t/L) - \eta_+(t, T)}{\sigma(t, T)},$$

$$\widehat{h}_2(F_t, T, t) = \frac{2 \ln K - \ln(LF_t) + \eta_-(t, T)}{\sigma(t, T)},$$

$$\widehat{h}_3(F_t, t, T) = \frac{\ln (L/F_t) + \eta_-(t, T)}{\sigma(t, T)},$$

$$\widehat{h}_4(F_t, t, T) = \frac{\ln (K/F_t) + \eta_-(t, T)}{\sigma(t, T)},$$

$$\widehat{h}_5(F_t, t, T) = \frac{2 \ln K - \ln(LF_t) + \eta_+(t, T)}{\sigma(t, T)},$$

$$\widehat{h}_6(F_t, t, T) = \frac{\ln(K/F_t) + \eta_+(t, T)}{\sigma(t, T)},$$

and for any fixed $0 \leq t < T$ and $F_t > 0$

$$J_{1,2}(F_t, t, T) = \int_t^T e^{\varkappa(u,T)} \, dN \left(\frac{\ln(K/F_t) + \varkappa(t, T) \pm \frac{1}{2}\sigma^2(t, u)}{\sigma(t, u)} \right).$$

Remarks. Let us assume that $\beta_2 = \beta_1 = 1$. It can be checked that if $b \equiv 0$ and the coefficients $\varkappa$ and σ are assumed to be constant, the term $J_{1,2}(F_t, t, T)$ can be evaluated explicitly, and the valuation formula of Proposition 2 reduces to the special case of the formula obtained in Proposition 1 with $\gamma = r - \varkappa$. It is worthwhile to stress that the choice of a barrier as in (2) instead of (5) does not lead to a closed-form solution, in general.

Before we proceed to the proof of Proposition 2, let us recall an auxiliary result (for the proof, see, e.g., Appendix B in Musiela and Rutkowski [10]). Assume that $\tilde{Y}_t, t \in [0, U]$, follows under $\tilde{\mathbb{P}}$ a generalized Brownian motion with drift with respect to the filtration $\tilde{\mathbb{F}}$. Specifically,

$$\tilde{Y}_t = Y_0 + \sigma \tilde{W}_t + \nu t, \quad \tilde{Y}_0 = \tilde{y}_0 > 0, \tag{6}$$

where $\tilde{W}_t, t \in [0, U]$, follows under $\tilde{\mathbb{P}}$ a standard one-dimensional Brownian motion with respect to $\tilde{\mathbb{F}}$, and the coefficients $\sigma > 0$ and $\nu \in \mathbb{R}$ are constants.

Lemma 1 Let $\tilde{\tau}$ be the first passage time to zero by the process $\tilde{Y}$ given by formula (6), specifically,

$$\tilde{\tau} = \inf \{ t < U : \tilde{Y}_t = 0 \}.$$

Then for any $0 < s \leq U$

$$\tilde{\mathbb{P}}\{\tilde{\tau} < s\} = N \left(\frac{-\tilde{y}_0 - \nu s}{\sigma\sqrt{s}} \right) + e^{-2\nu\sigma^{-2}\tilde{y}_0} N \left(\frac{-\tilde{y}_0 + \nu s}{\sigma\sqrt{s}} \right),$$

where N is the standard normal cumulative distribution function. Furthermore, for any $0 \leq u < s \leq U$ and any $y \geq 0$, we have, on the set $\{\tilde{\tau} > u\}$,

$$\tilde{\mathbb{P}}\{\tilde{Y}_s \geq y, \tilde{\tau} \geq s \mid \tilde{\mathcal{F}}_u\} = N \left(\frac{-y + \tilde{Y}_u + \nu(s - u)}{\sigma\sqrt{s - u}} \right)$$

$$- e^{-2\nu\sigma^{-2}\tilde{Y}_u} N \left(\frac{-y - \tilde{Y}_u + \nu(s - u)}{\sigma\sqrt{s - u}} \right).$$

Proof of Proposition 2. Under the present assumptions, a defaultable bond is formally equivalent to the contingent claim X which settles at the bond's maturity date T, and is given by the expression:

$$X := \beta_1 F_V^{\varkappa}(T,T)\mathbb{1}_{\{\bar{\tau} \geq T, V_T < L\}} + L\mathbb{1}_{\{\bar{\tau} \geq T, V_T \geq L\}} + \beta_2 v_{\bar{\tau}} B^{-1}(\bar{\tau},T)\mathbb{1}_{\{t < \bar{\tau} < T\}}.$$

Consequently, the forward price of a defaultable bond admits the following representation

$$F_D(t,T) = \mathbb{E}_{\mathbb{P}_T}\left(\beta_1 F_V^{\varkappa}(T,T)\mathbb{1}_{\{\bar{\tau} \geq T, V_T < L\}} + L\mathbb{1}_{\{\bar{\tau} \geq T, V_T \geq L\}} \,\Big|\, \mathcal{F}_t\right)$$
$$+ \beta_2 \,\mathbb{E}_{\mathbb{P}_T}\left(v_{\bar{\tau}} B^{-1}(\bar{\tau},T)\mathbb{1}_{\{t < \bar{\tau} < T\}} \,\Big|\, \mathcal{F}_t\right).$$

The representation above is an immediate consequence of the definition of the forward martingale measure $\mathbb{P}_T$ (see Section 13.2 in Musiela and Rutkowski [10]). We conclude that we have, on the set $\{\bar{\tau} > t\}$,

$$F_D(t,T) = L\,\mathbb{P}_T\{F_V^{\varkappa}(T,T) \geq L, \bar{\tau} \geq T \,|\, \mathcal{F}_t\}$$
$$+ \beta_1 \,\mathbb{E}_{\mathbb{P}_T}\left(F_V^{\varkappa}(T,T)\mathbb{1}_{\{F_V^{\varkappa}(T,T) < L, \bar{\tau} \geq T\}} \,\Big|\, \mathcal{F}_t\right)$$
$$+ \beta_2 K\,\mathbb{E}_{\mathbb{P}_T}\left(e^{\varkappa(\bar{\tau},T)}\mathbb{1}_{\{t < \bar{\tau} < T\}} \,\Big|\, \mathcal{F}_t\right) =: I_1(t) + I_2(t) + I_3(t),$$

where $\bar{\tau}$ equals (as usual, $\inf \varnothing = +\infty$)

$$\bar{\tau} = \inf\{t < T : F_V^{\varkappa}(t,T) \leq K\} = \inf\{t < T : Y_t \leq 0\},$$

where in turn $Y_t := \ln(F_V^{\varkappa}(t,T)/K)$ for $t \in [0,T]$. It is clear that

$$Y_t = Y_0 + \int_0^t (\sigma(u) - b(u,T))\, dW_u^T - \frac{1}{2}\int_0^t \|\sigma(u) - b(u,T)\|^2 \, du.$$

We consider the following deterministic time change $A : [0,T] \to \mathbb{R}_+$ associated with Y:

$$A_t = \int_0^t \|\sigma(u) - b(u,T)\|^2 \, du.$$

Let $A^{-1} : [0, A_T] \to [0,T]$ stand for the inverse time change. Then the time-changed process $\tilde{Y}_t := Y_{A_t^{-1}}$, $t \in [0, A_T]$, follows under $\mathbb{P}_T$ a one-dimensional Brownian motion with the drift coefficient $-1/2$, with respect to the time-changed filtration $\tilde{\mathbb{F}}$, where we set $\tilde{\mathcal{F}}_t = \mathcal{F}_{A_t^{-1}}$ for $t \in [0, A_T]$ (cf. Revuz and Yor [12]). More explicitly, $\tilde{Y}$ satisfies

$$\tilde{Y}_t = Y_0 + \tilde{W}_t - \tfrac{1}{2}t, \quad \forall t \in [0, A_T],$$

for a certain $(\mathbb{P}_T, \tilde{\mathbb{F}})$-standard Brownian motion $\tilde{W}$.

We shall first examine $I_1(t)$. Let us denote $\tilde{L} = \ln(L/K)$, and let us set $\tilde{\tau} := \inf\{t < A_T : \tilde{Y}_t \leq 0\}$. Notice that for any fixed $t < T$, we have, on the set $\{\bar{\tau} > t\} = \{\tilde{\tau} > A_t\}$,

$$\mathbb{P}_T\{F_V^{\varkappa}(T,T) \geq L, \bar{\tau} \geq T \,|\, \mathcal{F}_t\} = \mathbb{P}_T\{\tilde{Y}_{A_T} \geq \tilde{L}, \tilde{\tau} \geq A_T \,|\, \tilde{\mathcal{F}}_{A_t}\}.$$

Making use of Lemma 1, with $\tilde{\mathbb{P}} = \mathbb{P}_T$, $\sigma = 1$, $\nu = -1/2$, $u = A_t$ and $s = A_T$, we obtain

$$\mathbb{P}_T\{\tilde{Y}_{A_T} \geq \tilde{L}, \tilde{\tau} \geq A_T \,|\, \tilde{\mathcal{F}}_{A_t}\}$$

$$= N\left(\frac{\ln(K/L) + \tilde{Y}_{A_t} - \frac{1}{2}(A_T - A_t)}{\sqrt{A_T - A_t}}\right)$$

$$- e^{\tilde{Y}_{A_t}} N\left(\frac{\ln(K/L) - \tilde{Y}_{A_t} - \frac{1}{2}(A_T - A_t)}{\sqrt{A_T - A_t}}\right).$$

Consequently, we have

$$I_1(t) = L\mathbb{P}_T\{\tilde{Y}_{A_T} \geq \tilde{L}, \tilde{\tau} \geq A_T \mid \tilde{\mathcal{F}}_{A_t}\}$$

$$= LN\left(\frac{\ln(F_t/L) - \varkappa(t,T) - \frac{1}{2}\sigma^2(t,T)}{\sigma(t,T)}\right)$$

$$- e^{-\varkappa(t,T)}\frac{LF_t}{K} N\left(\frac{2\ln K - \ln(F_t L) + \varkappa(t,T) - \frac{1}{2}\sigma^2(t,T)}{\sigma(t,T)}\right).$$

This shows that

$$I_1(t) = L\left(N\left(\hat{h}_1(F_t,t,T)\right) - (F_t/K)e^{-\varkappa(t,T)}N\left(\hat{h}_2(F_t,t,T)\right)\right),$$

as expected.

To simplify the notation, we shall evaluate $I_2(t)$ and $I_3(t)$ for $t = 0$ only. The case of $t > 0$ follows by similar arguments as those used in the derivation of the formula for $I_1(t)$, and thus it presents no difficulties.

Let us focus on $I_2(0)$. In view of the definition of the processes $\tilde{Y}$ and A, we have

$$\mathbb{E}_{\mathbb{P}_T}\left(F_V^\varkappa(T,T)\mathbb{1}_{\{F_V^\varkappa(T,T)<L,\tilde{\tau}\geq T\}}\right) = K\mathbb{E}_{\mathbb{P}_T}\left(e^{\tilde{Y}_{A_T}}\mathbb{1}_{\{\tilde{Y}_{A_T}<\tilde{L},\tilde{\tau}\geq A_T\}}\right),$$

and thus we may re-express $I_2(0)$ as follows:

$$I_2(0) = \beta_1 K \int_0^{\tilde{L}} e^x \, d\mathbb{P}_T\{\tilde{Y}_{A_T} < x, \tilde{\tau} \geq A_T\}.$$

Using again Lemma 1, we obtain

$$d\mathbb{P}_T\{\tilde{Y}_{A_T} < x, \tilde{\tau} \geq A_T\}$$

$$= dN\left(\frac{x - \tilde{Y}_0 + \frac{1}{2}A_T}{\sqrt{A_T}}\right) + e^{\tilde{Y}_0} dN\left(\frac{-x - \tilde{Y}_0 - \frac{1}{2}A_T}{\sqrt{A_T}}\right)$$

$$= dN\left(\frac{x - \ln(F_0/K) + \varkappa(0,T) + \frac{1}{2}\sigma^2(0,T)}{\sigma(0,T)}\right)$$

$$+ e^{-\varkappa(0,T)}\frac{F_0}{K} dN\left(\frac{-x - \ln(F_0/K) + \varkappa(0,T) - \frac{1}{2}\sigma^2(0,T)}{\sigma(0,T)}\right).$$

Therefore, $I_2(0) = I_{21}(0) + I_{22}(0)$, where, by standard calculations

$$I_{21}(0) = \beta_1 K \int_0^{\tilde{L}} e^x \, dN\left(\frac{x - \ln(F_0/K) + \varkappa(0,T) + \frac{1}{2}\sigma^2(0,T)}{\sigma(0,T)}\right)$$

$$= \beta_1 F_0 e^{-\varkappa(0,T)} N\left(\frac{\ln(L/F_0) + \varkappa(0,T) - \frac{1}{2}\sigma^2(0,T)}{\sigma(0,T)}\right)$$

$$- \beta_1 F_0 e^{-\varkappa(0,T)} N \left(\frac{\ln(K/F_0) + \varkappa(0,T) - \frac{1}{2}\sigma^2(0,T)}{\sigma(0,T)} \right)$$

$$= \beta_1 F_0 e^{-\varkappa(0,T)} \left(N \big(\hat{h}_3(F_0,0,T) \big) - N \big(\hat{h}_4(F_0,0,T) \big) \right)$$

and

$$I_{22}(0) = \beta_1 e^{-\varkappa(0,T)} F_0 \int_0^{\tilde{L}} e^x \, dN \left(\frac{-x - \ln(F_0/K) + \varkappa(0,T) - \frac{1}{2}\sigma^2(0,T)}{\sigma(0,T)} \right)$$

$$= \beta_1 K N \left(\frac{2\ln K - \ln(LF_0) + \varkappa(0,T) + \frac{1}{2}\sigma^2(0,T)}{\sigma(0,T)} \right)$$

$$- \beta_1 K N \left(\frac{\ln(K/F_0) + \varkappa(0,T) + \frac{1}{2}\sigma^2(0,T)}{\sigma(0,T)} \right)$$

$$= \beta_1 K \left(N \big(\hat{h}_5(F_0,0,T) \big) - N \big(\hat{h}_6(F_0,0,T) \big) \right).$$

To establish the last two formulae, note that for any $c \neq 0$, and $a, b, d \in \mathbb{R}$, we have (we set here $\tilde{d} = d - c^{-1}$)

$$\int_a^b e^x \, dN(cx + d) = e^{\frac{1}{2}(\tilde{d}^2 - d^2)} \big(N(cb + \tilde{d}) - N(ca + \tilde{d}) \big).$$

Observe that $I_{21}(0) > 0$ and $I_{22}(0) < 0$; we always have $I_2(0) > 0$, though. It remains to evaluate $I_3(0)$, where

$$I_3(0) = \beta_2 K \, \mathbb{E}_{\mathbb{P}_T} \big(e^{\varkappa(\tilde{\tau}, T)} \, \mathbb{1}_{\{\tilde{\tau} < T\}} \big) = \beta_2 K \int_0^T e^{\varkappa(t,T)} \, d\mathbb{P}_T \{ \tilde{\tau} < t \}.$$

In view of Lemma 1, we have

$$\mathbb{P}_T \{ \tilde{\tau} < s \} = N \left(\frac{-\check{Y}_0 + \frac{1}{2}s}{\sqrt{s}} \right) + e^{\check{Y}_0} N \left(\frac{-\check{Y}_0 - \frac{1}{2}s}{\sqrt{s}} \right),$$

where $\check{Y}_0 = Y_0$, and, as before, $\tilde{\tau} = \inf \{ t < A_T : \check{Y}_t \leq 0 \}$. Since clearly $\mathbb{P}_T \{ \tilde{\tau} < t \} = \mathbb{P}_T \{ \tilde{\tau} < A_t \}$, we obtain

$$\mathbb{P}_T \{ \tilde{\tau} < t \} = N \left(\frac{-Y_0 + \frac{1}{2}A_t}{\sqrt{A_t}} \right) + e^{Y_0} N \left(\frac{-Y_0 - \frac{1}{2}A_t}{\sqrt{A_t}} \right)$$

$$= N \left(\frac{\ln \frac{K}{F_0} + \varkappa(0,T) + \frac{1}{2}A_t}{\sqrt{A_t}} \right) + e^{-\varkappa(0,T)} \frac{F_0}{K} N \left(\frac{\ln \frac{K}{F_0} + \varkappa(0,T) - \frac{1}{2}A_t}{\sqrt{A_t}} \right).$$

We conclude that $I_3(0) = I_{31}(0) + I_{32}(0)$, where

$$I_{31}(0) = \beta_2 K \int_0^T e^{\varkappa(t,T)} \, dN \left(\frac{\ln(K/F_0) + \varkappa(0,T) + \frac{1}{2}\sigma^2(0,t)}{\sigma(0,t)} \right)$$

$$= \beta_2 K J_1(F_0,0,T)$$

and

$$I_{32}(0) = \beta_2 F_0 e^{-\varkappa(0,T)} \int_0^T e^{\varkappa(t,T)} \, dN \left(\frac{\ln(K/F_0) + \varkappa(0,T) - \frac{1}{2}\sigma^2(0,t)}{\sigma(0,t)} \right)$$

$$= \beta_2 F_0 e^{-\varkappa(0,T)} J_2(F_0,0,T).$$

This completes the proof of Proposition 2. $\qquad\qquad\square$

To the best of our knowledge, explicit formulae for $J_1(F_t, t, T)$ and $J_2(F_t, t, T)$ are not available in the general time-dependent setup (even when, e.g., the dividend ratio $\varkappa$ is constant). Incidentally, quite simple expressions for these two terms can be obtained provided that we set $\varkappa = 0$; that is, in the absence of dividends. The following result is an immediate corollary to Proposition 2.

Corollary 1 *Under the assumptions of Proposition 2, if $\varkappa \equiv 0$ then*

$$F_D(t, T) = L\big(N\big(-d_1(F_t, t, T)\big) - (F_t/K)N\big(d_6(F_t, t, T)\big)\big)$$
$$+ \beta_1 F_t\big(N\big(d_2(F_t, t, T)\big) - N\big(d_4(F_t, t, T)\big)\big)$$
$$+ \beta_1 K\big(N\big(d_5(F_t, t, T)\big) - N\big(d_3(F_t, t, T)\big)\big)$$
$$+ \beta_2 K N\big(d_3(F_t, t, T)\big) + \beta_2 F_t N\big(d_4(F_t, t, T)\big),$$

where

$$d_1(F_t, t, T) = \frac{\ln(L/F_t) + \frac{1}{2}\sigma^2(t, T)}{\sigma(t, T)} = d_2(F_t, t, T) + \sigma(t, T),$$

$$d_3(F_t, t, T) = \frac{\ln(K/F_t) + \frac{1}{2}\sigma^2(t, T)}{\sigma(t, T)} = d_4(F_t, t, T) + \sigma(t, T),$$

$$d_5(F_t, t, T) = \frac{\ln(K^2/F_t L) + \frac{1}{2}\sigma^2(t, T)}{\sigma(t, T)} = d_6(F_t, t, T) + \sigma(t, T).$$

Proof. Since the inequality $F_t > K$ is satisfied on the set $\{\bar{\tau} > t\}$, we have

$$J_1(F_t, t, T) = \int_t^T dN\left(\frac{\ln(K/F_t) + \frac{1}{2}\sigma^2(t, u)}{\sigma(t, u)}\right)$$
$$= N\left(\frac{\ln(K/F_t) + \frac{1}{2}\sigma^2(t, T)}{\sigma(t, T)}\right)$$

and

$$J_2(F, t, T) = \int_t^T dN\left(\frac{\ln(K/F_t) - \frac{1}{2}\sigma^2(t, u)}{\sigma(t, u)}\right)$$
$$= N\left(\frac{\ln(K/F_t) - \frac{1}{2}\sigma^2(t, T)}{\sigma(t, T)}\right).$$

The formula now follows from simple calculations.   □

Let us observe that the formula of Corollary 1 covers as a special case the valuation result established by Briys and de Varenne [4]. In some other recent studies of first passage time models, in which the triggering barrier is assumed to be either a constant or an unspecified stochastic process, typically no closed-form solution for the value of a corporate debt is available, and thus a numerical approach is required (see, e.g., Collin-Dufresne and Goldstein [5], Kim et al. [7], Longstaff and Schwartz [8], Nielsen et al. [11], or Saá-Requejo and Santa-Clara [13]).

REFERENCES

[1] T.R. Bielecki and M. Rutkowski, *Credit risk: Modelling, valuation and hedging*, forthcoming in Springer-Verlag, 2001.

[2] F. Black and M. Scholes, The pricing of options and corporate liabilities, *J. Political Econom.* **81**, 637–654 (1973).

[3] F. Black and J.C. Cox, Valuing corporate securities: some effects of bond indenture provisions, *J. Finance* **31**, 351–367 (1976).

[4] E. Brys and F. de Varenne, Valuing risky fixed rate debt: an extension, *J. Finan. Quant. Anal.* **32**, 239–248 (1997).

[5] P. Collin-Dufresne and R.S. Goldstein, Do credit spread reflect stationary leverage ratios? Working paper, Carnegie Mellon University and Ohio State University, 2000.

[6] D. Heath, R. Jarrow and A. Morton, Bond pricing and the term structure of interest rates: a new methodology for contingent claim valuation, *Econometrica* **60**, 77–105 (1992).

[7] I.J. Kim, K. Ramaswamy and S. Sundaresan, The valuation of corporate fixed income securities, Working paper, University of Pennsylvania, 1993.

[8] F.A. Longstaff and E.S. Schwartz, A simple approach to valuing risky fixed and floating rate debt, *J. Finance* **50**, 789–819 (1995).

[9] R.C. Merton, On the pricing of corporate debt: the risk structure of interest rates, *J. Finance* **29** 449–470 (1974).

[10] M. Musiela and M. Rutkowski, *Martingale methods in financial modelling*, Springer-Verlag, Berlin, 1997.

[11] T.N. Nielsen, J. Saá-Requejo and P. Santa-Clara, Default risk and interest rate risk: the term structure of default spreads, Working paper, INSEAD, 1993.

[12] D. Revuz and M. Yor, *Continuous martingales and Brownian motion*, Springer-Verlag, New York, 1991.

[13] J. Saá-Requejo and P. Santa-Clara, Bond pricing with default risk, Working paper, 1999.

PRICING OPTIONS
FOR MARKOVIAN MODELS

GIANMARIO TESSITORE[1] and JERZY ZABCZYK[2]

[1] Università di Genova, Dipartimento di Matematica
via Dodecaneso 35, 16145 Genova, Italy
[2] Polish Academy of Sciences, Institute of Mathematics
Śniadeckich 8, 00-950 Warszawa, Poland
and Warwick University
Mathematics Institute, Coventry, England

Abstract The paper derives explicit formulae for the superprices of options. These superprices are given as value functions of certain stochastic control problems. The results hold for both finite and infinite dimensional Markovian models. The classical martingale methods have been replaced by a dynamic programming approach.

Key words Superprices, direct approach, dynamic programming, separation theorems.

1. INTRODUCTION

Let $(\Omega, \mathbb{P}, \mathcal{F})$ be a probability space on which a discrete time price process (S_n) is defined. The process (S_n) might be a general Markov chain on a set E^+. It is also assumed that the rate offered by the bank is $r > -1$, so the bank deposits B_n, $n = 0, 1, \ldots$ grow according to the equation

$$B_{n+1} = (1 + r)B_n, \quad B_0 = 1 . \tag{1}$$

A probability measure $\tilde{\mathbb{P}}$ is called an equivalent martingale measure if it is equivalent to $\mathbb{P}$ and the discounted sequence

$$(1 + r)^{-n} S_n, \quad n = 0, 1, \ldots,$$

is a martingale on $(\Omega, \tilde{\mathbb{P}}, \mathcal{F})$. It is well known (see [8], [12] and [17]) that, if the number K of assets is finite and if there exists an equivalent martingale measure $\tilde{\mathbb{P}}$, then the so called *superprices* $\mathcal{O}_E(N, s)$, $\mathcal{O}_A(N, s)$ of European and American options for time N, with the contingent claims $\varphi(S_n)$ at time n, are given by the formulae:

$$\mathcal{O}_E(N, s) = \sup_{\hat{\mathbb{P}}} \mathbb{E}_{\hat{\mathbb{P}}}[(1 + r)^{-N} \varphi(S_N^s)], \tag{2}$$

$$\mathcal{O}_A(N, s) = \sup_{\hat{\mathbb{P}}, \tau \leq N} \mathbb{E}_{\hat{\mathbb{P}}}[(1 + r)^{-\tau} \varphi(S_\tau^s)], \tag{3}$$

where the suprema are taken with respect to all equivalent martingale measures and with respect to all stopping times τ.

The present note is concerned with *related but different* formulae for European and American options derived, as in our previous papers [24], [20] and [21], by an application of dynamic programming and separation theorems. Contrary to the papers [8], [12] and [17], we do not use any decomposition results for semimartingales needed to derive (2) and (3). Today such approach is called *direct* (see [23]). It turns out that one can construct a stochastic control system:

$$\widehat{S}_{n+1} = \widehat{F}(\widehat{S}_n, \widehat{u}_n, \widehat{\xi}_{n+1}), \quad \widehat{S}_0 = s, \tag{4}$$

on E^+, where $\widehat{\xi}_1, \widehat{\xi}_2, \ldots$ is a sequence of independent random variables, uniformly distributed on the interval $[0, 1)$, such that

$$O_E(N, s) = \sup_{\widehat{\pi}} \widehat{\mathbb{E}} \left(\frac{1}{(1 + r)^N} \varphi(\widehat{S}_N^{s, \widehat{\pi}}) \right), \tag{5}$$

$$O_A(N, s) = \sup_{\widehat{\pi}, \widehat{\tau} \leq N} \widehat{\mathbb{E}} \left(\frac{1}{(1 + r)^{\widehat{\tau}}} \varphi(\widehat{S}_{\widehat{\tau}}^{s, \widehat{\pi}}) \right). \tag{6}$$

In the above formulae $\widehat{\pi}$ denotes any admissible control strategy and $\widehat{\tau}$ any stopping time for the new controlled system (4). Moreover $\widehat{S}_n^{s, \widehat{\pi}}$ denotes the position of the controlled system at moment n provided the system was controlled with $\widehat{\pi}$ and started from $s \in E^+$. In particular the option prices are solutions of the Bellman equations associated with the control problems which can be solved by an iterative method.

Our main results are formulated as Theorem 4, Theorem 5 and Theorem 2. In Theorem 4 the prices of options are expressed as iterations of some non-linear operators and Theorem 5 states that they are value functions of a stochastic control problem, with stopping for American options, for a *new control system*. An explicit description of the structure of the set of the control parameters for the new system, vital for calculations, is given in Theorem 4. The proofs are based on a combination of functional analytic results on convex sets and extremal points as well as of the dynamic programming ideas. General results are applied to specific multi-asset and multinomial models in Sections 3.1 – 3.3. The case when noise is continuously distributed is discussed in Section 3.4. We also show that the described approach is directly related to the so called viability theory, see [1]. In the final section we describe a connection between the martingale measure and the control theoretic approaches.

The reduction to a control problem, well known in continuous time (see, e.g., [11], [7], [2], [3], [16]), is usually done differently, and, in a sense, in a reverse direction. First one derives a dual representation of the prices in terms of the equivalent martingale measures (see (2) and (3)). Secondly one finds, usually under more restrictive assumptions, a stochastic control formulation and, finally, the prices are identified with solutions of the corresponding Bellman equation. Most results on option pricing in discrete time are obtained in a similar way or are concerned with the binomial model (see [18], [5] and [4]).

It follows from the discussions in the final section that in several situations the formulae of the type (5) and (6) can be derived from (2) and (3), however, this cannot always be done. The set of all equivalent martingale measures might be very large and therefore not easy to use. In addition there exist models for which there are no equivalent martingale measures but option prices can still be obtained using the approach presented here. Such situation was analyzed in [19], where the concept of a generalized martingale measure was introduced. As is shown here, if the price process is infinite dimensional, equivalent martingale measures rarely exist. Moreover, to cover infinite markets with martingale approach, new infinite dimensional decomposition theorems would be needed.

The present paper is a rewritten version of the working paper [22].

Acknowledgments We thank Rudiger Frey for his comments on an earlier version of the paper and S. Kwapien and M. Rutkowski for discussions on infinite dimensional models. Thanks go also to the reviewers for interesting remarks and useful suggestions.

2. GENERAL MODEL

We start from a general market assuming that the vector of stock prices is an element of a Borel set E^+ contained in a Banach space E. In particular, if the market is finite and consists of shares of d-types, then one usually takes $E^+ = (0, +\infty)^d$ and $E = \mathbb{R}^d$. A portfolio is any element σ of the dual space E^* and $\langle \sigma, s \rangle$ denotes the value of the functional σ on the element s. Thus if $s \in E^+$ then $\langle \sigma, s \rangle$ is the value of the portfolio σ calculated with respect to the price vector s.

Let Y be a compact metric space and $\xi_1, \xi_2, \ldots$ a sequence of independent, Y-valued, identically distributed random variables. The dynamics of the prices is determined by the equation,

$$S_{n+1} = F(S_n, \xi_{n+1}), \quad S_0 = s \in E^+, \tag{7}$$

where F is a continuous mapping from $E^+ \times Y$ into E^+.

Let β_n and σ_n denote, respectively, an amount of money deposited in a bank at time n and a portfolio of shares fixed at moment n and kept unchanged in the time interval $(n, n+1)$. Let $r > -1$ be the interest rate offered by the bank and X_n the capital of the agent at moment n. Assuming self-financing strategy one arrives at the following relations:

$$X_n = \beta_n + \langle \sigma_n, S_n \rangle, \tag{8}$$
$$X_{n+1} = (1+r)\beta_n + \langle \sigma_n, S_{n+1} \rangle, \quad n = 0, 1, \ldots. \tag{9}$$

Relations (7), (8) and (9) yield

$$\begin{cases} X_{n+1} = (1+r)X_n + \langle \sigma_n, F(S_n, \xi_{n+1}) - (1+r)S_n \rangle, \ X_0 = x \in \mathbb{R}, \\ S_{n+1} = F(S_n, \xi_{n+1}), \ S(0) = s \in E^+. \end{cases} \tag{10}$$

It follows from (8) and (9) that β_n is uniquely determined by σ_n, X_n, S_n. We therefore identify any investment strategy with the sequence of σ_n only. More precisely, as it is customary in stochastic control, a strategy π is defined as a sequence $(\sigma_0, \sigma_1, \ldots)$ of measurable transformations $\sigma_0 : (\mathbb{R} \times E^+) \to E^*, \ldots, \sigma_n : (\mathbb{R} \times E^+)^{n+1} \to E^*, \ldots$. Instead of

$$\sigma_n((X_0, S_0), \ldots, (X_n, S_n))$$

we write shortly σ_n. In mathematical finance one usually starts from an increasing family of σ-fields, $\mathcal{F}_n \subset \mathcal{F}$ such that, for each $n \in \mathbb{N}$, ξ_n is $\mathcal{F}_n$-measurable and ξ_{n+1} is independent of $\mathcal{F}_n$. Both definitions lead to the same results. To underline the dependence of the sequences (X_n), (S_n) on the initial conditions and on the strategy π we will often write $X_n^{(x,s),\pi}$ and S_n^s.

Let φ be a Borel function on E^+, which we will call the *contingent claim*. The *upper price* $O_E(s, N)$ of the *European option* with exercise time N, initial price of the stock $s \in E^+$ and the contingent claim φ is the infimum of all those $x \in \mathbb{R}$ for which there exists a self-financing strategy $\pi = (\sigma_0, \sigma_1, \sigma_2, \ldots)$ such that

$$\mathbb{P}(X_N^{(x,s),\pi} \geq \varphi(S_N^s)) = 1. \tag{11}$$

The infimum of all those $x \in \mathbb{R}$ for which there exists a self-financing strategy $\pi = (\sigma_0, \sigma_1, \ldots)$ such that

$$\mathbb{P}\left(X_n^{(x,s),\pi} \geq \varphi(S_n^s), \ n = 0, 1, \ldots, N\right) = 1, \tag{12}$$

is, by definition, the upper price $O_A(s, N)$ of the American option with the terminal time N, contingent claim φ and the initial price s of the stocks.

If G is a measurable transformation from a measurable space $(E, \mathcal{E})$ into $(E_1, \mathcal{E}_1)$, then $G(E)$ denotes the image of G and $G^*(\nu)$ the image of the probability measure ν:

$$G^*\nu(\Gamma) = \nu\{x \in E; G(x) \in \Gamma\} = \nu(G^{-1}(\Gamma)).$$

The convex hull of a set H will be denoted by $\mathrm{conv}\, H$. Thus for each $s \in E^+$ and any probability measure ν on Y, $F(s, Y)$, $F^*(s, \nu)$ and $\mathrm{conv}\, F(s, Y)$ are respectively the images of Y and ν by the transformation $F(s, \cdot)$ and the convex hull of $F(s, Y)$. We will always assume that

(A1)   For arbitrary $s \in E^+$, $(1 + r)s \in \mathrm{conv}\, F(s, Y)$.

If $(A1)$ does not hold then there exist arbitrage opportunities.

Proposition 1 *Assume that for some* $s \in E^+$, $(1 + r)s \notin \mathrm{conv}\, F(s, Y)$, *then* $O_E(s, 1) = -\infty$.

Proof. We define $D_s = \mathrm{conv}\, F(s, Y)$. Then D_s is a compact, convex subset of E. By the classical separation theorem (see [6], Chapter 5), there exists $\sigma \in E^*$ such that

$$\inf_{z \in D_s} \langle \sigma, z - (1 + r)s \rangle = \alpha > 0.$$

Let x be any real number and π a strategy that at time 0 chooses portfolio $\gamma\sigma$ where γ is a positive number. Then

$$X_1^{(x,s),\pi} = (1 + r)x + \langle \gamma\sigma, F(s, \xi_1) - (1 + r)s \rangle$$

and, with probability 1,

$$X_1^{(x,s),\pi} \geq (1 + r)x + \gamma\alpha.$$

Since $\varphi(F(s, \xi_1))$ is a bounded random variable one can find γ such that

$$(1 + r)x + \gamma\alpha \geq \varphi(F(s, \xi_1)) \quad \mathbb{P}\text{-a.s.}$$

Consequently $O_E(s, 1) \leq x$. Since x was an arbitrary number, $O_E(s, 1) = -\infty$, as required.
□

2.1. Associated Stochastic Control System

We now pass to a construction of the associated stochastic control system.

For any metric space Z, by $\mathcal{P}(Z)$ we denote the set of all probability measures defined on $(Z, \mathcal{B}(Z))$, equipped with the topology of weak convergence. To simplify we also use the following notation

$$U = \mathcal{P}(Y), \qquad P^{\nu}(s, \cdot) = F^*(s, \nu), \quad s \in E^+, \ \nu \in U,$$

$$M_s = \left\{ \mu \in \mathcal{P}(F(s, Y)) : \int_{F(s,Y)} z\mu(dz) = (1 + r)s \right\}.$$

If (A1) holds then for each $s \in E^+$ the set M_s is nonempty, convex and compact. By M_s^e we denote the set of all extremal points of M_s. Moreover U_s is any subset of U which is transformed by the mapping $F^*(s, \cdot)$ onto M_s and U_s^e is the subset of U_s which is transformed onto M_s^e.

The *state space* of the new control system will be E^+, the same as for the price process S_n. The set of *available control parameters*, at the state $s \in E^+$, will be U_s^e and *transition probabilities* will be

$$P^{\nu}(s, \cdot) = F^*(s, \nu), \quad s \in E^+, \ \nu \in U_s^e.$$

Despite of their rather complicated definition the sets U_s^e have in fact rather simple structure. Namely we have the following theorem.

Theorem 2 *Assume that the set $F(s, Y) = D$ spans a linear space L of a finite dimension d. Then any measure $\mu \in M_s^e$ is supported by at most $d + 1$ points of D.*

Proof. We can assume that $F(s, Y) \subset R^d$, $c = (1 + r)s \in \operatorname{conv} F(s, Y)$. Then M_s consists of all probability measures μ with the supports contained in D and such that

$$\int_D z\mu(dz) = c .\tag{13}$$

We define the functions $e_0(z) = 1$, $e_1(z) = z_1, \ldots, e_d(z) = z_d$ for $z = (z_1, \ldots, z_d)$ in $\mathbb{R}^d$. We first show that, if $e_0, e_1, \ldots, e_d$ do not span $H = L^2(D, \mu)$, then there exist two different measures μ_1, μ_2 in M_s such that

$$\mu = 1/2 \, (\mu_1 + \mu_2),$$

so μ cannot be an extremal measure. To prove this we define

$$\mu_1(dz) = f_1(z)\mu(dz), \quad \mu_2(dz) = f_2(z)\mu(dz),$$

where

$$f_1(z) = 1 - h(z), \quad f_2(z) = 1 + h(z), \quad z \in D.$$

It is clear that $\mu = 1/2 \, (\mu_1 + \mu_2)$. Measures μ_1, μ_2 are probability measures and belong to M_s if and only if

$$-1 \leq h(z) \leq 1, \quad z \in D, \quad < h, e_j > = 0, \quad j = 0, 1, \ldots, d,$$

where $< \cdot, \cdot >$ stands for the scalar product in H. Since the functions $e_0, \ldots, e_d$ do not span H, there exists a continuous function $g \in H$ which is not in the linear span of $e_0, \ldots, e_d$. Let g_0 be the orthogonal projection of g onto the linear span of $e_0, \ldots, e_d$. Define

$$h = \gamma(g - g_0),$$

where $\gamma > 0$ is sufficiently small. Then measures μ_1, μ_2 have the required properties. $\qquad \square$

Thus the sets M_s^e, U_s^e can explicitly be described in the finite dimensional situations. For specific applications see Section 3 on linear models.

The next result allows to give a dynamic interpretation of the controlled system. For the proof we can refer, for instance, to [25].

Theorem 3 *Assume that there exists an equivalent metric on E^+ with respect to which E^+ is a complete, separable, metric space and assume that the mapping $(\nu, s) \to P^\nu(s, \cdot)$ from $U \times E^+$ into $\mathcal{P}(E^+)$ is measurable. Then there exists a measurable mapping $\widehat{F} : (s, \nu, x) \to \widehat{F}(s, \nu, x)$ from $E^+ \times U \times [0, 1)$ into E^+ such that for an arbitrary, uniformly distributed random variable ξ with values in the interval $[0, 1)$ and for arbitrary s, ν the law of $\widehat{F}(s, \nu, \xi)$ is exactly $P^\nu(s, \cdot)$.*

It is well known (see [13]) that, if E^+ is an open subset of E or an intersection of a countable family of open subsets, then the required metric exists. In particular this is true if E^+ consists of all vectors with positive coordinates in R^k or in l^2.

It follows from the theorem that, given the family of transition kernels $P^\nu(s, \cdot)$, $s \in E^+$, and the family of admissible sets of control parameters $U_s^e \subset U_s$, one can construct (see, e.g., [25]) a new control system on $(\widehat{\Omega}, \widehat{\mathcal{F}}, \widehat{\mathbb{P}})$ and E^+ by:

$$\widehat{S}_{n+1} = \widehat{F}(\widehat{S}_n, \widehat{\nu}_n, \widehat{\xi}_{n+1}), \quad \widehat{S}_0 = s, \tag{14}$$

where $\widehat{\xi}_1, \widehat{\xi}_2, \ldots$ are independent random variables, identically and uniformly distributed on $[0, 1)$, and $\widehat{F}$ is a measurable mapping from $E^+ \times U_s \times [0, 1) \to E^+$ such that

$$\mathbb{P}(\widehat{F}(s, \nu, \widehat{\xi}_1,) \in \Gamma) = P^\nu(s, \Gamma)$$

for any $s \in E^+$ and $\nu \in U_s$.

Let V_n^1 and V_n^2 be the value functions corresponding to the following optimization problems:

$$V_n^1(s) = \sup_{\widehat{\pi}} \widehat{\mathbb{E}}\left(\frac{1}{(1+r)^n}\varphi(\widehat{S}_n^{s,\widehat{\pi}})\right), \quad s \in E^+,$$

$$V_n^2(s) = \sup_{\widehat{\pi}, \widehat{\tau} \le n} \widehat{\mathbb{E}}\left(\frac{1}{(1+r)^{\widehat{\tau}}}\varphi(\widehat{S}_{\widehat{\tau}}^{s,\widehat{\pi}})\right), \quad s \in E^+, n = 0, 1, \ldots,$$

where $\widehat{\pi}, \widehat{\tau}$ denote admissible control strategies and stopping times. They satisfy the following Bellman equations (see, e.g., [25])

$$V_{n+1}^1(s) = \sup_{\mu \in U_s} \frac{1}{1+r}P^\mu V_n^1(s), \quad V_0^1(s) = \varphi(s),$$

$$V_{n+1}^2(s) = \max\left(\varphi(s), \sup_{\mu \in U_s} \frac{1}{1+r}P^\mu V_n^2(s)\right), \quad V_0^2(s) = \varphi(s).$$

The following operators Q, R, acting on functions defined on E^+ will play an important role.

$$Q\varphi(s) = \frac{1}{1+r} \sup_{\nu \in U_s^e} P^\nu \varphi(s), \tag{15}$$

$$R\varphi(s) = \max\left(\varphi(s), \frac{1}{1+r} \sup_{\nu \in U_s^e} P^\nu \varphi(s)\right), \quad s \in E^+. \tag{16}$$

Taking into account the formulated Bellman equations we can claim that

$$V_n^1(s) = Q^n\varphi(s), \quad V_n^2(s) = R^n\varphi(s), \quad s \in E^+. \tag{17}$$

2.2. Main Results and Proofs

We will need also the following condition:

(A2)   The set $\{(s, \nu) \in E^+ \times U : \nu \in U_s, s \in E^+ \}$ is closed.

The following theorem is the main result of the paper.

Theorem 4 *Under the assumptions (A1), (A2) for an arbitrary s in E^+ and an arbitrary upper-semicontinuous function φ:*

$$
\begin{aligned}
O_E(s, N) &= Q^N \varphi(s), & (18) \\
O_A(s, N) &= R^N \varphi(s). & (19)
\end{aligned}
$$

Taking into account (17) we have the following Corollary.

Theorem 5 *Assume that there exists an equivalent metric on E^+ with respect to which E^+ is a complete, separable metric space. Under the assumptions (A1), (A2) and assuming that φ is upper-semicontinuous, the following formulae hold:*

$$
O_E(N, s) = \sup_{\widehat{\pi}} \widehat{\mathbb{E}} \left(\frac{1}{(1+r)^N} \varphi(\widehat{S}_N^{s, \widehat{\pi}}) \right),
$$

$$
O_A(N, s) = \sup_{\widehat{\pi}, \widehat{\tau} \leq N} \widehat{\mathbb{E}} \left(\frac{1}{(1+r)^{\widehat{\tau}}} \varphi(\widehat{S}_{\widehat{\tau}}^{s, \widehat{\pi}}) \right),
$$

where $\widehat{\pi}$ denotes any admissible control strategy and $\widehat{\tau}$ any stopping time for the new controlled system.

With upper-semicontinuous functions important examples of digital options can be covered, see [15]. Theorem 4 will be a consequence of some auxiliary results. We first consider the one step model for which $N = 1$.

Theorem 6 *Under the assumption (A1) for an arbitrary s in E^+ and an arbitrary upper-semicontinuous function φ:*

$$
O_E(s, 1) = \frac{1}{1+r} \sup_{u \in U_s^e} P^u \varphi(s), \tag{20}
$$

$$
O_A(s, 1) = \max \left(\varphi(s), \frac{1}{1+r} \sup_{u \in U_s^e} P^u \varphi(s) \right). \tag{21}
$$

We need the following separation result.

Proposition 7 *Let D be a compact subset of a Banach space E and φ a bounded Borel function on D.*

   i) *If there exists $\sigma \in E^*$ such that*

$$
\langle \sigma, z \rangle \geq \varphi(z) \quad \text{for all } z \in D, \tag{22}
$$

then for any probability measure μ concentrated on D such that

$$\int_D z\mu(dz) = 0,$$

(23)

one has

$$\int_D \varphi(z)\mu(dz) \leq 0.$$

(24)

ii) *If φ is upper-semicontinuous and for any probability measure μ concentrated on D satisfying (23) the inequality (24) holds then for every $\varepsilon > 0$ there exists $\sigma \in E^*$ such that*

$$\langle \sigma, z \rangle \geq \varphi(z) - \varepsilon \quad \text{for all } z \in D.$$

(25)

Proof. i) If (22) and (23) hold then

$$\int_D \varphi(z)\mu(dz) \leq \int_D \langle \sigma, z \rangle \mu(dz) = \left\langle \sigma, \int_D z\mu(dz) \right\rangle = 0$$

as required.

ii) For $\varepsilon > 0$ we define

$$\Lambda_\varepsilon = \left\{ \left(\int_D z\mu(dz), - \int_D (\varphi(z) - \varepsilon)\mu(dz) \right); \text{ supp}\,\mu \subseteq D, \ \mu \in \mathcal{P}(D) \right\}.$$

Then Λ_ε is a convex subset of $E \times \mathbb{R}$. Let $\tilde{\Lambda}_\varepsilon$ be the closure of the set of all vectors $\tau\lambda$ where $\tau \geq 0$ and $\lambda \in \Lambda_\varepsilon$. Then $\tilde{\Lambda}_\varepsilon$ is a closed convex cone that contains 0. We will show that $(0, -1) \notin \tilde{\Lambda}_\varepsilon$. Assume, to the contrary, that $(0, -1) \in \tilde{\Lambda}_\varepsilon$. Then there exists a sequence (τ_n, μ_n) such that $\tau_n \geq 0$, $\text{supp}\,\mu_n \subseteq D$, μ_n probability measures, $n = 1, \ldots$ such that

$$\tau_n \int_D z\mu_n(dz) \to 0 \quad \text{and} \quad \tau_n \int_D (\varphi(z) - \varepsilon)\mu_n(dz) \to 1.$$

(26)

Without any loss of generality we can assume that $\mu_n \to \mu_\infty$ (weakly) as $n \to +\infty$. Then

$$\int_D z\mu_n(dz) \to \int_D z\mu_\infty(dz) = a.$$

If $a \neq 0$, then by (26), $\tau_n \to 0$ and consequently $\tau_n \int_D (\varphi(z) - \varepsilon)\mu_n(dz) \to 0$, contrary to our assumption. If $a = 0$ then

$$\int_D z\mu_\infty(dz) = 0$$

and therefore, by (24) $\int_D (\varphi(z) - \varepsilon)\mu_\infty(dz) \leq -\varepsilon$. Consequently

$$\limsup_n \int_D (\varphi(z) - \varepsilon)\mu_n(dz) \leq \int_D (\varphi(z) - \varepsilon)\mu_\infty(dz) \leq -\varepsilon.$$

Thus for sufficiently large n,

$$\int_D (\varphi(z) - \varepsilon)\mu_n(dz) \leq -\frac{\varepsilon}{2} < 0,$$

and

$$\tau_n \int_D (\varphi(z) - \varepsilon)\mu_n(dz) \geq 0,$$

which is again a contradiction.

By the classical separation result (see [6], Chapter 5), applied to $\tilde{\Lambda}_\varepsilon$ and $\{(0, -1)\}$, there exists $\sigma_0 \in E^*$, $\varrho \in \mathbb{R}$, $c \in \mathbb{R}$ such that

$$\tau \left[\int_D \langle \sigma_0, z \rangle \mu(dz) - \varrho \int_D (\varphi(z) - \varepsilon)\mu(dz) \right] \geq c, \qquad -\varrho < c, \tag{27}$$

for all $\tau \in \mathbb{R}^+$ and all probability measures μ on D. Letting $\tau \to 0$, (27) yields $c \leq 0$ and $\varrho > 0$. Letting $\tau \to +\infty$ (27) yields

$$\int_D \langle \sigma_0, z \rangle \mu(dz) - \varrho \int_D (\varphi(z) - \varepsilon)\mu(dz) \geq 0.$$

In this way, choosing $\sigma = \varrho^{-1}\sigma_0$, the proof of the proposition is complete. □

Proof of Theorem 6. Assume that for some $s \in E^+$ and $x \in \mathbb{R}$ there exists $\sigma \in E^*$ such that

$$\begin{aligned} X_1^{(x,s),\sigma} &= (1+r)x + \langle \sigma, F(s, \xi_1) - (1+r)s \rangle \tag{28} \\ &\geq \varphi(F(s, \xi_1)) \quad \mathbb{P}\text{-a.s.} \end{aligned}$$

Applying Proposition 7 i) for $z = F(s, y) - (1+r)s$, $y \in Y$ and the function φ defined as

$$z \to \varphi(z + (1+r)s) - (1+r)x \tag{29}$$

one gets that if $\mathrm{supp}\,\mu \subseteq D_s = F(s, Y)$, $\int_{D_s} z\mu(dz) = (1+r)s$ then

$$\int_{D_s} \varphi(z)\mu(dz) - (1+r)x \leq 0 \tag{30}$$

and therefore

$$x \geq \frac{1}{1+r} \int_{D_s} \varphi(z)\mu(dz). \tag{31}$$

So

$$O_E(s, 1) \geq \frac{1}{1+r} \sup_{\nu \in U_s} P^\nu \varphi(s).$$

Conversely, we define

$$\bar{x} = \frac{1}{1+r} \sup_{\nu \in U_s} P^\nu \varphi(s).$$

Then, for all probability measures μ, $\mathrm{supp}\,\mu \subset D_s$, $\int_{D_s} z\mu(dz) = (1+r)s$, one has that

$$\int_{D_s} [\varphi(z) - (1+r)\bar{x}]\mu(dz) \leq 0$$

and by Proposition 7 ii) for arbitrary $\varepsilon > 0$ there exists $\sigma \in E^*$ such that

$$\begin{aligned} \langle \sigma, F(s, \xi_1) - (1+r)s \rangle &\geq \varphi(F(s, \xi_1)) - (1+r)\bar{x} - \varepsilon \tag{32} \\ &\geq \varphi(F(s, \xi_1)) - (1+r)\left[\bar{x} + \frac{\varepsilon}{1+r}\right]. \end{aligned}$$

Consequently

$$X_1^{(\bar{x} + \frac{\varepsilon}{1+r}, s), \sigma} \geq \varphi(S_1^s) \quad \mathbb{P}\text{-a.s.,} \tag{33}$$

so

$$O_E(s,1) \leq \bar{x} + \frac{\varepsilon}{1+r}.$$

In this way the identity (20) has been proved.

The proof of (21) easily follows from (20) because in the definition of $O_A(s,1)$ one requires additionally that $x = X_0^{(x,s),\pi} \geq \varphi(S_0^s) = \varphi(s)$. □

To treat the case of general N we need the following lemma.

Lemma 8 *For an arbitrary upper-semicontinuous function φ the functions $Q\varphi$ and $R\varphi$ are also upper-semicontinuous.*

Proof. We can assume that $r = 0$ and consider only the transformation Q. Let $s_n \to s \in E^+$ and f be a continuous function such that $\varphi \leq f$. For any $\varepsilon > 0$ and any n there exists ν_n such that

$$Q\varphi(s_n) - \varepsilon \leq \int_Y f(F(s_n,y))\nu_n(dy),$$

and consequently

$$\limsup_n Q\varphi(s_n) - \varepsilon \leq \int_Y f(F(s,y))\nu_\infty(dy),$$

where ν_∞ is the weak limit of ν_n whose existence can be assumed without any loss of generality. Taking into account that φ is the lower envelope of continuous functions f we have

$$\limsup_n Q\varphi(s_n) - \varepsilon \leq \int_Y \varphi(F(s,y))\nu_\infty(dy),$$

Under the assumption (A2), $\nu_\infty \in U_s$ and consequently

$$\limsup_n Q\varphi(s_n) - \varepsilon \leq Q\varphi(s).$$

This implies the result. □

Proof of Theorem 4 (Continuation). Let

$$Z_n^{z,\pi} = \begin{pmatrix} X_n^{z,\pi} \\ S_n^s \end{pmatrix}, \qquad z = (x,s) \in \mathbb{R} \times E^+$$

be the position of the system under considerations at moments $n = 0, 1, \ldots$ and K a Borel subset of $\mathbb{R} \times E^+$. It is convenient, at this stage, to introduce the concepts of *viability sets and capture basins* (see [1]). Namely let $\text{Viab}_N(K)$ be the set of all initial states z which can be transferred to K in N steps. Similarly let $\text{Capt}_N(K)$ be the set of all initial states which can be kept in K with probability one for all $n \leq N$, by a suitable choice of a strategy π. More precisely,

$$\begin{aligned} \text{Viab}_N(K) &= \{z \in \mathbb{R} \times E^+; \, \exists_\pi \mathbb{P}(Z_N^{z,\pi} \in K) = 1\}, \\ \text{Capt}_N(K) &= \{z \in \mathbb{R} \times E^+; \, \exists_\pi \mathbb{P}(Z_n^{z,\pi} \in K, \, n = 0, 1, \ldots, N) = 1\}. \end{aligned}$$

It is clear that if $K = \text{epi } \varphi = \{(x,s); \, x \geq \varphi(s)\}$ is the epigraph of φ, then

$$O_E(s,N) = \inf\{x; \, (x,s) \in \text{Viab}_N(\text{epi } \varphi)\},$$

$$O_A(s, N) = \inf\{x; (x, s) \in \text{Capt}_N(\text{epi}\,\varphi)\}.$$

It follows from (10) that if $x' \geq x$ then for any $N \in \mathbb{N}$, any strategy π and any $s \in E^+$

$$X_N^{(x',s),\pi} \geq X_N^{(x,s),\pi} \quad \mathbb{P}\text{-a.s.}$$

Consequently for each N there exist functions φ_N, ψ_N such that

$$\{(x,s); x > \varphi_N(s)\} \subseteq \text{Viab}_N(\text{epi}\,\varphi) \subseteq \{(x,s); x \geq \varphi_N(s)\},$$
$$\{(x,s); x > \psi_N(s)\} \subseteq \text{Capt}_N(\text{epi}\,\varphi) \subseteq \{(x,s); x \geq \psi_N(s)\}.$$

It is clear that

$$\varphi_N(s) = O_E(N, s),$$
$$\psi_N(s) = O_A(N, s).$$

We have the following characterizations.

Theorem 9 *Under the assumptions of Theorem 4, for each $N \in \mathbb{N}$ and each $s \in E^+$,*

$$\{(x,s); x \geq Q^N \varphi(s)\} \supseteq \text{Viab}_N(\text{epi}\,\varphi) \supseteq \{(x,s); x > Q^N \varphi(s)\}, \tag{34}$$

$$\{(x,s); x \geq R^N \varphi(s)\} \supseteq \text{Capt}_N(\text{epi}\,\varphi) \supseteq \{(x,s); x > R^N \varphi(s)\}. \tag{35}$$

Proof. We only prove (34) as the proof of (35) is completely analogous. Let $x > Q^N \varphi(s)$. We have to construct a strategy π such that

$$X_N^{(x,s),\pi} \geq \varphi(S_N^s) \quad \mathbb{P}\text{-a.s.}$$

It follows from (32) and (33) that for an arbitrary upper-semicontinuous function f and arbitrary $\varepsilon > 0$ there exists $\sigma_\varepsilon \in E^+$ such that

$$\langle \sigma, z \rangle \geq [f(z + (1 + r)s) + \varepsilon] - (1 + r)\left[\bar{x} + \frac{2\varepsilon}{1 + r}\right],$$

where $\bar{x} = Qf(s)$, for all $z \in D_s$. Clearly σ_ε can be chosen as a measurable function of s and $y = \bar{x} + \frac{2\varepsilon}{1+r}$. Moreover,

$$X_1^{(y,s),\sigma_\varepsilon} \geq f(S_1^s) \quad \mathbb{P}\text{-a.s.}$$

Choosing consecutively

$$f = Q^{N-1}\varphi + \varepsilon, \quad f = Q^{N-2}\varphi + \varepsilon, \quad \ldots, \quad f = \varphi + \varepsilon,$$

where $\varepsilon = \frac{1}{N}(x - Q\varphi(s)) > 0$, one can construct a strategy π such that

$$X_N^{(x,s),\pi} \geq \varphi(S_N^s) \quad \mathbb{P}\text{-a.s.}$$

Conversely, let N be the first natural number such that

$$\text{Viab}_N(\text{epi}\,\varphi) \not\subseteq \{(x,s); x \geq Q^N \varphi(s)\}$$

for an upper-semicontinuous function φ. It is clear that $N \geq 2$. Let

$$(x', s') \in \text{Viab}_N(\text{epi } \varphi) \quad \text{but} \quad x' < Q(Q^{N-1}\varphi)(s').$$

Then, for a set $K' \subset \{(y, t); y < Q^{N-1}\varphi(t)\}$,

$$\mathbb{P}((X_1^{(x',s'),\pi'}, S_1^{s'}) \in K') > 0.$$

However, for any strategy π and any $(y, t) \in K'$

$$\mathbb{P}((X_{N-1}^{(y,t),\pi} \geq \varphi(S_{N-1}^t)) < 1.$$

In particular

$$\begin{aligned}
\mathbb{P}(X_N^{(x',s'),\pi'} &\geq \varphi(S_N^{s'})) \\
&= \mathbb{P}(\{X_N^{(x',s'),\pi'} \geq \varphi(S_N^{s'})\} \cap \{(X_1^{(x',s'),\pi'}, S_1^{s'}) \in K'\}) \\
&\quad + \mathbb{P}(\{X_N^{(x',s'),\pi'} \geq \varphi(S_N^{s'})\} \cap \{(X_1^{(x',s'),\pi'}, S_1^{s'}) \notin K'\}) \\
&= I_1 + I_2.
\end{aligned}$$

But

$$\begin{aligned}
I_1 &< \mathbb{P}(X_1^{(x',s'),\pi'}, S_1^{s'}) \in K'), \\
I_2 &\leq \mathbb{P}(X_1^{(s',s'),\pi'}, S_1^{s'}) \notin K')
\end{aligned}$$

and therefore

$$\mathbb{P}(X_N^{(x',s'),\pi'} \geq \varphi(S_N^{s'})) < 1,$$

which is again a contradiction. $\square$

3. LINEAR MODELS

We assume that $K < +\infty$, $E = \mathbb{R}^K$, $E^+ = (0, +\infty)^K$, $Y \subset (-1, +\infty)^K$ and

$$F(s, y) = (\mathbb{I} + y)s, \quad s \in (0, +\infty)^K, \; y \in Y, \tag{36}$$

where $\mathbb{I}$ denotes a K-vector with all coordinates equal to 1 and the vectors in (36) are multiplied coordinatewise. Thus

$$S_{n+1} = F(S_n, \xi_{n+1}), \quad S_0 = s \tag{37}$$

and the prices of assets are governed by linear stochastic equations with correlated noises:

$$S_{n+1}^k = (1 + \xi_{n+1}^k)S_n^k, \quad S_0^k = s^k, \quad k = 1, \ldots, K; \quad n = 1, 2, \ldots. \tag{38}$$

Here $\xi_n = \begin{pmatrix} \xi_n^1 \\ \vdots \\ \xi_n^K \end{pmatrix}$ is a sequence of independent, identically distributed random variables

taking values in the set Y. The initial price vector $s = S_0$ is in $\mathbb{R}_+^K$ ($\mathbb{R}_+ = [0, +\infty)$).

From now on we will require the contingent claim φ to be an *upper-semicontinuous* function. Since

$$M_s = \left\{ F^*(s, \nu); \quad \nu \in \mathcal{P}(Y), \int_Y (\mathbb{I} + y)s\, \nu(dy) = (1 + r)s \right\}, \tag{39}$$

the sets U_s^e are identical with the set $\tilde{U}^e$ of extremal points of

$$\tilde{U} = \{\nu \in U : \int_Y z\nu(dz) = r\mathbb{I}\}. \tag{40}$$

If, in addition, the number of elements in Y is L and :

$$Y = \left\{ \begin{pmatrix} a_1^1 \\ \vdots \\ a_K^1 \end{pmatrix}, \dots, \begin{pmatrix} a_1^L \\ \vdots \\ a_K^L \end{pmatrix} \right\} = \{a^1, \dots, a^L\},$$

then

$$\tilde{U} = \left\{ (p_1, \dots, p_L) : \sum_{\ell=1}^{L} p_\ell a^\ell = r\mathbb{I}, \sum_{\ell=1}^{L} p_\ell = 1, \; p_\ell \geq 0, \; \ell = 1, \dots, L \right\}. \tag{41}$$

Since $\tilde{U}$ is a polyhedron, the set $\tilde{U}^e$ is *finite* and consists of all vertices of $\tilde{U}$. It follows from Theorem 2 that $p_j \neq 0$ for at most $K + 1$ of $j = 1, \dots, L$. If $u = (p_1, \dots, p_L) \in \tilde{U}^e$ we have:

$$P^u(s, (\mathbb{I} + a^\ell)s) = p_\ell, \quad \ell = 1, \dots, L, \quad s \in \mathbb{R}_+^K. \tag{42}$$

We will now discuss several special cases.

3.1. Cox-Ross-Rubinstein Model

By assumption only one type of assets exists and, moreover, its prices are assumed to have only two possible rates of change. Consequently we have $K = 1$, $L = 2$, $Y = \{a, b\}$ with $-1 < a < r < b$. Then the set:

$$\tilde{U} = \{(p_1, p_2) : p_1 \geq 0, \; p_2 \geq 0, \; p_1 + p_2 = 1, \; p_1 a + p_2 b = r\}$$

consists of only one vector $u = \left(\frac{b-r}{b-a}, \frac{r-a}{b-a} \right)$. The control problem introduced in Section 2 becomes trivial and the general formulae for prices of options reduce to the well known ones.

3.2. Multinomial Single-Asset Case

We again consider a single asset but allow its prices to have more then two possible rates of increment. For instance take $K = 1$ and $L = 3$; $Y = \{a^1, a^2, b\}$ with $-1 < a^1 < a^2 < r < b$. Then:

$$\tilde{U} = \left\{ (p_1, p_2, q) : p_1, p_2, q \geq 0, \; p_1 + p_2 + q = 1, \; p_1 a^1 + p_2 a^2 + qb = r \right\}$$
$$= \left\{ p_1, p_2 \geq 0, \; p_1 + p_2 \leq 1, \; p_1 \frac{b - a^1}{b - r} + p_2 \frac{b - a^2}{b - r} = 1, \; q = 1 - p_1 - p_2 \right\}.$$

Therefore the set $\tilde{U}$ is an interval and its extremal points are:

$$\tilde{U}^e = \left\{ \left(\frac{b-r}{b-a^2}, 0, \frac{r-a^1}{b-a^1} \right), \left(0, \frac{b-r}{b-a^2}, \frac{r-a^2}{b-a^2} \right) \right\} = \{u_1, u_2\}.$$

The control problem introduced in Section 2 has the state space $\mathbb{R}_+$ and two control parameters corresponding to the following transition probabilities:

$$P^1(s, (1 + a^1)s) = \frac{b - r}{b - a^1}, \qquad P^2(s, (1 + a^2)s) = \frac{b - r}{b - a^2}.$$

The following generalization was implicit in [21].

Proposition 10 *Assume that*

$$K = 1, \quad L = I + J, \quad Y = \{a^1, \ldots a^I, b^1, \ldots, b^J\},$$

where $-1 < a^1 < \ldots < a^I < r < b^1 < \ldots < b^J$, *then the set V of extremals of the corresponding set $\tilde{U}$ consists of $I \times J$ elements of the form:*

$$V = \{(0, \ldots, 0, \widehat{p}_{i,j}, 0, \ldots, \tilde{p}_{i,j}, 0, \ldots, 0) : (i, j) \in I \times J\}$$

where $\widehat{p}_{i,j} = \dfrac{b^j - r}{b^j - a^i}$ *is placed at the i coordinate and* $\tilde{p}_{i,j} = \dfrac{r - a^i}{b^j - a^i}$ *is placed at the coordinate* $I + j$. *The corresponding transition probabilities are:*

$$P^{i,j}(s, (1 + a^i)s) = \widehat{p}_{i,j}, \qquad P^{i,j}(s, (1 + b^j)s) = \tilde{p}_{i,j}.$$

3.3. Multi-Asset Case

We now allow the market to include more then one asset. We initially assume that two types of assets are present, each with two possible rates of price change. Therefore $K = 2, L = 4$ and

$$Y = \left\{ \begin{pmatrix} a \\ c \end{pmatrix}, \begin{pmatrix} b \\ d \end{pmatrix}, \begin{pmatrix} a \\ d \end{pmatrix}, \begin{pmatrix} b \\ c \end{pmatrix} \right\},$$

where $-1 < a < r < b$ and $-1 < c < r < d$. In this case $\tilde{U}$ is the set of all vectors (p_1, p_2, p_3, p_4) in $\mathbb{R}_+^4$ verifying

$$\begin{cases} p_1 a + p_2 b + p_3 a + p_4 b = r, \\ p_1 c + p_2 d + p_3 d + p_4 c = r, \\ p_1 + p_2 + p_3 + p_4 = 1, \\ p_i \geq 0, \quad i = 1, 2, 3, 4, \end{cases}$$

or equivalently, by easy computations:

$$\begin{cases} p_2 + p_4 = \varrho, \\ p_2 + p_3 = \vartheta, \\ p_1 = 1 + p_2 - \varrho - \vartheta, \\ p_i \geq 0, \quad i = 1, 2, 3, 4, \end{cases} \qquad \text{where } \varrho = \frac{r - a}{b - a}, \qquad \vartheta = \frac{r - c}{d - c}.$$

For the sake of brevity we focus only on the case where $\varrho \leq \vartheta$. The set $\tilde{U}$ is again a segment with the following extremal points:

$$(1 - \vartheta, \varrho, \vartheta - \varrho, 0) \text{ and } (0, \vartheta + \varrho - 1, 1 - \varrho, 1 - \vartheta) \quad \text{if } \varrho + \vartheta \geq 1,$$

$(1 - \vartheta, \varrho, \vartheta - \varrho, 0)$ and $(1 - \varrho - \vartheta, 0, \vartheta, \varrho)$ if $\varrho + \vartheta \geq 1$.

Note that one of the coordinates of the extremal points is zero in agreement with Theorem 2. The control problem introduced in Section 2 has the state space $\mathbb{R}_+ \times \mathbb{R}_+$, and two control parameters corresponding to the following transition probabilities:

$$P^1\left(\begin{pmatrix} s^1 \\ s^2 \end{pmatrix}, \begin{pmatrix} (1+a)s^1 \\ (1+c)s^2 \end{pmatrix}\right) = 1 - \vartheta,$$

$$P^1\left(\begin{pmatrix} s^1 \\ s^2 \end{pmatrix}, \begin{pmatrix} (1+b)s^1 \\ (1+d)s^2 \end{pmatrix}\right) = \varrho,$$

$$P^1\left(\begin{pmatrix} s^1 \\ s^2 \end{pmatrix}, \begin{pmatrix} (1+b)s^1 \\ (1+d)s^2 \end{pmatrix}\right) = \vartheta + \varrho - 1,$$

if $\vartheta + \varrho \geq 1$ or

$$P^2\left(\begin{pmatrix} s^1 \\ s^2 \end{pmatrix}, \begin{pmatrix} (1+a)s^1 \\ (1+d)s^2 \end{pmatrix}\right) = \vartheta,$$

$$P^2\left(\begin{pmatrix} s^1 \\ s^2 \end{pmatrix}, \begin{pmatrix} (1+b)s^1 \\ (1+c)s^2 \end{pmatrix}\right) = \varrho,$$

$$P^2\left(\begin{pmatrix} s^1 \\ s^2 \end{pmatrix}, \begin{pmatrix} (1+a)s^1 \\ (1+c)s^2 \end{pmatrix}\right) = 1 - \vartheta - \varrho,$$

if $\vartheta + \varrho < 1$.

3.4. Continuously Distributed Noise

For simplicity we assume that $K = 1$ and the set Y is an interval $[a, b]$ such that $-1 < a < r < b$. By Theorem 2 the measures in $\tilde{U}^e$ have two point supports and are of the form:

$$(y - z)(z - r)^{-1}\delta_y + (y - z)(r - y)^{-1}\delta_z, \quad a \geq y \geq r \geq z \geq b.$$

The transition probabilities of the new controlled Markov chain are very simple,

$$P^{y,z}(s, \{(1+y)s\}) = (y - z)(z - r)^{-1} = 1 - P^{y,z}(s, \{(1+z)s\}).$$

If, in addition, φ is a convex function, then the operator Q acts on φ as a linear operator:

$$Q\varphi(s) = \varphi((1+a)s)(b-r)(b-a)^{-1} + \varphi((1+b)s)(r-a)(b-a)^{-1}.$$

4. HEDGING WITH CONSTRAINTS

We now show that the theory can easily be adapted to the situation where some additional constraints on the self-financing strategies are imposed. We limit ourselves to the linear model described in Section 3. We assume, for instance, that, at any moment, the agent is obliged to possess a non-negative amount on his bank account and that his portfolio has to contain non-negative quantities of all types of the shares. Those are the so called *non borrowing* constraints. They can be expressed (see [21]) as constraints on the portfolio σ, depending on the current capital x and the current vector of share prices s in the following manner:

$$\langle \sigma, -(r+1)\mathbb{I}s \rangle \geq -(1+r)x, \qquad \langle \sigma, e_i s \rangle \geq 0$$

where $e_1, \ldots e_K$ is the canonical basis in $\mathbb{R}^K$.

We treat the case of more general constraints:

$$\langle \sigma, v_\nu s \rangle \geq -(1+r)x, \quad \langle \sigma, w_\mu \rangle \geq 0, \quad \nu = 1, \ldots, N_1, \quad \mu = 1, \ldots, N_2, \tag{43}$$

where v_ν, w_μ, $\nu = 1, \ldots, N_1, \mu = 1, \ldots, N_2$ are vectors in $\mathbb{R}^K$. We assume that the vectors w_μ, $\mu = 1, \ldots, N_2$, are linearly independent. We define:

$$\tilde{U} = \Big\{ (p_1, \ldots, p_L) \text{ with } p_\ell \geq 0, \ \ell = 1, \ldots, L \text{ such that}$$

$$\text{there exists } q_1, \ldots, q_{N_1}, \pi_1, \ldots, \pi_{N_2} \text{ with } q_\nu \geq 0, \ \pi_\mu \geq 0, \text{ verifying} \tag{44}$$

$$\sum_{\ell=1}^{L} p_\ell(a^\ell - r\mathbb{I}) + \sum_{\nu=1}^{N_1} q_\nu v_\nu + \sum_{\mu=1}^{N_2} \pi_\mu w_\nu = 0, \quad \sum_{\ell=1}^{L} p_\ell + \sum_{\nu=1}^{N_1} q_\nu = 1 \Big\}.$$

As it can easily be verified $\tilde{U}$ is a compact polyhedron. Let V be the finite set of its extremal points. If $u = (p_1, \ldots, p_L) \in V$ we can define a transition kernel $P^u(\cdot, \cdot)$ on $\mathbb{R}_+^K$ by the formula:

$$\tilde{P}^u(s, (\mathbb{I} + a^\ell)s) = p_\ell, \quad \ell = 1, \ldots, L, \quad s \in \mathbb{R}_+^K,$$

$$\tilde{P}^u(s, \partial) = 1 - \sum_{\ell=1}^{L} p_\ell; \qquad P^u(\partial, \partial) = 1, \tag{45}$$

where ∂ is an auxiliary absorbing state. Now, as in Section 2, we introduce a stochastic control model with the set V of control parameters, the state space $\mathbb{R}_+^K \cup \{\partial\}$ and with the dynamics determined by the kernels: $(P^u, u \in V)$. By $\tilde{S}_n^{s,\zeta}$, $n = 0, 1, \ldots$ we denote the controlled process starting from s and corresponding to the V-valued strategy $\zeta = (\zeta_1, \zeta_2, \ldots, \zeta_{N-1})$. Finally let $\tilde{\mathcal{O}}_E(N, s)$ and $\tilde{\mathcal{O}}_A(N, s)$ be defined as $\mathcal{O}_E(N, s)$ and $\mathcal{O}_A(N, s)$ but with the additional requirement that the hedging strategy $\pi = (\sigma_0, \sigma_1, \ldots)$ verifies the constraints (43). The following is the analogue of Theorem 5.

Theorem 11 *If the set $\tilde{U}$ given by (44) is non-empty, then*

$$\tilde{\mathcal{O}}_E(N, s) = \sup_\zeta \mathbb{E}\left(\frac{1}{(1+r)^N} \varphi(\tilde{Z}_N^{s,\zeta}) \right), \tag{46}$$

$$\tilde{\mathcal{O}}_A(N, s) = \sup_{\zeta, \tau} \mathbb{E}\left(\frac{1}{(1+r)^\tau} \varphi(\tilde{Z}_\tau^{s,\zeta}) \right). \tag{47}$$

If the set $\tilde{U}$ is empty then $\mathcal{O}_E(N, s) = \mathcal{O}_A(N, s) = -\infty$. In the above formula, $\varphi(\partial) = 0$.

Proof. If we take into account the constraints (43), we arrive at

$$\langle \sigma, (a^\ell - r\mathbb{I})s \rangle \geq \varphi((\mathbb{I} + a^\ell)s) - (1+r)x, \quad \ell = 1, \ldots, L,$$

$$\langle \sigma, v_\nu s \rangle \geq -(1+r)x, \langle \sigma, w_\mu \rangle \geq 0, \qquad \nu = 1, \ldots, N_1, \tag{48}$$

$$\langle \sigma, w_\mu \rangle \geq 0, \qquad \qquad \mu = 1, \ldots, N_2.$$

Using Proposition 7 in the same way as in the proof of Theorem 4 the above inequalities hold if and only if for all $p_i \geq 0$, $i = 1, \ldots, L$; $q_\nu \geq 0$, $\nu = 1, \ldots, N_1$; $\pi_\mu \geq 0$, $\mu = 1, \ldots, N_2$ verifying

$$\sum_{i=1}^{L} p_\ell(a^\ell + r\mathbb{I}) + \sum_{\nu=1}^{N_1} q_\nu v_\nu + \sum_{\mu=1}^{N_2} \pi_\mu w_\mu = 0,$$

it must hold

$$\sum_{i=1}^{L} p_\ell \varphi((\mathbb{I} + a^\ell)s) - (1+r) \left[\sum_{i=1}^{L} p_\ell + \sum_{\nu=1}^{N_1} q_\nu \right] x \leq 0.$$

Since by renormalization we can always assume that $\sum_{i=1}^{L} p_\ell + \sum_{\nu=1}^{N_1} q_\nu = 1$ the above relation can be rewritten

$$x \geq (1+r)^{-1} \sum_{i=1}^{L} p_\ell \varphi((\mathbb{I} + a^\ell)s),$$

the rest of the proof is identical to that of Theorem 4. Since $\sum_{i=1}^{L} p_\ell$ can be smaller than 1, we have to allow for the possibility that the system leaves the state space and falls into the auxiliary absorbing state ∂. $\qquad\qquad\square$

5. PRICING BY EQUIVALENT MARTINGALE MEASURES

5.1. Finite Number of Assets

We fix the number of steps $N \geq 1$ and define $\Omega = E^+ \times Y^N$. Thus any $\omega \in \Omega$ is of the form $\omega = (s_0, y_1, \ldots, y_N)$. For $\omega = (s_0, y_1, \ldots, y_N)$ and all $n = 1, \ldots, N$ we define

$$S_0(\omega) = s_0, \quad S_n(\omega) = F(S_{n-1}(\omega), y_n), \quad \xi_n(\omega) = y_n,$$
$$\mathcal{F}_n = \sigma(S_0, \xi_1, \ldots, \xi_n).$$

If μ_0 is the distribution of share prices at time 0 and ν is the distribution of the random variables ξ_n in the model (7) then one can take $(\Omega, \mathcal{F}_N, \mathbb{P})$ with $\mathbb{P} = \mu_0 \times \prod_1^N \nu$ as the basic probability space. A probability measure $\tilde{\mathbb{P}}$ on $\mathcal{F}_N$ is an equivalent martingale measure if it is equivalent to $\mathbb{P}$ and, for each $n = 1, \ldots, N - 1$,

$$E_{\tilde{\mathbb{P}}}(F(S_n, \xi_{n+1})|\mathcal{F}_n) = (1+r)S_n \quad \mathbb{P}\text{-a.s.} \tag{49}$$

An arbitrary probability measure $\tilde{\mathbb{P}}$ on $(\Omega, \mathcal{F}_N)$ is uniquely determined by an initial measure μ_0 and a sequence of transition kernels u_n, $n = 1, \ldots, N$, from $E^+ \times Y^n$ into the space $\mathcal{P}(Y)$ of probability measures on Y. The value $u_n(\Gamma; s_0, y_1, \ldots, y_n)$ is the conditional probability, with respect to $\tilde{\mathbb{P}}$, that ξ_n will be in the set Γ given that the random variables $S_0, \xi_1, \ldots, \xi_{n-1}$ took values $s_0, y_1, \ldots, y_{n-1}$. The measure $\tilde{\mathbb{P}}$ described in this way will be denoted by $P(u_0, u_1, \ldots, u_N)$ and its restriction to $E^+ \times Y^n$ by $P_n(u_0, u_1, \ldots, u_n)$.

A necessary and sufficient condition for the martingale property (49) to hold can be expressed in terms of the kernels u_n. Namely (49) holds if and only if, for any $n = 1, \ldots, N$,

$$\int_Y F(S_{n-1}, y_n) u_n(dy_n; s_0, y_1, \ldots, y_{n-1}) = (1+r)S_{n-1} \quad \mu_0 u_1 \ldots u_{n-1}\text{-a.s.,} \tag{50}$$

where in (50), S_{n-1} stands for $S_{n-1}(s_0, y_1, \ldots, y_{n-1})$. Moreover, we have that the measure $P(\mu_0, u_1, \ldots, u_N)$ is equivalent to $\mathbb{P}$ if and only if

$$\text{for } n \leq N, \text{ the measures } u_n \text{ are } P_n(\mu_0, u_1, \ldots, u_n)\text{-a.s. equivalent to } \nu. \tag{51}$$

It is not difficult to see that the set of all equivalent martingale measures can be identified with the set of all sequences $u_1, \ldots, u_N$ of transition kernels for which (50) and (51) hold. Some strategies introduced for (14) can therefore be identified with equivalent martingale measures, however, the converse statement is not always true. As we will show there are several finite dimensional cases in which equivalent martingale measures do not exist but the control system (14) is well defined. Thus the control theoretic formulae (5) and (6) might define the option prices, whereas the martingale measure type formulae (2) and (3) loose their validity.

Note that the condition (50) looks specially simple for linear models (compare (39) and (40)). In that case (50) is independent of S_{n-1} and reads

$$\int_Y y u_n(dy) = r\mathbb{I}.$$

However it might happen that in the set $\mathbb{M}$ consisting of all probability measures u with support Y and such that

$$\int_Y y u(dy) = r\mathbb{I}$$

there can be no measures equivalent to the measure ν defining $\mathbb{P}$. It might also happen that there are no equivalent measures to ν among extremals. Here are illustrative examples.

Example 12 Consider the situation of Section 3.3 with two assets and two possible prices. That is,

$$Y = \left\{ \begin{pmatrix} a \\ c \end{pmatrix}, \begin{pmatrix} b \\ d \end{pmatrix}, \begin{pmatrix} a \\ d \end{pmatrix}, \begin{pmatrix} b \\ c \end{pmatrix} \right\}$$

where $-1 < a < r < b$ and $-1 < c < r < d$. In this case $\mathbb{M}$ is the set of all vectors (p_1, p_2, p_3, p_4) in $\mathbb{R}_+^4$ verifying:

$$\begin{cases} p_2 + p_4 = \varrho, \\ p_2 + p_3 = \vartheta, \\ p_1 = 1 + p_2 - \varrho - \vartheta, \\ p_i > 0, \quad i = 1, 2, 3, 4, \end{cases} \qquad \text{where } \varrho = \frac{r-a}{b-a}, \qquad \vartheta = \frac{r-c}{d-c},$$

(where, for brevity, we again assume that $\varrho \leq \vartheta$). The set $\tilde{U}^e$ consists only of the two extremal points:

$$(1 - \vartheta, \varrho, \vartheta - \varrho, 0) \text{ and } (0, \vartheta + \varrho - 1, 1 - \varrho, 1 - \vartheta) \quad \text{if } \varrho + \vartheta \geq 1$$

$$(1 - \vartheta, \varrho, \vartheta - \varrho, 0) \text{ and } (1 - \varrho - \vartheta, 0, \vartheta, \varrho) \quad \text{if } \varrho + \vartheta < 1.$$

Example 13 Consider the same situation as above but now with $r = a$. In this case $\mathbb{M}$ is empty. On the contrary $\tilde{U}^e$ just trivializes to the point $(1 - \vartheta, 0, \vartheta, 0)$.

5.2. Infinite Number of Assets

If the number of assets is infinite then, as a rule, an equivalent martingale measure does not exist and if this is the case the classical formulae for option prices (2) and (3) are meaningless.

Consider, for instance, an infinite dimensional linear model on the space $E = l^2$ with $E^+ = l_+^2$ consisting of all positive sequences in E and

$$F(s, y) = (\mathbb{I} + y)s, \quad s \in l_+^2, \ y \in Y,$$

where Y is the infinite Cartesian product of two point sets $\{a_n, b_n\}$,

$$a_n = r + \gamma_n, \quad b_n = r - \gamma_n.$$

The numbers γ_n are positive and smaller than r. Then, by finite dimensional considerations, the martingale measure should be an infinite product of $1/2(\delta_{\{a_n\}} + \delta_{\{b_n\}})$. If, however, the objective measure describing the one step noise is a product of measures which associate with a_n, b_n probabilities p_n, $1 - p_n$, respectively then, by the Kakutani theorem, the martingale measure is equivalent to the objective one if and only if

$$\sum_n^{+\infty} (1/2 - p_n)^2 < +\infty,$$

which is a very restrictive condition.

REFERENCES

[1] J.P. Aubin, *Viability theory*, Birkhäuser, 1996.

[2] J. Cvitanic, Optimal trading under constraints, *Proceedings CIME Conference, Bressanone 1996*, edited by W. Runggaldier, Lecture Notes in Math. **1656**, Springer-Verlag, 1997.

[3] J. Cvitanic, H. Pham and N. Touzi, A closed-form solution to the problem of super-replication under transaction costs, *Finance Stoch.*, to appear.

[4] M.V. Dothan, *Prices in financial markets*, Oxford University Press, 1990.

[5] D. Duffie, *Asset pricing theory*, Princeton University Press, 1992.

[6] N. Dunford and J.T. Schwartz, *Linear operators*, Part I, Interscience Publishers, New York and London, 1958.

[7] N. El Karoui and M.-C. Quenez, Dynamic programming and pricing of contingent claims in an incomplete market, *SIAM J. Control Optimization* **33** (1995), 27–66.

[8] H. Föllmer and Yu.M. Kabanov, Optional decomposition and Lagrange multipliers, *Finance Stoch.* **2** (1998), 69–81.

[9] R. Frey, *Superreplication in stochastic volatility models and optimal stopping*, Manuscript.

[10] J. Harrison and D. Kreps, Martingales and arbitrage in multiperiod security markets, *J. Econ. Theory* **20** (1979), 381–408.

[11] I. Karatzas, *Lectures on the mathematics of finance*, CRM Monograph Series **8**, American Mathematical Society, Providence (Rhode Island), 1996.

[12] D.O. Kramkov, Optional decomposition of supermartingales and hedging contingent claims in incomplete security markets, *Probab. Theory Relat. Fields* **105** (1996), 459–479.

[13] K. Kuratowski, *Topology*, Volumes I and II, New York, 1966 and 1968.

[14] M. Motoczynski and L. Stettner, On option pricing in multidimensional Cox-Ross-Rubinstein model, *Applicationes Mathematicae* **25** (1998), 55–72.

[15] M. Musiela and M. Rutkowski, *Martingale methods in financial modelling*, Springer-Verlag, 1997.

[16] H. Pham, *Imperfection de marché et méthodes d'évaluation et couverture d'options*, Notes de cours, Scuola Normale Superiore, Pisa, 1998.

[17] A.N. Shiryaev, *Essentials of stochastic finance*, World Scientific, Hongkong, 1999.

[18] A.N. Shiryaev, J.M. Kabanov, D.O. Kramkov, A.B. Melnikow, On the theory of pricing European and American options I. Discrete time, *Theory Probab. Appl.* **39** (1994), 23–79 (in Russian).

[19] M. Taqqu & W. Willinger, The analysis of finite security markets using martingales, *Adv. Appl. Probab.* **19** (1987), 1–25.

[20] G. Tessitore and J. Zabczyk, Pricing options and Bellman's principle, Preprint della Scuola Normale Superiore, Pisa, 1995.

[21] G. Tessitore and J. Zabczyk, Pricing options for multinomial models, *Bull. Pol. Acad. Sci., Math.* **44** (1996), 363–380.

[22] G. Tessitore and J. Zabczyk, Pricing options for multinomial models and multiasset models, Preprint of the Institute of Mathematics, Polish Academy of Sciences, Warszawa, 1999.

[23] N. Touzi, *Problèmes de control optimal en finance*, Habilitation, Paris IX Dauphin, 1999.

[24] J. Zabczyk, Pricing options by dynamic programming, *Stochastic Processes and Related Topics*, 153–160, H.J. Engelbert, H. Föllmer and J. Zabczyk (Eds.), Gordon and Breach, 1996.

[25] J. Zabczyk, *Chance and decision*, Quaderni della Scuola Normale Superiore, Pisa, 1996.

THREE INTERTWINED BROWNIAN TOPICS: EXPONENTIAL FUNCTIONALS, WINDING NUMBERS, AND RAY–KNIGHT THEOREMS ON LOCAL TIME

MARC YOR

Université Pierre et Marie Curie
Laboratoire de Probabilités
F-75252 Paris, France

RESUMÉ OF LECTURES

The following notes describe very succinctly the contents of the lectures I gave during the Winter school. I discussed a number of results about Brownian motion, concerning (in that order):

a) **Exponential Functionals**, i.e., functionals of the type

$$A_t^{a,b} = \int_0^t ds \exp(aB_s + bs),$$

where B is a Brownian motion and a and b are real constants.

b) **Ray–Knight Theorems for Local Times**, i.e., the description of the law of

$$\{\ell_T^x; x \in \mathbf{R}\},$$

where $\{\ell_t^x\}$ denotes the family of Brownian local times and T is a particular random time.

c) **Some Asymptotic Laws of Brownian Windings** around a finite number of points.

In each case, I tried to present the results in a form which could be suitably extended to an adequate class of Lévy processes, in replacement of Brownian motion.

The discussion of *Topic* **a)** was largely motivated by the problem of computing the payoff of a continuously averaged Asian option in a Black–Scholes setting, that is, to obtain an explicit expression of $E[(A_t^{a,b} - k)^+]$.

A related problem, which, during the nineties, has been solved by a number of authors, among whom Paulsen [2] (1993), and his coauthors Gjessing and Nilsen, is the computation of the laws of perpetuities of the form

$$Z \overset{\text{def}}{=} \int_0^\infty \exp(\xi_{s-})d\eta_s,$$

where (ξ, η) is a two-dimensional Lévy process such that Z is well-defined.

Introducing the generalized Ornstein–Uhlenbeck process

$$X_t \equiv X_t^x(\xi, \eta) \overset{\text{def}}{=} \exp(\xi_t)\left(x + \int_0^t \exp(-\xi_{s-})d\eta_s\right), \quad t \geq 0,$$

which is a homogeneous Markov process, allows to view the distribution of Z as the unique invariant probability measure of X. See, e.g., Carmona, Petit and Yor [1] for some applications.

To introduce *Topic* **b)** in a manner which is akin to the above discussion of *Topic* **a)**, one may remark that the generalized process

$$\tau_t^{(\xi,\eta)} : \quad f \longrightarrow \int_0^t f(\xi_t - \xi_{s-})d\eta_s \quad (f \quad \text{Borel bounded, ...})$$

is an homogeneous Markov process. The particular case when $\eta_s \equiv s$ is of special interest since, assuming that the local times $\{\ell_t^a\}$ of ξ exist, for the corresponding measure-valued process one has:

$$\tau_t^{(\xi)}(dx) = \ell_t^{\xi_t - x}dx.$$

Furthermore, time-changing $\tau^{(\xi)}$ with either

$$\lambda_u = \inf\{t : \ell_t^0 > u\}, \ u \geq 0, \quad \text{or} \quad \sigma(a) = \inf\{t : \xi_t \geq a\}, \ a \geq 0,$$

(in which case we assume that ξ does not have positive jumps) yields two interesting Markov processes $(\tau_{\lambda_u}^{(\xi)}, u \geq 0)$ and $(\tau_{\sigma(a)}^{(\xi)}, a \geq 0)$, for which the Ray–Knight theorems (in the case ξ is Brownian motion, say) describe the laws of their marginals for fixed u and a.

By now, extensive generalisations of the Ray–Knight theorems for Brownian local times have been obtained, e.g.:

– For one-dimensional diffusions, using time and space changes of variables.

– For perturbed Brownian motions (see Carmona, Petit and Yor [3]) which although not Markovian by themselves retain enough of the Markov property to transfer it to their local time processes.

– The genealogical structure of a continuous state branching process with, or without, immigration (in short, CB(I)) may be coded by a so-called height process, whose local time process is itself a CB(I) process; see the works of Le Gall and Le Jan [4] and Lambert [5].

– In [6], the authors show in particular that for a regular Lévy process with local times $\{L_t^x\}$, denoting $\lambda_u = \inf\{t : L_t^0 > u\}$, there exists a centered, stationary Gaussian process

$\{G_x, x \in \mathbf{R}\}$ independent of L such that

$$\{L_{\lambda_u}^x + G_x^2; x \in \mathbf{R}\} \overset{(law)}{=} \{(G_x + \sqrt{u})^2; x \in \mathbf{R}\}.$$

In the case when X is Brownian motion, this result yields the classical Ray–Knight theorem, since then $(G_x, x \in \mathbf{R})$ is a Brownian motion; more generally, if X is the symmetric stable Lévy process with parameter $\alpha > 1$, then $\{G_x; x \in \mathbf{R}\}$ satisfies

$$E[(G_x - G_y)^2] = c_\alpha \mid x - y \mid^{\alpha - 1}, \qquad (x, y \in \mathbf{R}),$$

i.e., G is a fractional Brownian motion.

 – Hu and Warren [7] consider the stochastic bifurcation model of Bass and Burdzy, which consists of the family, indexed by $x \in \mathbf{R}$, of solutions

$$X_t(x) = x + B_t + \int_0^t ds\, b(X_s(x)),$$

where $b(y) = \beta_1 1_{(y \le 0)} + \beta_2 1_{(y > 0)}$.
 They obtain Ray–Knight theorems for the family $(\{L_t^x\}, x \in \mathbf{R})$ of semimartingale local times, each of them being defined *at level* 0, of the solution $(X_t(x); t \ge 0)$.

To introduce *Topic* c), I first go back to Topic a), only to mention that, a priori, exponential functionals and winding numbers of, respectively, linear Brownian motion and planar Brownian motion seem to be two very far removed subjects; in fact, this is not really so, as may be argued by considering the skew-product representation of planar Brownian motion $(Z_t, t \ge 0)$:

$$Z_t = \mid Z_t \mid \exp(i\vartheta_t) \equiv \exp\{\beta_u + i\gamma_u\}\Big|_{u = \int_0^t \frac{ds}{|Z_s|^2}},$$

where $(\vartheta_t, t \ge 0)$ denotes the (continuous) winding number of Z, and $\beta + i\gamma$ is another planar Brownian motion; one sees that the clock

$$\left(\int_0^t \frac{ds}{\mid Z_s \mid^2}, \; t \ge 0\right)$$

is the inverse process of the exponential

$$\left(\int_0^u ds\, \exp(2\beta_s), \; u \ge 0\right).$$

As discussed in, e.g., [12], many consequences of this remark may be drawn for either *Topic* a) or c), including in particular the asymptotic distribution for $(\vartheta_t, \; t \to \infty)$, due originally to Spitzer (1958).
 Adequate extensions of this result to a finite number of points are the subject of [8] (see also [9] for a bigger picture, related to polymer studies).
 S. Watanabe [10] obtained further generalizations to stochastic line integrals along Brownian paths on Riemann surfaces.
 However, these studies only concern the homological aspects of Brownian behaviour, whereas Gruet [11] obtains some asymptotic results about the length of the word associated to the homotopy class of the Brownian path.

REFERENCES

Topic a):

[1] Ph. Carmona, F. Petit, M. Yor, Exponential Functionals of Lévy processes, in: *Lévy Processes: Theory and Applications*, O. Barndorff-Nielsen, T. Mikosch, S. Resnick (Eds.), Birkhäuser, 2001.

[2] J. Paulsen, Risk theory in a stochastic economic environment, *Stochastic Processes Appl.* **46** (1993), 327–361.

Topic b):

[3] Ph. Carmona, F. Petit, M. Yor, Some extensions of the arc-sine law as partial consequences of the scaling property of Brownian motion, *Probab. Theory Relat. Fields* **100** (1994), 1–29.

[4] J.F. Le Gall, Y. Le Jan, Branching processes in Lévy processes: the exploration process, *Ann. Probab.* **26** (1998), 213–252.

[5] A. Lambert, The genealogy of branching processes with immigration, to appear in *Probab. Theory Relat. Fields* (2001).

[6] N. Eisenbaum, H. Kaspi, M.B. Marcus, J. Rosen, Z. Shi, A Ray–Knight theorem for symmetric Markov processes, to appear in *Ann. Probab.* (2001).

[7] Y. Hu, J. Warren, Ray–Knight theorems related to a stochastic flow, *Stochastic Processes Appl.* **86**, n°2 (2000), 287–306.

Topic c):

[8] J. Pitman, M. Yor, (i) Asymptotic laws of planar Brownian motion, *Ann. Probab.* **14** (1986), 733–799; (ii) Further asymptotic laws of planar Brownian motion, *Ann. Probab.* **17** (1989), 965–1011.

[9] Y. Hu, M. Yor, Asymptotic studies of Brownian functionals, *Bolyai Society Mathematical Studies*, Vol. **9**, "Random Walks", 187–217, Budapest, 1999.

[10] S. Watanabe, Asymptotic windings of Brownian motion paths on Riemann surfaces, *Preprint* (1998), to appear in a volume dedicated to Professor T. Hida.

[11] J.-C. Gruet, On the length of the homotopic Brownian word in the thrice punctured sphere, *Probab. Theory Relat. Fields* **111** (1998), 489–516.

[12] M. Yor, From planar Brownian windings to Asian options, *Insurance (Math. and Economics)* **13** (1993), 23–34.

LIST OF PARTICIPANTS

R. Buckdahn
Université de Bretagne Occidentale, Facultés des Sciences et Techniques, Départment de Mathématiques, 6, Av. le Gorgeu, B.P. 809, F-29285 Brest Cedex, France
e-mail: rainer.buckdahn@univ-brest.fr

A. Cherny
Moscow State University, Faculty of Mechanics and Mathematics, Department of Probability Theory, 119899, Moscow, Russia
e-mail: a_cherny@mail.ru

F. Coquet
Université de Rennes 1, Institut Mathématique, Campus de Beaulieu, F-35042 Rennes, France
e-mail: coquet@univ-rennes1.fr

J. Creutzig
Friedrich-Schiller-Universität Jena, Fakultät für Mathematik und Informatik, Institut für Stochastik, Ernst-Abbe-Platz 1-4, D-07743 Jena, Germany
e-mail: creutzig@minet.uni-jena.de

P. Dencker
Universität Rostock, Fachbereich Mathematik, Universitätsplatz 1, D-18055 Rostock, Germany
e-mail: peter.dencker@stud.uni-rostock.de

H.-J. Engelbert
Friedrich-Schiller-Universität Jena, Fakultät für Mathematik und Informatik, Institut für Stochastik, Ernst-Abbe-Platz 1-4, D-07743 Jena, Germany
e-mail: engelbert@minet.uni-jena.de

S. Geiß
Department of Mathematics, University of Jyväskylä, P.O. Box 35 (MAD), FIN-40351 Jyväskylä, Finland
e-mail: geiss@math.jyu.fi

T. Goll
Universität Freiburg, Institut für Mathematische Stochastik, Eckerstr. 1, D-79104 Freiburg i. Br., Germany
e-mail: goll@stochastik.uni-freiburg.de

B. Grigelionis
Vilnius University, Institute of Mathematics and Informatics, Akademijos 4, Naugarduko 24, LT-2600 Vilnius, Lithuania
e-mail: jurgita@ktl.mii.lt

Yu. Kabanov
Université de Franche-Comté Besançon, Faculté de Sciences et des Techniques, Laboratoire de Mathématiques, 16 Route de Gray, F-25030 Besançon Cedex, France
e-mail: Kabanov@math.univ-fcmte.fr

A. Komisarski
Warsaw University, Faculty of Mathematics, Informatics and Mechanics, ul. Banacha 2, 02-097 Warsaw, Poland
e-mail: andkom@mimuw.edu.pl

W. Kurenok
Belarus State University, Department of Mathematics and Mechanics, 220050, Minsk-50, av. F. Skoriny, 4, Belarus
e-mail: Kurenok@mmf.bsu.unibel.by

S. Laue
Universität Kaiserslautern, Fachbereich Mathematik, Erwin-Schrödinger-Str., D-67663 Kaiserslautern, Germany
e-mail: Laue@mathematik.uni-kl.de

J. Leitner
Universität Konstanz, Fachbereich für Mathematik und Statistik, D-78457 Konstanz
e-mail: Johannes.Leitner@uni-konstanz.de

M. Mania
Georgian Academy of Sciences, A. Ramadze Mathematical Institute, M. Alexidze St. 1, 380093 Tbilisi, Georgian Republic
e-mail: mania@imath.acnet.ge

B. Markussen
University of Copenhagen, Department of Theoretical Statistics, Universitetsparken 5, 2100 Copenhagen OE, Denmark
e-mail: markusb@math.ku.dk

A. Matoussi
TU Berlin, Fachbereich Mathematik, Str. des 17. Juni 136, D-10623 Berlin, Germany
e-mail: matoussi@math.TU-Berlin.de

K. Mittmann
Friedrich-Schiller-Universität Jena, Fakultät für Mathematik und Informatik, Institut für Stochastik, Ernst-Abbe-Platz 1-4, D-07743 Jena, Germany
e-mail: mittmann@minet.uni-jena.de

E. Novak
Universität Erlangen, Mathematisches Institut, Bismarckstraße 1 1/2, D-91054 Erlangen, Germany (until September 2000) and
Friedrich-Schiller-Universität Jena, Fakultät für Mathematik und Informatik, Institut für Stochastik, Ernst-Abbe-Platz 1-4, D-07743 Jena, Germany
e-mail: novak@minet.uni-jena.de

I. Pavljukevitch
Humboldt-Universität zu Berlin, Fachbereich Mathematik, Institut für Stochastik, Unter den Linden 6, D-10099 Berlin, Germany
e-mail: Ilja.Pavljukevitsch@rz.hu-berlin.de

M. Röckner
Universität Bielefeld, Fakultät für Mathematik, Universitätsstr. 25, D-33615 Bielefeld
e-mail: cdraeger@mathematik.uni-bielefeld.de

F. Russo
Université Paris 13, Département de Mathématiques, Institut Galilée, av. Jean-Baptiste Clement, F-93430 Villetaneuse, France
e-mail: russo@zeus.math.univ-paris13.fr

M. Rutkowski
Politechnika Warszawska, Institute of Mathematics, pl. Politechniki 1, 00-661 Warszawa, Poland
e-mail: markrut@alpha.mini-pw.edu.pl

W. M. Schmidt
Deutsche Bank, T & S (Z), OTC-Derivative, Taunusanlage 12, D-60325 Frankfurt a. M., Germany
e-mail: wolfgang-michael.schmidt@db.com

H.-J. Starkloff
Technische Universität Chemnitz, Fakultät für Mathematik, Lehrstuhl für Stochastik, D-09107 Chemnitz, Germany
e-mail: h.starkloff@mathematik.tu-chemnitz.de

Ch. Stricker
Université de Franche-Comté Besançon, Faculté de Sciences et des Techniques, Laboratoire de Mathématiques, 16 Route de Gray, F-25030 Besançon Cedex, France
e-mail: stricker@math.univ-fcomte.fr

A. Thalmaier
Universität Bonn, Institut für Angewandte Mathematik, Wegelerstraße 6, D-53115 Bonn, Germany
e-mail: anton@wiener.iam.uni-bonn.de

G. Tittel
Friedrich-Schiller-Universität Jena, Fakultät für Mathematik und Informatik, Institut für Stochastik, Ernst-Abbe-Platz 1-4, D-07743 Jena, Germany
e-mail: tittel@minet.uni-jena.de

P. Vallois
Université Henri Poincaré Nancy 1, Institut Elie Cartan, B.P. 239, 54506 Vandoevre les Nancy Cedex, France
e-mail: Pierre.Vallois@antares.iecn.u-nancy.fr

T. Volz

TU Darmstadt, Fachbereich Mathematik, Schloßgartenstr. 7, D-64289 Darmstadt, Germany

e-mail: volz@mathematik.tu-darmstadt.de

M. Walther

Friedrich-Schiller-Universität Jena, Fakultät für Mathematik und Informatik, Institut für Stochastik, Ernst-Abbe-Platz 1-4, D-07743 Jena, Germany

e-mail: mail-mario@gmx.de

H. von Weizsäcker

Universität Kaiserslautern, Fachbereich Mathematik, Erwin-Schrödinger-Str., D-67663 Kaiserslautern, Germany

e-mail: weizsaecker@mathematik.uni-kl.de

J. Wolf

Friedrich-Schiller-Universität Jena, Fakultät für Mathematik und Informatik, Institut für Stochastik, Ernst-Abbe-Platz 1-4, D-07743 Jena, Germany (until August 2000) and *Bundesaufsichtsamt für das Versicherungswesen, Graurheindorfer Str. 108, D-53117 Bonn, Germany*

e-mail: jochen.wolf@bav.bund.de

R. Wunderlich

Technische Universität Chemnitz, Fakultät für Mathematik, Lehrstuhl für Stochastik, D-09107 Chemnitz, Germany

e-mail: wunderlich@mathematik.tu-chemnitz.de

M. Yor

Université Paris VI, Laboratoire de Probabilités, 4, Place Jussieu, Tour 56, F- 75252 Paris Cedex 05, France

e-mail: deaproba@proba.jussieu.fr

J. Zabczyk

Institute of Mathematics, Polish Academy of Sciences, P.O. Box 137, Śniadeckich 8, 00-950 Warsaw, Poland

e-mail: zabczyk@panim.impan.gov.pl

Index

Milton Keynes UK
Ingram Content Group UK Ltd.
UKHW020024071024
449327UK00032B/2912